BOLYAI SOCIETY
MATHEMATICAL STUDIES

12

Springer
Berlin
Heidelberg
New York
Hong Kong
London
Milan
Paris
Tokyo

BOLYAI SOCIETY MATHEMATICAL STUDIES

1. **Combinatorics, Paul Erdős is Eighty, Vol. 1**
   D. Miklós, V.T. Sós, T. Szőnyi (Eds.)
2. **Combinatorics, Paul Erdős is Eighty, Vol. 2**
   D. Miklós, V.T. Sós, T. Szőnyi (Eds.)
3. **Extremal Problems for Finite Sets**
   P. Frankl, Z. Füredi, G. Katona, D. Miklós (Eds.)
4. **Topology with Applications**
   A. Császár (Ed.)
5. **Approximation Theory and Function Series**
   P. Vértesi, L. Leindler, Sz. Révész, J. Szabados, V. Totik (Eds.)
6. **Intuitive Geometry**
   I. Bárány, K. Böröczky (Eds.)
7. **Graph Theory and Combinatorial Biology**
   L. Lovász, A. Gyárfás, G. Katona, A. Recski (Eds.)
8. **Low Dimensional Topology**
   K. Böröczky, Jr., W. Neumann, A. Stipsicz (Eds.)
9. **Random Walks**
   P. Révész, B. Tóth (Eds.)
10. **Contemporary Combinatorics**
    B. Bollobás (Ed.)
11. **Paul Erdős and His Mathematics I+II**
    G. Halász, L. Lovász, M. Simonovits, V. T. Sós (Eds.)
12. **Higher Dimensional Varieties and Rational Points**
    K. Böröczky, Jr., J. Kollár, T. Szamuely (Eds.)

Károly Böröczky, Jr.
János Kollár
Tamás Szamuely (Eds.)

Higher Dimensional Varieties and Rational Points

Springer

JÁNOS BOLYAI MATHEMATICAL SOCIETY

Károly Böröczky, Jr.
Alfréd Rényi Institute of Mathematics
Hungarian Academy of Sciences
Reáltanoda utca 13–15
1053 Budapest, Hungary
e-mail: carlos@renyi.hu

János Kollár
Department of Mathematics
Princeton University
Princeton NJ-08544-1000, USA
e-mail: kollar@math.princeton.edu

Tamás Szamuely
Alfréd Rényi Institute of Mathematics
Hungarian Academy of Sciences
Reáltanoda utca 13–15
1053 Budapest, Hungary
e-mail: szamuely@renyi.hu

Mathematics Subject Classification (2000): 14Jxx, 14G05, 14F35, 11G50

Cataloging-in-Publication Data applied for
A catalog record for this book is available from the Library of Congress.
Bibliographic information published by Die Deutsche Bibliothek
Die Deutsche Bibliothek lists this publication in the Deutsche Nationalbibliografie; detailed bibliographic data is available in the Internet at http://dnb.ddb.de

ISSN 1217-4696
ISBN 978-3-642-05644-4

Springer-Verlag Berlin Heidelberg New York
a member of BertelsmannSpringer Science+Business Media GmbH

http://www.springer.de

Printed in Hungary

Cover design: Erich Kirchner, Heidelberg

Printed on acid-free paper 44/3142/db – 5 4 3 2 1 0

CONTENTS

Introduction

It has been understood for a long time that there is a close connection between the geometric and the arithmetic properties of algebraic varieties. Results of this type have been firmly established for curves for some time and substantial progress was also made for surfaces. The higher dimensional versions of these problems are largely wide open, and there does not seem to be full agreement about even the main conjectures.

The aim of the Summer School and Conference on "Higher Dimensional Varieties and Rational Points" organized at the Alfréd Rényi Institute of Mathematics in Budapest, Hungary between 3 and 21 September 2001 was to bring together students and experts from the arithmetic and geometric sides in order to get a better understanding of the current problems and advances.

This volume brings together most of the lecture series and several of the conference lectures to present a comprehensive view of the subject. The aim of these notes is to give an introduction to advanced students and to specialists in related fields by emphasizing the main ideas of the subjects.

The paper by Carolina Araujo and János Kollár gives an introduction to the study of rational curves on varieties, with special emphasis on rationally connected varieties. Rationally connected varieties are the higher dimensional analogs of rational curves and surfaces. One of the main problems is to establish the higher dimensional versions of basic arithmetic properties of rational curves on surfaces. The study of rational curves leads to several satisfactory answers over local fields, but a substantial gap still remains in general.

The paper by János Kollár applies the methods of the previous survey to the study of the fundamental group of open subsets of rationally connected varieties. These results have already been applied to inverse Galois problems and to rationally connected varieties over finite fields.

Olivier Debarre's lectures emphasize the methods that have been very efficient in the study of Fano varieties. Fano varieties are the basic building blocks of rationally connected varieties. Many of the intricacies of the general cases are absent, but they seem to retain all the essential geometric and arithmetic complexity of the full theory. The first half of the lectures describes the best understood case, toric Fano varieties, and the second half provides a general overview.

Sándor Kovács addresses the higher dimensional version of the Shafarevich conjecture for families of smooth varieties. We know since Faltings' proof of the Mordell conjecture that the Shafarevich conjecture for curves is very closely related to finding rational points. In higher dimensions there does not seem to be such a connection, instead these problems center on the geometry, arithmetic and curvature properties of moduli spaces. The arithmetic significance of such results is only beginning to emerge.

Jean-Louis Colliot-Thélène explores the local-to-global method of finding rational points through examples. Given a variety over $\mathbb{Q}$, one can first find solutions over p-adic fields (called local solutions) and then hope to patch these together into a rational solution (called global solution). The second step is usually quite difficult or even impossible. The theory of descent and the fibration method are the most successful approaches. As a consequence, one gets very nice answers for diagonal cubic equations over number fields and over function fields of curves over finite fields.

The notes of Brendan Hassett provide one of the best examples of mixing geometry and arithmetic. It is expected that rationally connected varieties have many rational points over a suitable finite degree field extension, but currently very few higher dimensional cases are settled. The best current method uses delicate geometric constructions to produce families of elliptic curves (or more generally abelian varieties) on the varieties under study and then exploits the fact that over a finite extension of the base elliptic curves have many rational points thanks to the group law. The paper also clarifies several points that have not been adequately treated in the literature.

Yuri Tschinkel studies the relationship between the geometry of the anticanonical class of a variety and the distribution of its rational points. General conjectures by Batyrev and Manin (refined by Peyre) predict asymptotic formulas for the number of points of bounded height in which the constants are related to geometric properties of the variety. Aside from some sporadic examples, these are known only for varieties with a large algebraic group acting on them. The case when the group is not reductive leads to very interesting geometric and analytic problems, many of which have not been treated earlier even in the simplest cases. These examples cover many of the most frequently occurring special cases and they provide the best test case for several of the general conjectures.

Acknowledgments

This volume is the fruit of a conference and workshop hosted by the Alfréd Rényi Institute of Mathematics under the auspices of the Center of Excellence program. Financial support from the High Level Scientific Conferences project of the European Union (contract No. HPCF-CT-2001-00062) and the Paul Erdős Summer Research Center is gratefully acknowledged. The editors are indebted to Emmanuel Peyre for the cover illustration. Last but not least, they wish to express their thanks to Jean-Louis Colliot-Thélène, Dezső Miklós, András Némethi, Veronika Pándi and Miles Reid for their help in organizing the conference and arranging the publication of the volume.

Part I: Geometry

BOLYAI SOCIETY
MATHEMATICAL STUDIES, 12

Higher Dimensional Varieties and Rational Points, pp. 13–68.

Rational Curves on Varieties

CAROLINA ARAUJO and JÁNOS KOLLÁR

The aim of these notes is to give an introduction to the ideas and techniques of handling rational curves on varieties. The main emphasis is on varieties with many rational curves, which are the higher dimensional analogs of rational curves and surfaces.

Sections one through five and seven closely follow the lectures of J. Kollár given at the summer school "Higher dimensional varieties and rational points". The notes were written up by C. Araujo and substantially edited later.

Terminology. We follow the terminology of [9]. Thus a variety is an integral separated scheme of finite type over a field (not necessarily algebraically closed). Somewhat inconsistently, by a curve we mean a geometrically reduced (not necessarily irreducible) separated 1-dimensional scheme of finite type over a field.

1. Rational Curves on Surfaces

In this section we study rational curves on surfaces. Our main point is that rational surfaces contain plenty of rational curves (at least when we are over an algebraically closed field).

In the following example we analyze the possible genera of curves lying on smooth surfaces in $\mathbb{P}^3$. This illustrates our point: while surfaces of degree $d \leq 3$ (which are rational) contain many rational curves, those of degree $d \geq 5$ tend not to contain rational curves at all.

Example 1 (Curves on Surfaces in $\mathbb{P}^3$). Let $S = S_d$ be a smooth surface of degree d in $\mathbb{P}^3_{\mathbb{C}}$. By the Noether–Lefschetz Theorem, if $d \geq 4$ and S_d is very general, any curve $C \subset S_d$ is a complete intersection. This means that there is a surface H_m of degree m in $\mathbb{P}^3$, for some m, such that $C = S_d \cap H_m$ scheme-theoretically. (See [20, I] for a survey of different proofs of this result.)

Let $\tilde{C} \to C$ be the normalization of C. We ask the following question: for fixed d and m, what are the possible values of the genus $g(\tilde{C})$ of $\tilde{C}$?

The biggest possible value of $g(\tilde{C})$ is attained when $C = S_d \cap H_m$ is already nonsingular. In this case $g(\tilde{C}) = g(C) = \frac{1}{2}md(m + d - 4) + 1$.

Lower bounds for $g(\tilde{C})$ are much more interesting and difficult to get.

Let H denote the hyperplane class in $\mathbb{P}^3$, so that $H_m \sim mH$.

By Riemann–Roch we have:

$$h^0\big(S, \mathcal{O}_S(mH)\big) = m^2\frac{d}{2} - m\frac{d(d-4)}{2} + \chi(\mathcal{O}_S).$$

By the adjunction formula we have:

$$p_a(C) = m^2\frac{d}{2} + m\frac{d(d-4)}{2} + 1.$$

The "expected" maximal number of singular points of C is

$$h^0\big(S, \mathcal{O}_S(mH)\big) - 1.$$

By "expected" we mean that this is not a theorem but it is frequently correct. Here is an argument to support it.

Requiring that the curve has a singularity at a given point imposes 3 conditions. We have a 2-dimensional family of points. Thus, requiring that the curve has a singularity at some point imposes 1 condition.

Putting these formulas together, we obtain that the expected smallest genus of $\tilde{C}$ is

$$md(d-4) + 2 - \chi(\mathcal{O}_S).$$

If $d \geq 5$, this grows linearly with m. So we expect that, given g, there are only finitely many values of m for which $\big|\mathcal{O}_S(m)\big|$ contains a curve of geometric genus g. It is important to keep in mind that this "conclusion" may not hold for every surface S_d. However, this is essentially correct for very general surfaces (although this is not at all easy to prove). In fact, by [21, Theorem 2.1], on a very general $S_d \subset \mathbb{P}^3_{\mathbb{C}}$, any irreducible curve of type $S_d \cap H_m$ has geometric genus $\geq \frac{1}{2}dm(d-5) + 2$.

If $d = 2$ or 3, not every curve on S_d is a complete intersection (for instance, the lines on smooth quadrics and cubics are not complete intersections). One can show that for $d \leq 3$ the linear system $|\mathcal{O}_S(m)|$ contains rational curves for every m. Instead of proving this fact here, we only prove that, for $d \leq 3$, S_d contains plenty of rational curves (this argument works over any algebraically closed field).

$d = 1$. In this case $S_1 \cong \mathbb{P}^2$, and for every m there are plenty of rational curves of degree m (i.e., rational curves that can be obtained as $C = S_1 \cap H_m$). These are given by maps $\phi : \mathbb{P}^1 \to \mathbb{P}^2$ sending $(s : t)$ to $(\phi_0(s,t) : \phi_1(s,t) : \phi_2(s,t))$, where $\phi_0(s,t)$, $\phi_1(s,t)$ and $\phi_2(s,t)$ are homogeneous polynomials of degree m.

$d = 2$. In this case S_2 is a smooth quadric surface in $\mathbb{P}^3$, and the stereographic projection from a point in S_2 gives a birational equivalence $S_2 \sim \mathbb{P}^2$.

$d = 3$. In this case S_3 is a smooth cubic surface in $\mathbb{P}^3$, which, again, is birational to $\mathbb{P}^2$ (it can be realized as $\mathbb{P}^2$ blown up at 6 points).

Example 2 (Rational curves on $\mathbb{P}^n$). Given finitely many points $P_1, \dots, P_m$ in $\mathbb{P}^n$, let us write down a rational curve $C \subset \mathbb{P}^n$ passing through these points. Set $P_i = (a_{i0} : \cdots : a_{in})$. Pick any points $c_1, \dots, c_m \in \mathbb{A}^1$. The Lagrange interpolation formula gives degree $m-1$ polynomials $f_0, \dots, f_n$ such that $f_j(c_i) = a_{ij}$ for every i, j. These determine a morphism $F : \mathbb{A}^1 \to \mathbb{P}^n$ such that $F(c_i) = P_i$.

Note that if the P_i and c_i are defined over a field k then F has coordinate functions defined over k.

Assume next that we fix not only the points P_i but also tangent directions at each P_i. It is not hard to see that there is a rational curve $C \subset \mathbb{P}^n$ through these points with the preassigned tangent directions.

More generally, we can even specify finitely many terms of the Taylor expansion of $F : \mathbb{A}^1 \to \mathbb{P}^n$ at each point c_i.

The next example shows that any smooth rational variety has an even stronger property.

Example 3 (Constructing curves on rational varieties). Let X be a smooth projective rational variety of dimension n, and $\Phi : \mathbb{P}^n \dashrightarrow X$ a birational map given by polynomial coordinate functions $\phi_0, \dots, \phi_N$.

Let C be a smooth projective curve with finitely many points $c_i \in C$. For each i let us specify finitely many terms of the Taylor expansion near c_i of a map from C to X. (That is, we choose morphisms $g_i : \operatorname{Spec} \mathcal{O}_{c_i,C}/m_i^{e_i} \to X$, where $\mathcal{O}_{c_i,C}$ is the local ring of C at c_i with maximal ideal m_i.)

Our aim is to find a morphism $C \to X$ that has the specified Taylor expansion at the points c_i. Moreover, if the data are defined over a field k we want $C \to X$ to be defined over k.

The first step is to reduce the problem to $X = \mathbb{P}^n$. In order to do this, we extend each g_i to something bigger. If we are over $\mathbb{C}$, we extend each g_i to a small analytic neighborhood of c_i. Over general fields this corresponds to extending each g_i to the completion $\hat{\mathcal{O}}_{c_i,C}$. The completion of a 1-dimensional smooth local ring is always the power series ring in one variable, thus these can be best given by the pull back maps

$$G_i^* : \mathcal{O}_{P_i,X} \to \hat{\mathcal{O}}_{c_i,C} \cong k[[t]].$$

By choosing the extension general enough, we may assume that the image of $\operatorname{Spec} \hat{\mathcal{O}}_{c_i,C}$ is not contained in the exceptional locus of Φ^{-1}. Thus G_i can be lifted to a morphism $H_i : \operatorname{Spec} k[[t]] \to \mathbb{P}^n$ such that $G_i = \Phi \circ H_i$. At the function level this means that

$$G_i^*(u_j) = W_i(t) \cdot H_i^*(\phi_j) \quad \text{for every } j,$$

where the u_j are homogeneous coordinates on $X \subset \mathbb{P}^N$, $G_i^*(u_j) \in k[[t]]$ and the extra factor $W_i(t)$ comes from the nonuniqueness of the projective coordinates. Let w_i be the order of poles of W_i.

Assume now that we choose a different map $H_i' : \operatorname{Spec} k[[t]] \to \mathbb{P}^n$ that agrees with H_i up to order d_i. By composing with Φ we get $G_i' : \operatorname{Spec} k[[t]] \to X$ and the above formula implies that G_i' agrees with G_i up to order $d_i - w_i$.

This implies that if we solve our original problem for maps $C \to \mathbb{P}^n$ for every e_i then we also obtain a solution for the problem $C \to X$.

Now let us consider the problem for $C \to \mathbb{P}^n$. Choose a line bundle L on C of degree at least $2g(C) - 1 + \sum e_i$. Then the restriction map

$$H^0(C, L) \to \sum_i H^0(C, L \otimes \mathcal{O}_{c_i,C}/m_i^{e_i}) \cong \sum_i \mathcal{O}_{c_i,C}/m_i^{e_i}$$

is surjective. Thus for every j there is a section $\sigma_j \in H^0(C, L)$ such that the restriction of σ_j to $\mathcal{O}_{c_i,C}/m_i^{e_i}$ agrees with $g_i^*(x_j)$. The map $C \to \mathbb{P}^n$ given by all these σ_j has the required properties.

We should keep this in mind in Section 7, when we define rationally connected varieties, in order to characterize the higher dimensional analogs of rational curves and rational surfaces.

2. Deformation of Morphisms I

In this section we are interested in the following problem: suppose we have a rational curve C on a smooth variety X defined over some base field k. How can we produce more rational curves out of C?

More generally, let X be a smooth quasi-projective variety, Y a projective variety, and $f_0 : Y \to X$ a morphism. We would like to deform the morphism f_0 in order to produce more morphisms from Y to X.

That is, we would like to find a map

$$F : Y \times D \to X \quad \text{such that} \quad F|_{Y \times \{0\}} = f_0,$$

where $(D, 0)$ is a smooth pointed algebraic curve. For $t \in D$ we set $f_t = F|_{Y \times \{t\}} : Y \to X$, and we also require that F be nontrivial, that is, $f_t \neq f_0$ (as morphisms) for general t.

It turns out that the existence of F depends on the choice of D in a very unpredictable manner. To avoid this problem, we need to replace D with a "very small" neighborhood of 0 in D. In general, no affine neighborhood is small enough. Technically, the best is to work with the spectrum of the formal power series ring $\operatorname{Spec} k[[t]]$, but this is not very intuitive.

Assume for simplicity that we are over $\mathbb{C}$. Then we get a geometrically appealing picture by working with $D = \Delta$, where Δ denotes the unit disc in $\mathbb{C}$. So we want to find a map

$$F : Y \times \Delta \to X \quad \text{such that} \quad F|_{Y \times \{0\}} = f_0.$$

Here of course F is complex analytic. We also allow Δ to be replaced by a smaller disc at any time. (All discs in $\mathbb{C}$ are biholomorphic to each other, so this is only a matter of convenience.)

Suppose we have such a map. For each $p \in Y$, $\{f_t(p)\}_{t \in \Delta}$ is an analytic arc on X. Let $v_p \in T_{f_0(p)}X$ be the derivative of this arc at the point $f_0(p) \in X$ (see Fig. 1). We "expect" that this correspondence

$$p \in Y \mapsto v_p \in T_{f_0(p)}X$$

determines a nonzero element $\partial F/\partial t \in H^0(Y, f_0^* T_X)$. (This is again an expectation only. Given any $F(y,t) : Y \times \Delta \to X$ we can replace it by

$$G(y, s) := F(y, s^2) : Y \times \Delta \to X$$

to get another morphism with $\partial G/\partial s = 0$.) In this case, however, the expectation is completely correct, and we get the following criterion.

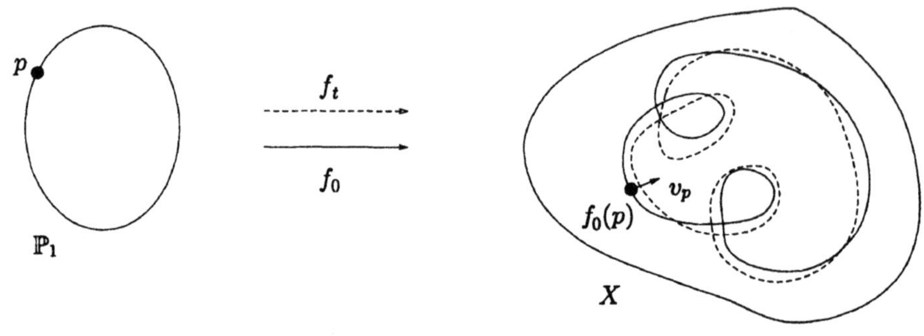

Fig. 1. Deforming a rational curve

If $H^0(Y, f_0^*T_X) = 0$, then f_0 cannot be deformed in a nontrivial way.

4 (The space of morphisms $Y \to X$). Let X and Y be varieties over some base field k. Suppose we have a morphism $f_0 : Y \to X$, and we want to study deformations of f_0. In this situation, it is very useful to consider a space parametrizing all morphisms from Y to X.

For instance, when $Y = X = \mathbb{P}^1$, we would like to say that isomorphisms $\mathbb{P}^1 \to \mathbb{P}^1$ are parametrized by the group PGL_2. More precisely, by this we mean that for any field k, the points of $PGL_2(k)$ are in a "natural" one-to-one correspondence with the k-isomorphisms $\mathbb{P}^1 \to \mathbb{P}^1$.

Similarly, when trying to understand morphisms $Y \to X$, we are looking for a scheme $\mathrm{Hom}\,(Y, X)$, together with a morphism, called *universal morphism*,

$$u : \mathrm{Hom}\,(Y, X) \times Y \to X,$$

both defined over k, such that for every field extension K/k and every K-morphism $f : Y_K \to X_K$ there is a unique point $p \in \mathrm{Hom}\,(Y, X)(K)$ such that $f = u|_{\{p\} \times Y}$. The point p is then denoted by $[f]$.

It turns out that when X is quasi-projective and Y is projective there is such a scheme $\mathrm{Hom}\,(Y, X)$ (see [6] or [12, I.1.10]).

It is also very useful to consider morphisms $Y \to X$ sending a specified point of Y to a specified point of X. More generally, given a closed subscheme B of Y, and $g : B \to X$ a morphism, we want to study morphisms $Y \to X$ extending g. Under the same assumptions as before, there exists a scheme parametrizing such morphisms. We denote it by $\mathrm{Hom}\,(Y, X, g)$. It is clearly a subscheme of $\mathrm{Hom}\,(Y, X)$.

The scheme $\mathrm{Hom}(Y,X)$ is not always a variety. Usually it has countably many components. However, each irreducible component of $\mathrm{Hom}(Y,X)$ is of finite type over k.

These spaces are very complicated in general. However, if

$$H^1(Y, f^*T_X) = 0,$$

$\mathrm{Hom}(Y,X)$ has a very simple description near the point $[f]$. This is the content of the second part of the next theorem.

Theorem 5. *Let Y and X be schemes over a field k, Y projective and without embedded points and X smooth and quasi-projective. Let $f : Y \to X$ be a morphism. Then*

1. *We have* $T_{[f]}\mathrm{Hom}(Y,X) \cong H^0(Y, f^*T_X)$.
2. (Kodaira–Spencer) *If* $H^1(Y, f^*T_X) = 0$, *then* $\mathrm{Hom}(Y,X)$ *is smooth at* $[f]$ *and has dimension* $h^0(Y, f^*T_X)$.
3. (Mumford) *In general,* $\dim_{[f]}\mathrm{Hom}(Y,X) \geq h^0(Y, f^*T_X) - h^1(Y, f^*T_X)$.

Proof. Statements (1) and (3) are special cases of [12, I.2.16], statements (2.16.1) and (2.16.2) respectively.

Now assume that $H^1(Y, f^*T_X) = 0$. It follows from (1) and (3) above that $\dim_{[f]}\mathrm{Hom}(Y,X) = h^0(Y, f^*T_X)$. Therefore, by [12, I.2.17], $\mathrm{Hom}(Y,X)$ is a local complete intersection. Since $\dim_{[f]}\mathrm{Hom}(Y,X) = \dim T_{[f]}\mathrm{Hom}(Y,X)$, we conclude that $\mathrm{Hom}(Y,X)$ is smooth at $[f]$. ∎

The theorem says that when $H^1(Y, f^*T_X) = 0$, there is an irreducible curve through $[f]$ on $\mathrm{Hom}(Y,X)$ in every direction of $T_{[f]}\mathrm{Hom}(Y,X)$. This means that any global section of f^*T_X comes from a deformation $F : Y \times \Delta \to X$.

This is not the case if $\mathrm{Hom}(Y,X)$ is not smooth at $[f]$.

3. Deforming Rational Curves

Now we apply the theory presented in the previous section to study deformations of curves on smooth varieties. We are mainly interested in deforming rational curves.

Like before, let X be a smooth quasi-projective variety over a field k, but now assume that $Y = C$ is a proper curve over k. A crucial simplification in this case is that $h^0(C, f^*T_X) - h^1(C, f^*T_X)$ is just the Euler characteristic $\chi(C, f^*T_X)$, which can be computed by Riemann–Roch:

$$\begin{aligned}\chi(C, f^*T_X) &= \deg(f^*T_X) + \chi(\mathcal{O}_C)\dim X \\ &= -\big(f_*[C]\cdot K_X\big) + \big(1 - p_a(C)\big)\dim X.\end{aligned}$$

Hence we obtain that

$$\dim_{[f]} \operatorname{Hom}(C, X) \geq -\big(f_*[C]\cdot K_X\big) + \big(1 - p_a(C)\big)\dim X.$$

Example 6. A smooth projective variety X is called *Fano* if its anticanonical class $-K_X$ is ample (for instance, we can take for X a smooth hypersurface in $\mathbb{P}^n$ of degree $d \leq n$). In this case $(C \cdot K_X) < 0$ for every curve $C \subset X$, and we expect curves on X to move a lot. For instance, if $f : \mathbb{P}^1 \to X$ is a nonconstant morphism, then

$$\dim_{[f]} \operatorname{Hom}(\mathbb{P}^1, X) \geq -\big(f_*[\mathbb{P}^1]\cdot K_X\big) + \dim X \geq 1 + \dim X.$$

The automorphisms of $\mathbb{P}^1$ account for a 3-dimensional family of maps. Thus, if $\dim X \geq 3$, every rational curve on X moves. (We really need $\dim X \geq 3$ here. Lines on a smooth cubic surface in $\mathbb{P}^3$ provide an example of nonmoving rational curves on a Fano variety of dimension 2.) In general, the larger the degree of the curve, the more it moves.

When X is a Fano variety defined over an algebraically closed field, a result of Mori shows that through any point of X there passes a rational curve (the next exercise shows this in the hypersurface case, when $d \leq n-1$).

Exercise 7. Let $X_d \subset \mathbb{P}^n$ be a smooth hypersurface of degree $d \leq n-1$ over an algebraically closed field. Show that through every point of X_d there is an $(n-1-d)$-dimensional family of lines. Compute the dimension of the family of all lines on X_d and compare it with the above estimate.

Let us study the case $C \cong \mathbb{P}^1$ in more detail. Every vector bundle on $\mathbb{P}^1$ splits into a direct sum of line bundles (see [9, V. Exercise 2.6]). Therefore we can write, for suitable integers a_i,

$$f^*T_X \cong \sum_{i=1}^{\dim X} \mathcal{O}_{\mathbb{P}^1}(a_i).$$

Definition 8. We say that $f : \mathbb{P}^1 \to X$ is *free* if all the a_i are ≥ 0.

We say that $f : \mathbb{P}^1 \to X$ is *very free* if all the a_i are ≥ 1.

Remark 9. Intuitively, a free curve can be moved in every direction, while a very free curve still moves freely in any direction with one point fixed.

In particular, if $Z \subset X$ has codimension ≥ 2, any free curve $f : \mathbb{P}^1 \to X$ has nearby deformations that are disjoint from Z. If $f : \mathbb{P}^1 \to X$ is very free, then it has nearby deformations f_t such that $f_t(0) = f(0)$ and $f_t(\mathbb{P}^1 \setminus \{0\})$ is disjoint from Z. Moreover, the tangent spaces of the images $f_t(\mathbb{P}^1)$ at $f(0)$ sweep out an open subset of the tangent space of X at $f(0)$.

These results are easy to obtain from Propositions 10 and 13. See [12, II.3] for detailed proofs.

Notice that a morphism $f : \mathbb{P}^1 \to X$ is free (resp. very free) if and only if $H^1(\mathbb{P}^1, f^*T_X(-1)) = 0$ (resp. $H^1(\mathbb{P}^1, f^*T_X(-2)) = 0$). By the Semicontinuity Theorem (see [9, III.12.8]), nearby deformations of a free (resp. very free) rational curve are still free (resp. very free).

Proposition 10. *Let X be an n-dimensional smooth quasi-projective variety over an algebraically closed field k. Let $f : \mathbb{P}^1 \to X$ be a nonconstant morphism.*

1. *If f is free, there exists a variety Y of dimension $n-1$, a point $y_0 \in Y$, and a dominant morphism $F : \mathbb{P}^1 \times Y \to X$ such that $F|_{\mathbb{P}^1 \times \{y_0\}} = f$.*
2. *Assume there exists a variety Y and a dominant morphism $F : \mathbb{P}^1 \times Y \to X$. If k has characteristic 0, $f_y := F|_{\mathbb{P}^1 \times \{y\}}$ is free for general $y \in Y$.*

Proof. (1) Let Z be a suitable neighborhood of $[f]$ in $\mathrm{Hom}(\mathbb{P}^1, X)$. Since f is free, any given point of $f(\mathbb{P}^1)$ can be moved in arbitrary direction. This implies that the evaluation morphism $F : \mathbb{P}^1 \times Z \to X$ is submersive at every point of $\mathbb{P}^1 \times \{[f]\}$, thus it is also dominant. (See [12, II.3.5.3] for a rigorous proof.) If $Y \subset Z$ is a general subvariety of dimension $n-1$ then $F|_{\mathbb{P}^1 \times Y} : \mathbb{P}^1 \times Y \to X$ is still dominant.

(2) Since F is dominant, there is a dense open subset $U \subset \mathbb{P}^1 \times Y$ such that $F : U \to X$ is smooth (here we use the characteristic 0 assumption). Therefore, the induced map $dF : T_{\mathbb{P}^1 \times Y} \to F^*T_X$ is surjective on U.

Let $\pi_2 : \mathbb{P}^1 \times Y \to Y$ denote the second projection, and assume that $y \in \pi_2(U)$. Then

$$\mathcal{O}_{\mathbb{P}^1}(2) \oplus \mathcal{O}_{\mathbb{P}^1}^{\dim Y} \cong T_{\mathbb{P}^1 \times Y}\big|_{\mathbb{P}^1 \times \{y\}} \to f_y^*T_X \cong \sum_{i=1}^{n} \mathcal{O}_{\mathbb{P}^1}(a_i)$$

is surjective on an open set. Therefore all the a_i are ≥ 0. ■

Corollary 11. *Let X be a smooth quasi-projective variety over an algebraically closed field of characteristic 0. Let $x \in X$ be a very general point, i.e., x is outside a countable union of proper closed subvarieties of X. Then any morphism $f : \mathbb{P}^1 \to X$ whose image passes through x is free.*

Proof. For each (of countably many) irreducible component Z of $\mathrm{Hom}(\mathbb{P}^1, X)$, consider the evaluation morphism $F_Z : \mathbb{P}^1 \times Z \to X$. Since being free is an open condition, the points of Z parametrizing nonfree morphisms form a closed subscheme $Z' \subset Z$ (notice that we may have $Z' = Z$). By Proposition 10, the restriction $F_Z|_{\mathbb{P}^1 \times Z'} : \mathbb{P}^1 \times Z' \to X$ is not dominant, and thus $\overline{F_Z(\mathbb{P}^1 \times Z')} \neq X$.

Set $V_Z = \overline{F_Z(\mathbb{P}^1 \times Z')} \subsetneq X$. Then any morphism $f : \mathbb{P}^1 \to X$ whose image passes through a point in $X \setminus \bigcup_Z V_Z$ is free. ■

Remark 12. This corollary says little when X is a variety over a countable field k. Indeed, in this case we may have $X(k) = \bigcup_Z V_Z(k)$ (notation as in the proof above). No such example is known, and this leads to the following open problem: *Is there a smooth projective variety X over $\bar{\mathbb{Q}}$ such that through every point of X there is a nonfree rational curve?*

A related question is the following: *Is there a smooth projective variety X over $\bar{\mathbb{Q}}$ such that through every point of X there is a rational curve, but rational curves do not cover $X(\mathbb{C})$?*

Variants of the latter are also interesting (and possibly easier) over finite fields.

We have an analogue of Proposition 10 for very free morphisms. The proof is very similar to the proof of Proposition 10, and we leave it to the reader.

Proposition 13. *Let X be a n-dimensional smooth quasi-projective variety over an algebraically closed field k. Fix $x_0 \in X$. Let $f : \mathbb{P}^1 \to X$ be a nonconstant morphism such that $f(0) = x_0$.*

1. *If f is very free, there exists a variety Y of dimension $n-1$, a point $y_0 \in Y$, and a dominant morphism $F : \mathbb{P}^1 \times Y \to X$ such that $F|_{\mathbb{P}^1 \times \{y_0\}} = f$ and $F|_{\{0\} \times Y} \equiv x_0$.*
2. *Assume there exists a variety Y and a dominant morphism $F : \mathbb{P}^1 \times Y \to X$ such that $F|_{\{0\} \times Y} \equiv x_0$. If k has characteristic 0 then $f_y := F|_{\mathbb{P}^1 \times \{y\}}$ is very free for general $y \in Y$.* ■

Example 14. Let X be a smooth Fano variety over a field k. We mentioned in Example 6 that through any point of $X_{\bar{k}}$ there passes a rational curve, defined over $\bar{k}$. When k has characteristic 0, Theorem 10.2 implies that through a very general point of $X_{\bar{k}}$ there passes a free rational curve. In fact, there is a free rational curve through every point, as we see in Theorem 29. On the other hand, so far we know nothing about the existence of rational curves on X that are defined over k. In general this is a very difficult problem that depends on the arithmetic nature of the field k. In the simplest case, when $k = \mathbb{R}$, we consider this problem in Application 19. The more general case of local fields is studied in Section 9.

4. Deformation of Morphisms II

Sometimes it is useful to consider a wider class of deformations than the one considered in the last two sections. Let $f : Y \to X$ be a morphism from a projective variety Y to a smooth quasi-projective variety X. Now, instead of changing only the morphism f, we also want to change Y and X in a predetermined way.

In order to get this setup, let $(R, 0)$ be an irreducible pointed scheme, and $\mathcal{X}$ and $\mathcal{Y}$ schemes over R, with $\mathcal{X} \to R$ smooth and $\mathcal{Y} \to R$ proper and flat. Let $\mathcal{X}_0$ and $\mathcal{Y}_0$ be the fibers over $0 \in R$. Starting with a given morphism $f_0 : \mathcal{Y}_0 \to \mathcal{X}_0$, we would like to obtain deformations $f_r : \mathcal{Y}_r \to \mathcal{X}_r$.

Like before, there exists an R-scheme $\operatorname{Hom}(\mathcal{Y}/R, \mathcal{X}/R)$ parametrizing morphisms $\mathcal{Y} \to \mathcal{X}$. This means that there is a universal morphism

$$u : \operatorname{Hom}(\mathcal{Y}/R, \mathcal{X}/R) \times_R \mathcal{Y} \to \mathcal{X}$$

such that for every R-scheme $S \to R$ and for every S-morphism $f : S \times_R \mathcal{Y} \to S \times_R \mathcal{X}$ there is a unique R-morphism $[f] : S \to \operatorname{Hom}(\mathcal{Y}/R, \mathcal{X}/R)$ such that f is the composition

$$f : S \times_R \mathcal{Y} \stackrel{(\mathrm{id}_S, [f], \mathrm{id}_{\mathcal{Y}})}{\longrightarrow} S \times_R \operatorname{Hom}(\mathcal{Y}/R, \mathcal{X}/R) \times_R \mathcal{Y} \stackrel{(\mathrm{id}_S, u)}{\longrightarrow} S \times_R \mathcal{X}.$$

The fiber of the map $\mathrm{Hom}\,(\mathcal{Y}/R, \mathcal{X}/R) \to R$ over the point $0 \in R$ is precisely $\mathrm{Hom}\,(\mathcal{Y}_0, \mathcal{X}_0)$:

$$\begin{array}{ccc} \mathrm{Hom}\,(\mathcal{Y}_0, \mathcal{X}_0) & \longrightarrow & \mathrm{Hom}\,(\mathcal{Y}/R, \mathcal{X}/R) \\ \downarrow & & \downarrow \\ \mathrm{Spec}\, k(0) & \longrightarrow & R \end{array}$$

(Here $k(0)$ denotes the residue field of 0 on R.)

The next result is an infinitesimal description of the space $\mathrm{Hom}\,(\mathcal{Y}/R, \mathcal{X}/R)$ near a point $[f_0]$ for which $H^1(\mathcal{Y}_0, f_0^* T_{\mathcal{X}_0}) = 0$. (For a proof see [12, I.2.17] and the proof of Theorem 5.)

Theorem 15. *Let* $(R, 0)$ *be an irreducible pointed scheme and* $\mathcal{X}$, $\mathcal{Y}$ *schemes over* R, *with* $\mathcal{X} \to R$ *smooth and* $\mathcal{Y} \to R$ *proper and flat. Assume that* $\mathcal{Y}_0$ *has no embedded points and let* $f_0 : \mathcal{Y}_0 \to \mathcal{X}_0$ *be a morphism. Then*

1. *If* $H^1(\mathcal{Y}_0, f_0^* T_{\mathcal{X}_0}) = 0$ *then* $\mathrm{Hom}\,(\mathcal{Y}/R, \mathcal{X}/R) \to R$ *is smooth in a neighborhood of* $[f_0]$, *and all fibers in this neighborhood have dimension* $h^0(\mathcal{Y}_0, f_0^* T_{\mathcal{X}_0})$.
2. *In general,*

$$\dim_{[f_0]} \mathrm{Hom}(\mathcal{Y}/R, \mathcal{X}/R) \geq h^0(\mathcal{Y}_0, f_0^* T_{\mathcal{X}_0}) - h^1(\mathcal{Y}_0, f_0^* T_{\mathcal{X}_0}) + \dim R. \qquad \blacksquare$$

Here is an example of how this theorem can be applied. Let X be a smooth variety over a field k. Suppose we have a connected union of rational curves $D \subset X$, and we want to deform it into a single rational curve. We represent D as the image of a tree of rational curves C under a morphism $f : C \to X$. Since C itself is reducible, $\mathrm{Hom}\,(C, X)$ does not give us any new integral curve on X. So we look for an irreducible k-scheme R and an R-scheme $\mathcal{Y}$ (proper and flat over R) such that the general fiber $\mathcal{Y}_r$ is an irreducible rational curve, while the special fiber $\mathcal{Y}_0$ is isomorphic to C. (We can usually take R to be $\mathbb{A}^1_k$ and $\mathcal{Y}$ to be the blow up of $\mathbb{P}^1_k \times \mathbb{A}^1_k$ at suitable points lying above $0 \in \mathbb{A}^1_k$. A concrete example is worked out in Application 17.) We also set $\mathcal{X} = X \times_k R$. If $H^1(C, f^* T_X) = 0$, by Theorem 15, $\mathcal{Z} = \mathrm{Hom}\,(\mathcal{Y}/R, \mathcal{X}/R) \to R$ is smooth at $[f]$, and nearby fibers have dimension $h^0(C, f^* T_X) > 0$. By choosing a point in one such fiber, $[f_r] \in \mathcal{Z}_r$, we obtain a deformation of f

$$f_r : \mathcal{Y}_r \to \mathcal{X}_r = X \times \{r\},$$

which is a morphism from a rational curve to X.

There is one thing we should be careful about: if k is not algebraically closed, it is possible that nearby fibers $\mathcal{Z}_r$ do not have k-points, and this method does not yield any deformation of D defined over k. We encounter this situation in Application 19.

Warning 16. It is important to keep in mind that in general the deformations obtained in this way do *NOT* give maps from $\mathcal{Y}$ to $\mathcal{X}$, not even over any open subscheme of R. Usually, the only thing we can hope to get is that $\mathrm{Hom}(\mathcal{Y}/R, \mathcal{X}/R) \to R$ is smooth at $[f]$. If this is the situation, we claim that there is a subscheme $[f] \in T \subset \mathrm{Hom}(\mathcal{Y}/R, \mathcal{X}/R)$ such that

1. the projection $T \to R$ is quasifinite and étale, mapping $[f]$ to 0;
2. the map f extends to a morphism

$$F_T : \mathcal{Y} \times_R T \to \mathcal{X} \times_R T.$$

To see this, let $t_1, \ldots, t_m$ be a regular sequence of generators for the maximal ideal in the local ring of $[f]$ on $\mathrm{Hom}(\mathcal{Y}_0, \mathcal{X}_0)$. (Such a sequence exists iff $[f]$ is a smooth point.) By flatness, these can be lifted to $T_1, \ldots, T_m$, in the local ring of $[f]$ on $\mathrm{Hom}(\mathcal{Y}/R, \mathcal{X}/R)$. Then $(T_1 = \cdots = T_m = 0)$ define T in a neighborhood of $[f]$.

Thus, if for instance $R = \mathbb{A}^1_{\mathbb{C}}$, we have deformations over an Euclidean neighborhood of 0 in $\mathbb{A}^1_{\mathbb{C}}$, realized as an open subset (in the Euclidean topology) of some $T \to \mathbb{A}^1_{\mathbb{C}}$. However, there may be no Zariski neighborhood of 0 in $\mathbb{A}^1_{\mathbb{C}}$ over which a nontrivial deformation exists.

Application 17 (Moving the union of free rational curves). Let X be a smooth variety over an algebraically closed field, and suppose we have two rational curves D_1 and D_2 on X such that $D_1 \cap D_2 \neq \varnothing$. When can we deform $D_1 \cup D_2$ into a rational curve?

Let $R = \mathbb{A}^1$, $\mathcal{X} = X \times \mathbb{A}^1$, and $\mathcal{Y}$ be the blow up of $\mathbb{P}^1 \times \mathbb{A}^1$ at some point on the fiber over $0 \in \mathbb{A}^1$. Then $C = \mathcal{Y}_0$ is the union of two rational curves C_1 and C_2, meeting transversely in a point P, while all other fibers are smooth rational curves. We view $D_1 \cup D_2$ as the image of a morphism $f_0 = f_1 \vee f_2 : C \to \mathcal{X}_0$. (Notice that we do not require that D_1 and D_2 meet transversely, nor that they intersect in a single point. If they meet in more than one point, we choose one of them to be the image of P.)

By Theorem 15, $D_1 \cup D_2$ can be deformed into a rational curve provided that $H^1(C, (f_1 \vee f_2)^* T_{\mathcal{X}_0}) = 0$. We claim that this is the case whenever f_1 and f_2 are free.

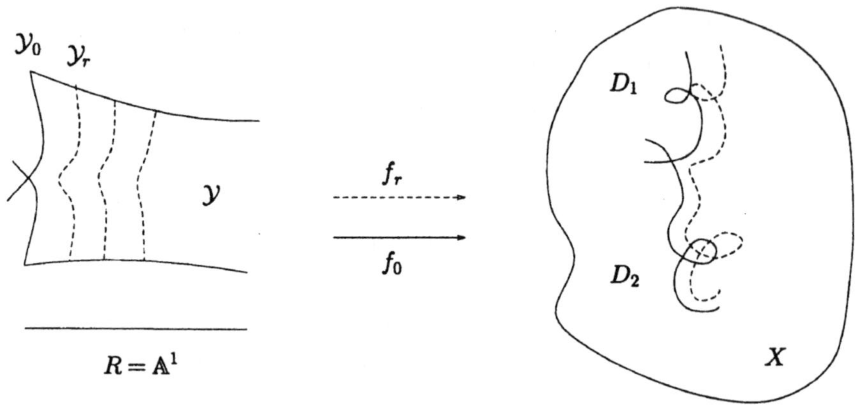

Fig. 2. Deforming the union of 2 free rational curves

Indeed, by taking $E = (f_1 \vee f_2)^* T_{\mathcal{X}_0}$ in exercise 18, we see that if both f_1 and f_2 are free, then $H^1\big(C, (f_1 \vee f_2)^* T_{\mathcal{X}_0}\big) = 0$, and hence $D_1 \cup D_2$ can be deformed into a rational curve. We actually have more than that: by taking $E = (f_1 \vee f_2)^* T_{\mathcal{X}_0}(-Q)$, where Q is any point in $C_1 \setminus C_2$, we see that $H^1\big(C, (f_1 \vee f_2)^* T_{\mathcal{X}_0}(-Q)\big) = 0$. By the Semicontinuity Theorem, $H^1\big(\mathbb{P}^1, f_r^* T_X(-1)\big) = 0$ for nearby deformations $f_r : \mathbb{P}^1 \to X$. Hence:

The union of two free rational curves with nonempty intersection can be deformed into a free rational curve.

Exercise 18. Let $C = \cup_{i=0}^n C_i$ be a reduced projective curve of genus zero (i.e., with $h^1(C, \mathcal{O}_C) = 0$) and E a vector bundle on C. Assume that

1. $H^1\big(C_i, E|_{C_i}(-1)\big) = 0$ for $i = 1, \ldots, n$, and
2. $H^1\big(C_0, E|_{C_0}\big) = 0$.

Then $H^1(C, E) = 0$.

Application 19 (Real Varieties). Let X be a smooth variety defined over $\mathbb{R}$, and assume that $X_{\mathbb{C}}$ contains free rational curves (defined over $\mathbb{C}$). We investigate the existence of free rational curves (defined over $\mathbb{R}$) on $X = X_{\mathbb{R}}$.

Assume there exists a point $P \in X(\mathbb{R})$ and a free rational curve $C \subset X_{\mathbb{C}}$ passing through P. If C is defined over $\mathbb{R}$, we are done. Otherwise, let $\bar{C}$ be its conjugate curve. Then $\bar{C}$ is also a free rational curve, and their union is defined over $\mathbb{R}$.

Set $R = \mathbb{A}^1_{\mathbb{R}}$, $\mathcal{X} = X_{\mathbb{R}} \times \mathbb{A}^1_{\mathbb{R}}$ and $\mathcal{Y} = (x_0^2 + x_1^2 = tx_2^2) \subset \mathbb{P}^2_{\mathbb{R}} \times \mathbb{A}^1_{\mathbb{R}}$. Then $\mathcal{Y}_0$ is the union of two conjugate rational curves meeting transversely in one point, while all other real fibers are smooth rational curves defined over $\mathbb{R}$. The curve $\mathcal{Y}_t$ contains $\mathbb{R}$-points for $t > 0$, while $\mathcal{Y}_t(\mathbb{R}) = \varnothing$ for $t < 0$. We view $C \cup \bar{C}$ as the image of a morphism $f_0 : \mathcal{Y}_0 \to \mathcal{X}_0$.

We have $H^1(\mathcal{Y}_0, f_0^* T_{\mathcal{X}_0}) = 0$. Let $\mathcal{Z} = \operatorname{Hom}(\mathcal{Y}/R, \mathcal{X}/R)$. By Theorem 15, $\mathcal{Z} \to R$ is smooth at $[f_0]$, and nearby fibers have dimension $h^0(\mathcal{Y}_0, f_0^* T_{\mathcal{X}_0}) > 0$. Now comes a subtle point: it is possible that nearby fibers $\mathcal{Z}_r$ have no $\mathbb{R}$-points. We must show that this is not the case. As in Warning 16, let $T \subset \mathcal{Z}$ be a curve through $[f_0]$ such that $T \to R$ is smooth at $[f_0]$. We apply the Inverse Function Theorem for real functions, and conclude that $T(\mathbb{R}) \to \mathbb{R}^1$ is a local homeomorphism at $[f_0]$ (in the Euclidean topology). Thus we obtain many $\mathbb{R}$-points in a neighborhood of $[f_0]$ in T. These points correspond to morphisms $f_t : \mathcal{Y}_t \to X \times \{t\}$ that are defined over $\mathbb{R}$.

Notice that we need that $t > 0$ in order to get a curve on $X_{\mathbb{R}}$ containing real points. The fact that $T \to R$ is étale ensures that we can pick such t.

By taking t near 0, we get that f_t is free.

Remarks 20.

1. The argument above requires the existence of a free rational curve $C \subset X_{\mathbb{C}}$ through a real point. When $X_{\mathbb{C}}$ contains a free rational curve, some dense open subset of $X_{\mathbb{C}}$ is covered by free rational curves. The set of real points, if nonempty, is Zariski dense, and so we can start the argument.
   Problems arise when X has no real points. One might suppose that in this case the question of existence of free rational curves on $X_{\mathbb{R}}$ does not even make sense. There can be no real maps $\mathbb{P}^1 \to X$ if X has no real points. There is, however, a quite interesting variant.
   Even without real points, X may contain empty real conics (i.e., curves obtained from morphisms $\mathbb{P}^2_{\mathbb{R}} \supset (x_0^2 + x_1^2 + x_2^2 = 0) \to X_{\mathbb{R}}$ defined over $\mathbb{R}$). If $X(\mathbb{R}) \neq \varnothing$, the above deformations corresponding to $t < 0$ are such. If $X(\mathbb{R}) = \varnothing$, the problem is open.
2. The argument in Application 19 can be generalized to show that if X is a variety defined over a p-adic field k, and $X_{\bar{k}}$ contains a free rational curve passing through a k-point x, then X_k contains free rational curves (defined over k) through x (see [14]). This is because a p-adic field also admits an Inverse Function Theorem. We return to this in Section 9.

5. Deforming Nonfree Rational Curves I

In Section 3 we studied deformations of free rational curves, and in Section 4 we saw how to deform unions of free rational curves into irreducible ones. Now we are interested in the following problem.

How can we deform a nonfree rational curve on X
when X contains many free rational curves?

One possible way of dealing with this problem was introduced in [17, 1.1.6]. In this section we explain this method. In the next section we describe a different approach to the same problem. The main trick is to use combs, which we define now.

Definition 21. Let k be an arbitrary field. A *comb* with *n teeth* over k is a projective curve with $n+1$ irreducible components $C_0, C_1, \ldots, C_n$ over $\bar{k}$ satisfying the following conditions:

1. The curve C_0 is defined over k.
2. The union $C_1 \cup \cdots \cup C_n$ is defined over k. (Each individual curve may not be defined over k.)
3. The curves $C_1, \ldots, C_n$ are smooth rational curves disjoint from each other, and each of them meets C_0 transversely in a single smooth point of C_0 (which may not be defined over k).

The curve C_0 is called the *handle* of the comb, and $C_1, \ldots, C_n$ are called the *teeth*. A *rational comb* is a comb whose handle is a smooth rational curve. A comb can be pictured as in Fig. 3.

In Section 9 we study combs over nonclosed fields in more detail. For the rest of this section we work over a fixed algebraically closed field $k = \bar{k}$ of arbitrary characteristic.

We start with an arbitrary rational curve D on a smooth quasi-projective variety X over k. Since D is not free, it is possible that we cannot move D on X. Assume that there are several free rational curves $D_1, \ldots, D_p \subset X$ meeting D in distinct points $d_i \in D$. Then there is a chance that we can deform their union $D \cup D_1 \cup \cdots \cup D_p$ into a rational curve.

Set $R = \mathbb{A}^p$, with affine coordinates y_i, and $\mathcal{X} = X \times \mathbb{A}^p$. Fix a birational morphism $\mathbb{P}^1 \to D$ and distinct points $z_1, \ldots, z_p \in \mathbb{P}^1$ mapping to $d_1, \ldots, d_p$. Let $\mathcal{Y}$ be the blowup of $\mathbb{P}^1 \times \mathbb{A}^p$ along the subvarieties

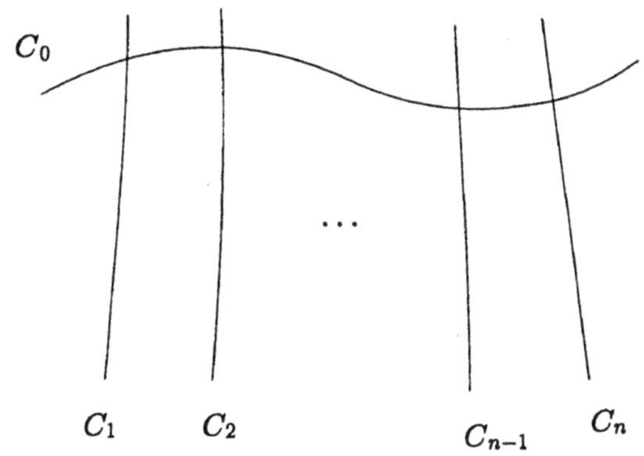

Fig. 3. Comb with n teeth

$\{z_i\} \times \{y_i = 0\}$, $i = 1, \ldots, p$. Then $\mathcal{Y}_0$ is a rational comb with handle C_0 and teeth $C_1, \ldots, C_p$. The general fiber is a smooth rational curve, but there are also fibers $\mathcal{Y}_r$, $r \in \mathbb{A}^p$ that are subcombs of $C_0 \cup C_1 \cup \cdots \cup C_p$. The number of teeth in this subcomb is precisely the number of coordinates of r that equal 0. We view $D \cup D_1 \cup \cdots \cup D_p$ as the image of a morphism $f_0 : \mathcal{Y}_0 \to \mathcal{X}_0$ mapping C_0 birationally onto D and each C_i birationally onto D_i, $1 \leq i \leq p$.

Notice that $\operatorname{Hom}(\mathcal{Y}_0, X)$ only gives us reducible or nonreduced curves on X. However, if $\dim_{[f_0]} \operatorname{Hom}(\mathcal{Y}/R, \mathcal{X}/R) > \dim_{[f_0]} \operatorname{Hom}(\mathcal{Y}_0, X)$, then we can deform some subcurve $D \cup D_{i_1} \cup \cdots \cup D_{i_l}$ of $D \cup D_1 \cup \cdots \cup D_p$ into a rational curve on X. We show that this is the case if p is large enough.

Remark 22. One limitation of this method is that we only know that *some* subcurve of $D \cup D_1 \cup \cdots \cup D_p$ deforms into an irreducible rational curve. In general, we do not even know the number of irreducible components of such a subcurve.

By Theorem 15 and Riemann–Roch, we have

$$\text{(5.1)}\quad \dim_{[f_0]} \operatorname{Hom}(\mathcal{Y}/R, \mathcal{X}/R) \geq -\big((D \cup D_1 \cup \cdots \cup D_p) \cdot K_X\big) + \dim X + p.$$

Now consider the restriction morphism $\operatorname{Hom}(\mathcal{Y}_0, X) \to \operatorname{Hom}(\mathbb{P}^1, X)$ sending $[h]$ to $\left[h|_{C_0}\right]$. The fiber over $[f] \in \operatorname{Hom}(\mathbb{P}^1, X)$ is isomorphic to

$$\prod_{i=1}^{p} \operatorname{Hom}\left(\mathbb{P}^1, X, 0 \mapsto f(z_i)\right).$$

By Theorem 5, when $[f] \in \operatorname{Hom}(\mathbb{P}^1, X, 0 \mapsto x)$ is free,

$$\dim_{[f]} \operatorname{Hom}(\mathbb{P}^1, X) = -\left(f_*[\mathbb{P}^1] \cdot K_X\right) + \dim X.$$

Since we have to impose $\dim X$ conditions in order to fix the image of the point $0 \in \mathbb{P}^1$, we get

$$\dim_{[f]} \operatorname{Hom}(\mathbb{P}^1, X, 0 \mapsto x) = \dim_{[f]} \operatorname{Hom}(\mathbb{P}^1, X) - \dim X = -\left(f_*[\mathbb{P}^1] \cdot K_X\right).$$

Therefore we obtain that

$$(5.2) \quad \dim_{[f_0]} \operatorname{Hom}(\mathcal{Y}_0, X) \leq \dim_{[f_0|C_0]} \operatorname{Hom}(\mathbb{P}^1, X) + \\ -\left((D \cup D_1 \cup \cdots \cup D_p) \cdot K_X\right) + (D \cdot K_X).$$

By inequalities (5.1) and (5.2), we just need to take

$$p > \dim_{[f_0|C_0]} \operatorname{Hom}(\mathbb{P}^1, X) - \dim X + (D \cdot K_X)$$

in order to have $\dim_{[f_0]} \operatorname{Hom}(\mathcal{Y}/R, \mathcal{X}/R) > \dim_{[f_0]} \operatorname{Hom}(\mathcal{Y}_0, X)$.

We have proved:

Theorem 23. *Let X be a smooth projective variety over an algebraically closed field. Let D be a rational curve on X (possibly nonfree), and represent it by a morphism $f : \mathbb{P}^1 \to X$. Assume there are p free rational curves $D_1, \ldots, D_p$ meeting D in distinct points, with $p > \dim_{[f]} \operatorname{Hom}(\mathbb{P}^1, X) - \dim X + (D \cdot K_X)$. Then there is an integer l, $1 \leq l \leq p$, and distinct indices $\{i_1, \ldots i_l\} \subset \{1, \ldots, p\}$ such that $D \cup D_{i_1} \cup \cdots \cup D_{i_l}$ can be deformed into a rational curve.* ∎

Remark 24. Given finitely many points on D, we can actually deform some curve $D \cup D_1 \cup \cdots \cup D_m$ keeping these points fixed. In this case, however, we may need to attach more free rational curves to D before we deform it.

We leave it to the reader to work out the necessary small modifications when D is not a rational curve.

In Section 7, we will see some applications of this theorem.

6. Deforming Nonfree Rational Curves II

Let X be a smooth projective variety over an algebraically closed field, and assume X has very free rational curves. Let $D \subset X$ be a curve.

In the last section we explained the approach of [17] to the problem of moving a curve D on X. First we attach several free rational curves $D_1, \ldots, D_p$ to D, and represent $D \cup D_1 \cup \cdots \cup D_p$ as the image of a morphism $f : C \to X$, where C is a comb with p teeth. Then we study deformations of f. There are two limitations of this method.

1. As we noted in Remark 22, by taking p large enough, we can only guarantee that some subcurve $D \cup D_{i_1} \cup \cdots \cup D_{i_l}$ deforms into an irreducible curve. We do not know which subcurve it is.
2. If we look at deformations of the morphism f, the obstruction space $H^1(C, f^*T_X)$ usually never vanishes, no matter how large p is.

Recently, Graber, Harris and Starr [5] introduced a variant to this picture which is very useful in some applications, especially in arithmetic questions.

By attaching very free rational curves $D_1, \ldots, D_p$ and by looking at a modified deformation setting (in the Hilbert scheme of X), we can guarantee that, for p large enough, the relevant deformation problem is unobstructed. We devote the rest of this section to explaining this new method.

First of all, instead of looking at the space of morphisms to X, we look at the Hilbert scheme of X. Below we gather some facts about Hilbert schemes. We refer to [6] and [12, I.1 and I.2] for details and proofs.

25 (Hilbert schemes). Let X be a projective scheme over a field k. There is a k-scheme Hilb(X) parametrizing closed subschemes of X. Like for the Hom schemes, this implies that there is a universal subscheme $U \subset \text{Hilb}(X) \times X$, flat over Hilb$(X)$, such that for every field extension K/k and every K-subscheme Z of X_K there is a unique point $[Z] \in \text{Hilb}(X)(K)$ such that the fiber of U over $[Z]$ is exactly Z.

Now assume that X is smooth, and let Z be a closed subscheme of X given by the ideal sheaf $\mathcal{I}_Z$. Assume that Z is a local complete intersection, that is, $\mathcal{I}_Z$ is locally generated by codim(Z, X) elements. Then $\mathcal{I}_Z/\mathcal{I}_Z^2$ is a locally free sheaf on Z and its dual $\text{Hom}(\mathcal{I}_Z/\mathcal{I}_Z^2, \mathcal{O}_Z)$ is the *normal bundle* N_Z of Z in X. We have:

1. $T_{[Z]}\,\text{Hilb}(X) \cong H^0(Z, N_Z)$.

2. If $H^1(Z, N_Z) = 0$, then $\mathrm{Hilb}\,(X)$ is smooth at $[Z]$.
3. In general, $\dim_{[Z]} \mathrm{Hilb}\,(X) \geq h^0(Z, N_Z) - h^1(Z, N_Z)$.

(See [6] or [12, I.2.8].)

Now we go back to our problem.

Let X be a smooth projective variety over an algebraically closed field and $D \subset X$ an irreducible curve. Our goal is to attach rational curves $D_1, \ldots, D_p$ to D so that the resulting comb $D \cup D_1 \cup \cdots \cup D_p$ deforms into an irreducible curve.

The first problem we face is that D may be quite singular, in which case we prefer to work with its normalization $\tilde{D} \to D$. On the other hand, here it is essential that D be embedded in X. To achieve both, we take an embedding $\tilde{D} \hookrightarrow \mathbb{P}^3$ and consider the diagonal map $\delta : \tilde{D} \to X \times \mathbb{P}^3$. This is an embedding with image $D' \cong \tilde{D}$ whose projection to the first factor is D. Any moving of D' can be projected to X to give a deformation of D. For most applications this reduction to the smooth curve D' is satisfactory. There may be, however, delicate geometric questions for which one needs to be more careful.

So assume $D \subset X$ is a smooth irreducible curve. Suppose there are disjoint smooth rational curves $D_1, \ldots, D_p$, each meeting D transversely in a single point. Then we can deform their union $C = D \cup D_1 \cup \cdots \cup D_p$ into an irreducible curve C' provided that the following conditions hold:

1. The sheaf N_C is generated by global sections.
2. $H^1(C, N_C) = 0$.

Indeed, we have an exact sequence

$$0 \to N_D \to N_C|_D \to Q \to 0,$$

where Q is a torsion sheaf supported at the points $P_i = D \cap D_i$. If condition 1 holds, we can find a global section $s \in H^0(C, N_C)$ such that, for each i, the restriction of s to a neighborhood of P_i is not in the image of N_D. This means that s corresponds to a first-order deformation of C that smoothes the nodes P_i of C. If condition 2 holds, there is no obstruction, and we can find a global deformation of C that smoothes its nodes P_i.

Remark 26. In many situations, we would like the resulting irreducible curve C' to have some extra properties. For instance, if we start with a smooth rational curve D, we may want C' to be a free (resp. very free) curve. In this case, by the Semicontinuity Theorem, we can just fix a point $P \in D$ (resp. points $P, Q \in D$) and replace condition (2) above by

2′. $H^1(C, N_C(-P)) = 0$ (resp. $H^1(C, N_C(-P-Q)) = 0$).

Theorem 27. *Let X be a smooth projective variety of dimension at least 3 over an algebraically closed field. Let $D \subset X$ be a smooth irreducible curve and M a line bundle on D. Let $C \subset X$ be a very free rational curve intersecting D and let $\mathcal{C}$ be a family of rational curves on X parametrized by a neighborhood of $[C]$ in* Hom $(\mathbb{P}^1, X)$.

Then there are curves $C_1, \dots, C_p \in \mathcal{C}$ such that $D^* = D \cup C_1 \cup \dots \cup C_p$ is a comb and satisfies the following conditions:

1. *The sheaf N_{D^*} is generated by global sections.*
2. *$H^1(D^*, N_{D^*} \otimes M^*) = 0$, where M^* is the unique line bundle on D^* that extends M and has degree 0 on the C_i.*

We have a lot of freedom in choosing curves from $\mathcal{C}$. Namely, as in Remark 9, for a general point $P \in D$ and for a general tangent direction $\ell \subset T_{P,X}$ there is a curve $C = C(P, \ell) \in \mathcal{C}$ passing through P and having tangent direction ℓ at P. Moreover, given a 1-dimensional subscheme $B \subset X$ we can also assume that C intersects $D \cup B$ only at P. (This is the only point where we need $\dim X \geq 3$.) By looking at the proof, we see that it is sufficient that there is a curve in $\mathcal{C}$ having a general normal direction $\bar{\ell} \subset N_{P,D}$ at P (instead of a general tangent direction $\ell \subset T_{P,X}$). This is very useful in the proof of Theorem 32.

For arithmetic applications, one has to be even more careful with the choice of the teeth C_i (see [16]).

Proof. *Step 1.* Here we produce a comb $D' \subset X$ with handle D such that $H^1(D, N_{D'}|_D \otimes M) = 0$.

Set $D_0 = D$. Suppose we have a comb $D_i = D \cup C_1 \cup \dots \cup C_i$ such that $d_i = h^1(D, N_{D_i}|_D \otimes M) \neq 0$. We show how to choose a curve $C_{i+1} \in \mathcal{C}$, disjoint from $C_1, \dots, C_i$, forming a comb $D_{i+1} = D_i \cup C_{i+1}$ so that $d_{i+1} < d_i$. Then we just repeat this procedure finitely many times until we reach $d_p = 0$.

By Serre duality, $H^1(D, N_{D_i}|_D \otimes M)$ is dual to

$$\operatorname{Hom}\left(\omega_D^{-1} \otimes M, (\mathcal{I}_{D_i}/\mathcal{I}_{D_i}^2)\big|_D\right),$$

so we would like to understand how $(\mathcal{I}_{D_i}/\mathcal{I}_{D_i}^2)\big|_D$ changes with i.

Pick any $C_{i+1} \in \mathcal{C}$ disjoint from $C_1, \ldots, C_i$ and intersecting D transversely in a single point P. The tangent direction of C_{i+1} at P corresponds to a normal direction $l_{C_{i+1}} \subset N_{P,D}$. By explicit computation we get an exact sequence

$$0 \to (\mathcal{I}_{D_{i+1}}/\mathcal{I}^2_{D_{i+1}})|_D \to (\mathcal{I}_{D_i}/\mathcal{I}^2_{D_i})|_D \to l^\vee_{C_{i+1}} \to 0,$$

where the last map is the composition of the restriction map $(\mathcal{I}_{D_i}/\mathcal{I}^2_{D_i})|_D \to N^\vee_{P,D}$ with the dual map $N^\vee_{P,D} \to l^\vee_{C_{i+1}}$.

In particular, we get an inclusion of groups

$$\operatorname{Hom}\left(\omega_D^{-1} \otimes M, (\mathcal{I}_{D_{i+1}}/\mathcal{I}^2_{D_{i+1}})|_D\right) \hookrightarrow \operatorname{Hom}\left(\omega_D^{-1} \otimes M, (\mathcal{I}_{D_i}/\mathcal{I}^2_{D_i})|_D\right).$$

So we always have $d_{i+1} \le d_i$.

Since $d_i \neq 0$, there is a nonzero element

$$\phi \in \operatorname{Hom}\left(\omega_D^{-1} \otimes M, (\mathcal{I}_{D_i}/\mathcal{I}^2_{D_i})|_D\right).$$

We are done if we can choose C_{i+1} so that ϕ does not come from an element $\tilde{\phi}$ of $\operatorname{Hom}\left(\omega_D^{-1} \otimes M, (\mathcal{I}_{D_{i+1}}/\mathcal{I}^2_{D_{i+1}})|_D\right)$.

For each $P \in D$, let ϕ_P denote the composition of ϕ with the restriction map $(\mathcal{I}_{D_i}/\mathcal{I}^2_{D_i})|_D \to N^\vee_{P,D}$. Since ϕ is nonzero, there is a dense open subset $V \subset D$ such that ϕ_P has rank 1 for every $P \in V$. Let $\xi_P \in N^\vee_{P,D}$ span the image of ϕ_P.

Choose a general point $P \in V$ not in $C_1 \cup \cdots \cup C_i$ and a curve $C_{i+1} \in \mathcal{C}$ through P such that

1. C_{i+1} intersects $D \cup C_1 \cup \cdots \cup C_i$ only at P, and
2. The projection of the tangent direction of C_{i+1} at P to $N_{P,D}$, denoted by $l_{C_{i+1}}$, is not contained in the kernel of $\xi_P : N_{P,D} \to k$.

It follows from our exact sequence that $\phi \in \operatorname{Hom}\left(\omega_D^{-1} \otimes M, (\mathcal{I}_{D_i}/\mathcal{I}^2_{D_i})|_D\right)$ comes from an element $\tilde{\phi}$ of $\operatorname{Hom}\left(\omega_D^{-1} \otimes M, (\mathcal{I}_{D_{i+1}}/\mathcal{I}^2_{D_{i+1}})|_D\right)$ if and only if $\xi_P|_{l_{C_{i+1}}} = 0$. This does not hold by the choice of C_{i+1}, and we are done.

Notice that, once we have a comb D' such that $H^1\big(D, N_{D'}|_D \otimes M\big) = 0$, any other comb D^* obtained from D' by adding extra teeth also satisfies $H^1\big(D, N_{D^*}|_D \otimes M\big) = 0$.

Step 2. Now we extend the comb D' obtained in Step 1 to a comb D^* so that $N_{D^*}|_D$ is generated by global sections.

Let D' be the comb obtained in Step 1. Let M' be any line bundle on D with degree $-(g+1)$, where g is the genus of the curve D. We use the method described above to attach more teeth to D' so that the resulting comb D^* also satisfies $H^1(D, N_{D^*}|_D \otimes M') = 0$. By Riemann–Roch, M' embeds in $\mathcal{O}_D(-P)$ for every $P \in D$. So we get that $H^1(D, N_{D^*}|_D(-P)) = 0$ for every $P \in D$, and this implies global generation of $N_{D^*}|_D$.

Step 3. Finally, we use the fact that the C_i are free to show that N_{D^*} is generated by global sections and that $H^1(D^*, N_{D^*} \otimes M^*) = 0$.

Let $D^* = D \cup C_1 \cup \cdots \cup C_p$ be the comb obtained in Step 2, so that $N_{D^*}|_D$ is generated by global sections and $H^1(D, N_{D^*}|_D \otimes M) = 0$.

For each i we have an exact sequence

$$0 \to N_{C_i} \to N_{D^*}|_{C_i} \to Q_i \to 0,$$

where Q_i is a torsion sheaf supported at $P_i = D \cap C_i$. Since C_i is free, $H^1(C_i, N_{C_i}(-P_i)) = 0$, and hence $H^1(C_i, N_{D^*}|_{C_i}(-P_i)) = 0$.

We also have the exact sequence

$$0 \to \sum_{i=1}^{p} N_{D^*}|_{C_i}(-P_i) \to N_{D^*} \to N_{D^*}|_D \to 0.$$

Therefore, the vanishing of $H^1(D^*, N_{D^*} \otimes M^*)$ follows from the vanishing of $H^1(D, N_{D^*}|_D \otimes M)$ and $H^1(C_i, N_{D^*}|_{C_i}(-P_i))$, $1 \le i \le p$.

Similarly, global generation of N_D^* follows from global generation of $N_{D^*}|_D$ and N_{C_i}, $1 \le i \le p$, plus the vanishing of $H^1(C_i, N_{D^*}|_{C_i}(-P_i))$. ■

7. Rationally Connected Varieties

In this section we introduce the concept of rationally connected varieties. After some motivation, we present possible definitions of the concept and show that they are equivalent. Then we study some properties of the class of rationally connected varieties. This is all done over $\mathbb{C}$ (or any uncountable algebraically closed field of characteristic 0). At the end of the section we consider subtleties that arise in positive characteristic.

28 (Motivation). It is rather clear that rational curves form a distinguished class among all algebraic curves. They are the simplest algebraic curves in many aspects (number theoretical, topological, etc.).

In dimension two, smooth rational surfaces are the best analogs of rational curves. Their class enjoys several nice properties:

1. (Castelnuovo's Criterion) A smooth projective surface S over $\mathbb{C}$ is rational if and only if $H^0(S, \omega_S^{\otimes 2}) = H^1(S, \mathcal{O}_S) = 0$.
2. (Deformation Invariance) Let $S \to T$ be a family of smooth projective surfaces over $\mathbb{C}$, T connected. If S_0 is rational, any S_t is rational.
3. (Noether's Theorem) Let S be a smooth projective surface. If there is a morphism $S \to \mathbb{P}^1$ whose general fibers are rational curves, then S itself is rational.

When we move to higher dimensions, and try to classify the "simplest" algebraic varieties, things get much more complicated. As a first try, one could take rational varieties as higher dimensional analogs of rational curves and rational surfaces. However, there are higher dimensional varieties that behave like rational varieties but are not rational. "Behaving like rational varieties" should be understood as having the relevant properties of rational varieties. Smooth cubic 3-folds, for instance, behave like rational varieties in many aspects, but they are not rational (see [3, Theorem 13.12]). Moreover, the analogs of properties 1–3 above all seem to fail for rational varieties of dimension 3 and up. Smooth cubic 3-folds satisfy $H^i(S, \mathcal{O}_S) = H^0(S, \omega_S^{\otimes m}) = 0$ for every $i, m \geq 1$. There are examples of conic bundles over $\mathbb{P}^2$ which are not rational. It is conjectured that deformation invariance also fails.

Another possibility would be to consider *unirational varieties.* A variety of dimension n is unirational if there is a dominant map $\mathbb{P}^n \dashrightarrow X$. However, the class of unirational varieties still seems to be too small. It is conjectured that deformation invariance fails for the class of unirational varieties.

A new concept was proposed in [17], namely the concept of *rationally connected varieties.* The definition was based on the fact that $\mathbb{P}^n$ has many rational curves, and on the expectation that a variety X behaves like $\mathbb{P}^n$ if and only if it has plenty of rational curves. The precise notion is given in Definition-Theorem 29 below.

It seems rather clear by now that in characteristic 0 rationally connected varieties are the right higher dimensional analogs of rational curves and rational surfaces. This class of varieties enjoys many nice properties:

0. (Birational invariance) The class of rationally connected varieties is closed under birational equivalence.
1. (Castelnuovo's Criterion) A smooth projective rationally connected variety X satisfies

$$H^0(X, (\Omega_X^1)^{\otimes m}) = 0 \quad for \ \ every \ \ m \geq 1,$$

and the converse is conjectured to hold. (This is Theorem 30.)
2. (Deformation Invariance) Being rationally connected is deformation invariant for smooth projective varieties over. (This is Theorem 31.)
3. (Noether's Theorem) If X is a smooth variety, and there is a dominant morphism $X \to Y$ to a rationally connected variety Y such that the general fibers are rationally connected, then X itself is rationally connected. (This is Corollary 33.)

The class of rationally connected varieties contains the class of unirational varieties. It should be said, however, that it is still an open problem to show that these classes are in fact distinct. That is, there is not a single rationally connected variety over $\mathbb{C}$ that is known not to be unirational. Potential candidates abound, for instance general quartics in $\mathbb{P}^4$, any smooth hypersurface of degree n in $\mathbb{P}^n$ for $n \geq 5$, smooth hypersurfaces of bidegree $(n, 2)$ in $\mathbb{P}^k \times \mathbb{P}^2$ for $n \gg k \geq 2$.

Definition-Theorem 29. Let X be a smooth projective variety over $\mathbb{C}$ (or any uncountable algebraically closed field of characteristic 0). We say that X is *rationally connected* if it satisfies the following equivalent conditions.

(1) There is a dense open set $X^0 \subset X$ such that, for every $x_1, x_2 \in X^0$, there is a chain of rational curves connecting x_1 and x_2.

(2) There is a dense open set $X^0 \subset X$ such that, for every $x_1, x_2 \in X^0$, there is a rational curve through x_1 and x_2.

(3) For every $x_1, x_2 \in X$, there is a rational curve through x_1 and x_2.

(4) For every integer $m > 0$, and every $x_1, \ldots, x_m \in X$, there is a rational curve through $x_1, \ldots, x_m$.

(5) For every integer $m > 0$, and every $x_1, \ldots, x_m \in X$, there is a free rational curve through $x_1, \ldots, x_m$.

(6) There is a very free morphism $f : \mathbb{P}^1 \to X$, that is,

$$f^* T_X = \sum_{i=1}^{\dim X} \mathcal{O}_{\mathbb{P}^1}(a_i) \quad \text{with} \quad a_i \geq 1.$$

If X is defined over an arbitrary field k of characteristic 0, we say that X is *rationally connected* if X_K satisfies the conditions above for some (or equivalently for all) uncountable algebraically closed field K extending k.

It is possible that the six conditions above are equivalent over any algebraically closed field of characteristic 0, but this is not known for countable fields, cf. Remark 12.

We of course need to prove that the six versions above are equivalent. The proof is somewhat more transparent if we introduce two other variants:

(1^+) For every $x_1, x_2 \in X$, there is a chain of rational curves connecting x_1 and x_2.

(6^+) For every $x \in X$ there is a very free morphism $f : \mathbb{P}^1 \to X$ through x.

The implications (5) $\Rightarrow$ (4) $\Rightarrow$ (3) $\Rightarrow$ (2) $\Rightarrow$ (1) are obvious, and so are (1^+) $\Rightarrow$ (1) and (6^+) $\Rightarrow$ (6).

Assume (1) and let us prove (6). The argument also shows (1^+) $\Rightarrow$ (6^+).

Let X^{free} be the set obtained in Corollary 11, so that any rational curve meeting X^{free} is free. It is the intersection of countably many dense open subsets. We repeatedly use two of its properties over $\mathbb{C}$: X^{free} is dense and for every irreducible subvariety $Z \subset X$, the intersection $X^{\text{free}} \cap Z$ is either empty or dense in Z. Thus, if a curve $C \subset X$ intersects X^{free}, a dense set of its deformations also intersect X^{free}.

Pick a point $x \in X^{\text{free}} \cap X^0$ and let x' be any point in X^0. By hypothesis, there is a chain of rational curves $C_1 \cup \cdots \cup C_n$ connecting x and x'. We show that there is in fact a chain of *free* rational curves connecting them.

Set $x_0 = x$, $x_i = C_i \cap C_{i+1}$ for $0 < i < n$, and $x_n = x'$. We replace each curve C_i by a free rational curve C_i' passing through x_i and x_{i+1}.

The curve C_0 is itself free because it intersects X^{free}, so we set $C_0' = C_0$.

We can deform C_0' to produce many free rational curves $D_1, \ldots, D_k$ meeting C_1. Indeed, view C_0' as the image of a morphism $f_0 : \mathbb{P}^1 \to X$, with $f_0(0) = x_1$. Let U be a neighborhood of $[f_0]$ in $\operatorname{Hom}(\mathbb{P}^1, X)$ parametrizing free morphisms and consider the evaluation map $F_U : \mathbb{P}^1 \times U \to X$. Since f_0 is free, the point $x_1 \in C_0'$ can be moved in any direction, i.e., $F_U(\{0\} \times U)$ contains a neighborhood of x_1 in X (see [12, II.3.5.3] for a rigorous argument). In particular $F_U(\{0\} \times U)$ contains a neighborhood of x_1 in C_1. Since C_0 meets X^{free}, we can choose deformations $D_1, \ldots, D_k$ that intersect C_1 (in distinct points) and also meet X^{free}. By Theorem 23, if we take k large enough, we can deform some subcurve $C_1 \cup D_{i_1} \cup \cdots \cup D_{i_m}$

of $D \cup D_1 \cup \cdots \cup D_k$ into a rational curve C_1', keeping x_1 and x_2 fixed. Since the D_i meet X^{free}, we can take C_1' to meet X^{free} as well. (See Fig. 4.)

We do the same with C_0' replaced with C_1', and C_1 replaced with C_2, and so on.

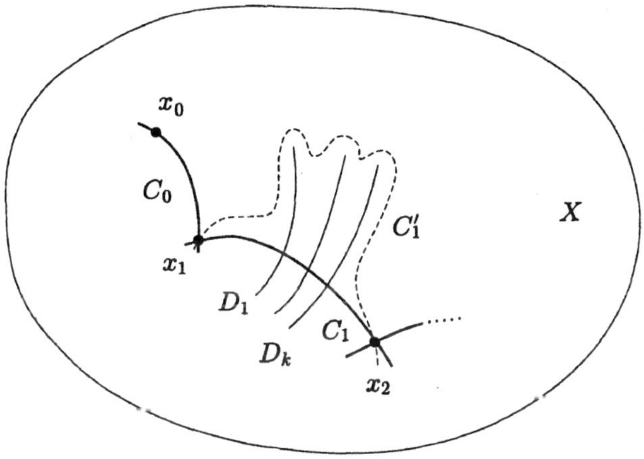

Fig. 4. Obtaining a free rational curve through x_1 and x_2

At the end we get a chain of free rational curves connecting x and x'. As in Application 17, we can deform this chain into a free rational curve, keeping x fixed. (We really would like to deform this chain into a free rational curve, keeping x and x' both fixed, but this is not automatic.)

Now consider the evaluation map $F : \mathbb{P}^1 \times \text{Hom}(\mathbb{P}^1, X, 0 \mapsto x) \to X$ sending $(t, [f])$ to $f(t)$. We have proved that for every $x' \in X^0$ there is an irreducible component $Z = Z(x') \subset \text{Hom}(\mathbb{P}^1, X, 0 \mapsto x)$ such that x' is in the closure of $F(\mathbb{P}^1 \times Z(x'))$.

The scheme $\text{Hom}(\mathbb{P}^1, X, 0 \mapsto x)$ has only countably many irreducible components, hence there is a single irreducible component $Z^* \subset \text{Hom}(\mathbb{P}^1, X, 0 \mapsto x)$ such that $F|_{\mathbb{P}^1 \times Z^*}$ is still dominant. By Proposition 13 we conclude that f is very free for general $[f] \in Z^*$.

Now let us prove (6) $\Rightarrow$ (1$^+$).

Let $f : \mathbb{P}^1 \to X$ be a very free morphism. By Proposition 13, there exists a variety Y of dimension $\dim X - 1$, a point $y \in Y$ and a dominant morphism

$$F : \mathbb{P}^1 \times Y \to X \quad \text{such that} \quad F|_{\mathbb{P}^1 \times \{y\}} = f \quad \text{and} \quad F|_{\{0\} \times Y} = x_0.$$

If x_1 and x_2 are in the image of $F : \mathbb{P}^1 \times Y \to X$, we have a length 2 chain of rational curves connecting x_1 to x_0 and then to x_2. Otherwise, since F is dominant, we still have a 1-parameter family $\{f_t\}$ of deformations of f fixing x_0 that degenerates into a tree of rational curves T_1 connecting x_1 and x_0 (see Fig. 5). Similarly there is a tree of rational curves T_2 connecting x_2 and x_0.

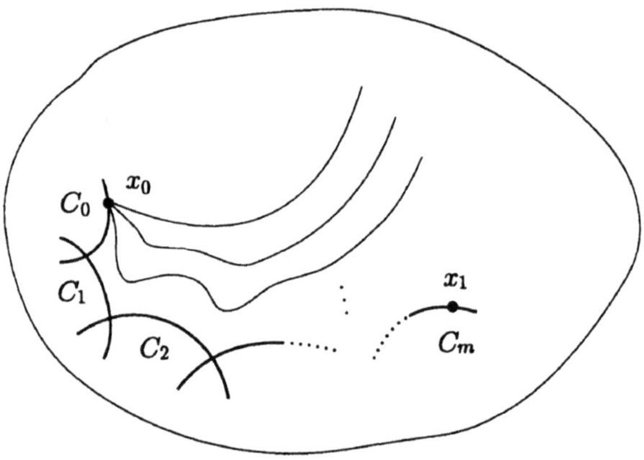

Fig. 5. A rational curve moving with one point fixed, and then degenerating into a union of rational curves.

To see that a 1-dimensional family of rational curves on X can only degenerate into a union of rational curves, we represent this family by a morphism $\phi : S \to X$, where S is a $\mathbb{P}^1$-bundle over a smooth curve T. Let $\bar{T}$ be a smooth compactification of T, and let $\bar{S} \to \bar{T}$ be a $\mathbb{P}^1$-bundle extending $S \to T$. Let $\bar{\phi} : \bar{S} \dashrightarrow X$ be a rational map extending ϕ. If $\bar{\phi}$ is a morphism, then we are done. Otherwise we perform a sequence of blowups of points $\tilde{S} \to \bar{S}$ in order to resolve the indeterminacy of $\bar{\phi}$, and obtain a morphism $\tilde{\phi} : \tilde{S} \to X$ extending ϕ. The result follows from the fact that blowups of points only add copies of $\mathbb{P}^1$ to the fibers of $\bar{S} \to \bar{T}$.

We complete the proof by showing $(6^+) \Rightarrow (5)$.

Assume that for every x_i there is a very free morphism $f_i : \mathbb{P}^1 \to X$ with $f_i(0) = x_i$. By Proposition 13, there are pointed varieties (Y_i, y_i) of dimension $\dim X - 1$, and dominant morphisms

$$F_i : \mathbb{P}^1 \times Y_i \to X \quad \text{such that} \quad F_i|_{\mathbb{P}^1 \times \{y_i\}} = f_i \quad \text{and} \quad F_i|_{\{0\} \times Y_i} = x_i.$$

We may also assume that $F_i|_{\mathbb{P}^1 \times \{y'\}}$ is very free for every $y' \in Y_i$.

Let $X^* \subset X$ be a dense open set contained in all the images $F_i(\mathbb{P}^1 \times (Y_i \setminus \{y_i\}))$. Let $g_0 : C_0 \cong \mathbb{P}^1 \to X$ be a very free morphism intersecting X^* and pick points $c_1, \ldots, c_m \in C_0$ such that $g_0(c_i) \in X^*$. By the choice of X^*, there are $z_i \in Y_i$ such that $F_i((\infty, z_i)) = g_0(c_i)$. Set $g_i := F_i|_{\mathbb{P}^1 \times \{z_i\}}$ (so $g_i(0) = x_i$ and $g_i(\infty) = g_0(c_i)$).

Let $C_0 \cup C_1 \cup \cdots \cup C_m$ be a rational comb such that $C_0 \cap C_i = \{c_i\}$ for $i = 1, \ldots, m$. Choose identifications $C_i \cong \mathbb{P}^1$ such that c_i is identified with ∞. Define a morphism

$$G : C \to X \quad \text{by} \quad G|_{C_i} = g_i,\ 0 \leq i \leq m.$$

Notice that it passes through the points $x_1, \ldots, x_m$.

Like in Application 17, the fact that all the g_i are very free implies that $G : C \to X$ can be deformed into a single very free morphism $\mathbb{P}^1 \to X$ passing through all the x_i. ∎

Now that we have established our definition, we can show that the class of rationally connected varieties enjoys several good properties.

Let us start with one half of Castelnuovo's theorem.

Theorem 30 [17]. *Let X be a smooth, projective, rationally connected variety over $\mathbb{C}$. Then*

$$H^0(X, (\Omega^1_X)^{\otimes m}) = 0 \quad \textit{for all } m \geq 1.$$

Proof. Choose $f : \mathbb{P}^1 \to X$ such that f^*T_X is ample (i.e., f is very free). Then $f^*\Omega^1_X$ is a direct sum of line bundles with negative degree, and so is every tensor power. Thus any section of $(\Omega^1_X)^{\otimes m}$ vanishes along $f(\mathbb{P}^1)$. Since X is covered by very free rational curves (this is characterization (6^+) of Definition-Theorem 29), we cannot have a nonzero section. ∎

It would be really interesting to know that the converse also holds, but this is known only in dimension 3 (see [17, 3.2]).

Theorem 31 [17]. *Let X be a variety over $\mathbb{C}$, S a connected $\mathbb{C}$-scheme and $p : X \to S$ a smooth projective morphism. If a fiber X_s is rationally connected for some $s \in S$, then every fiber of p is rationally connected.*

This is again not hard to prove. We can use characterization (6) of Definition-Theorem 29 to prove that being rationally connected is an open condition. Then we use characterization (1^+) to prove that it is also a closed condition. For a complete proof we refer to [17, 2.4] or [12, IV.3.11].

The analog of Noether's theorem was recently established by Graber, Harris and Starr.

Theorem 32 [5]. *Let X be a smooth projective variety over $\mathbb{C}$ and $f : X \to \mathbb{P}^1$ a morphism. If the general fiber of f is rationally connected, f has a section.* ∎

As a consequence of Theorem 32 we have:

Corollary 33. *Let $f : X \to Y$ be a dominant morphism of smooth projective varieties over $\mathbb{C}$. If Y and the general fiber of f are rationally connected, so is X.*

Proof. Let $x_1, x_2 \in X$ be two points such that f is smooth with rationally connected fibers over $f(x_1)$ and $f(x_2)$. We show that they can be connected by a chain of rational curves. Since Y is rationally connected, there is a rational curve joining $f(x_1)$ and $f(x_2)$. Represent it by a morphism $\mathbb{P}^1 \to Y$ birational onto its image, let Z' be the unique irreducible component of $\mathbb{P}^1 \times_Y X$ that dominates $\mathbb{P}^1$ and let $Z \to Z'$ be a resolution of singularities. The fibers of $Z' \to \mathbb{P}^1$ containing x_1, x_2 are already smooth, so we do not need to change them to get Z. Let $f_Z : Z \to \mathbb{P}^1$ denote the projection.

By Theorem 32, f_Z has a section $D \subset Z$. Let $C \subset X$ be the projection of D to X and x_1' (resp. x_2') be the intersection point of C with the fiber $X_{f(x_1)}$ (resp. $X_{f(x_2)}$). By assumption, there are rational curves $C_1 \subset X_{f(x_1)}$ (resp. $C_2 \subset X_{f(x_2)}$) joining x_1 and x_1' (resp. x_2 and x_2'). Then $C_1 \cup C \cup C_2$ is a chain of rational curves through x_1 and x_2 (see Fig. 6). ∎

34 (Rational connectedness in positive characteristic). In characteristic p, the conditions of Definition-Theorem 29 are not equivalent, and we have two distinct concepts, rationally connected varieties, and *separably rationally connected varieties*, which we define below.

Let X be a smooth projective variety defined over an arbitrary uncountable algebraically closed field. The following two conditions are equivalent.

1. Through two general points of X there passes a rational curve.
2. There exists a variety Z and a morphism $U = Z \times \mathbb{P}^1 \to X$ such that the induced morphism $U \times_Z U \to X \times X$ is dominant. (A point $(x_1, x_2) \in X \times X$ is in the image of this morphism if and only if there is a point $z \in Z$ such that the image of $\{z\} \times \mathbb{P}^1$ in X contains x_1 and x_2.)

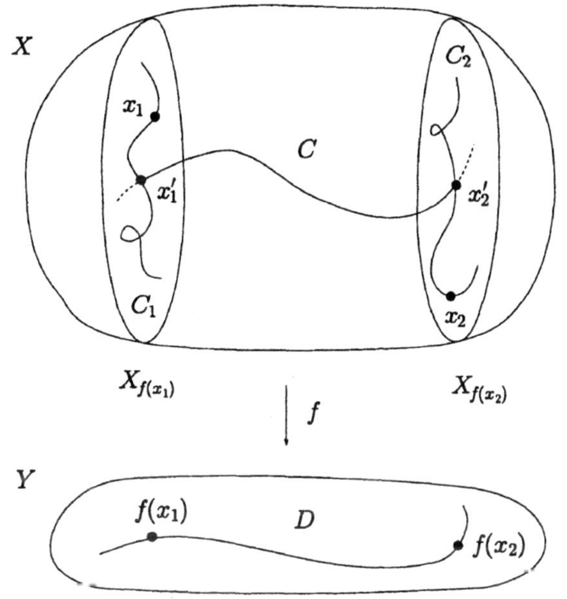

Fig. 6. Joining x_1 and x_2 by a chain of rational curves

In positive characteristic these conditions do not imply the existence of a very free morphism $f : \mathbb{P}^1 \to X$.

It turns out that the following two conditions are equivalent.

3. There is a morphism $f : \mathbb{P}^1 \to X$ such that f^*T_X is ample.
4. There exists a variety Z and a morphism $U = Z \times \mathbb{P}^1 \to X$ such that the induced morphism $U \times_Z U \to X \times X$ is dominant and smooth at the generic point.

This motivates the following definition.

Definition 35. Let X be a smooth projective variety over an arbitrary field k. We say that X is *separably rationally connected* if there is a morphism (defined over $\bar{k}$) $f : \mathbb{P}^1_{\bar{k}} \to X_{\bar{k}}$ such that f^*T_X is ample.

The condition above implies conditions (1) through (6) and also conditions (1^+) and (6^+) in Definition-Theorem 29.

Over fields of characteristic 0, the notions of rationally connected and separably rationally connected varieties coincide. However, there are examples of varieties in positive characteristic satisfying condition 1 above but not 3 (see for instance [12, V.5.19]).

We study separably rationally connected varieties over local fields in Section 9.

8. Genus Zero Stable Curves

In the last sections we saw that it is very useful to be able to deform reducible curves into irreducible ones. Frequently, we also need to degenerate rational curves so that they become reducible. For all of these processes it is very convenient to have a good theory that encompasses all genus zero curves in one object.

One possibility is to work with pairs (C, g), where C is a connected and possibly reducible curve of genus zero (that is, with $h^1(C, \mathcal{O}_C) = 0$) and $g : C \to X$ is a finite morphism. Geometrically these objects are easy to visualize. A technical problem arises, however, when C has $m \geq 3$ irreducible components meeting in one point. Here locally C looks like the m coordinate axes in $\mathbb{A}^m$. There are at least two problems.

1. The dualizing sheaf ω_C is not locally free, making duality less useful.
2. The deformation theory of this object is rather complicated.

[19] observed that it is better to insist that C have only nodes. The price we pay is that we have to allow $g : C \to X$ to map some irreducible components of C to points. This also means that even a reasonably nice curve on X does not correspond to a unique pair (C, g).

Example 36. First let $D \subset \mathbb{P}^3$ consist of 3 lines D_1, D_2, D_3 through a point P with independent directions. Then $h^1(D, \mathcal{O}_D) = 0$. The way we represent this object is by a morphism $g : C \to D$, where C is a rational comb with handle C_0 and teeth C_1, C_2, C_3. The morphism g maps C_i isomorphically onto D_i for $i = 1, 2, 3$ and contracts C_0 to P. We have 3 nodes on C_0.

Next consider the case when $D \subset \mathbb{P}^4$ consists of 4 lines $D_1, \ldots, D_4$ through a point P with independent directions. Again $h^1(D, \mathcal{O}_D) = 0$. As before, we can represent this object by a morphism $g : C \to D$, where C is a rational comb with handle C_0 and teeth C_1, C_2, C_3, C_4. The morphism g maps C_i isomorphically onto D_i for $i = 1, 2, 3, 4$ and contracts C_0 to P. This time we have 4 nodes on C_0, and these four points have moduli, the cross ratio.

We also have 3 degenerate cases when 2 of the nodes on C_0 come together. We represent them by replacing C_0 with two rational curves C_0' and C_0'' meeting transversely in a single point. The curves C_1, C_2, C_3, C_4 are still disjoint from each other. Two of them meet C_0' transversely and do not meet C_0'', while the others meet C_0'' transversely and do not meet C_0'. Each C_i maps isomorphically onto D_i, while C_0' and C_0'' are contracted to P (see Fig. 7).

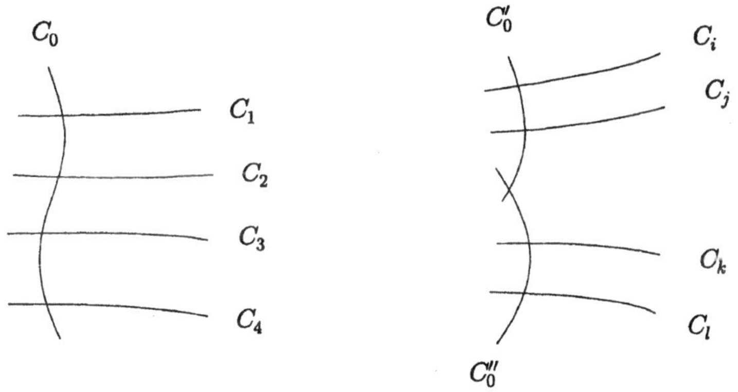

Fig. 7. Stable curves representing 4 lines through a point

Next we define stable curves. Then we give some basic properties of the space parametrizing such objects.

Definition 37. Let X be a scheme over a field k. A *stable curve over* X *(defined over* k*)* is a pair (C, f) where

1. C is a proper connected curve defined over k having only nodes,
2. $f : C \to X$ is a k-morphism, and
3. C has only finitely many automorphisms that commute with f. Equivalently, the following 2 situations are excluded:
   (a) The curve C has an irreducible component $C' \cong \mathbb{P}^1$ that contains at most 2 nodes of C and $f(C')$ is a point.
   (b) The curve C is a smooth elliptic curve and $f(C)$ is a point.

When X is a point, this is the Deligne-Mumford definition of a stable curve [4].

We also need to study cases when all the curves have marked points or they pass through certain points of X. Over an algebraically closed field these are just points, but in the nonclosed case we should keep track of the scheme structure on the points. This is the role of the scheme P in the definition below.

Definition 38. Let X be a scheme over a field k, P a smooth zero dimensional k-scheme and $t : P \to X$ a k-morphism. A *stable curve over X (defined over k) with base point $t(P)$* is a triple (C, f, σ) where

1. C is a proper connected curve defined over k having only nodes,
2. $\sigma : P \to C$ is a k-embedding into the smooth locus of C,
3. $f : C \to X$ is a k-morphism such that $t = f \circ \sigma$, and
4. C has only finitely many automorphisms that commute with f and σ.

We usually work with families of curves.

Definition 39. A *family of stable curves over X with base point $t(P)$* is a triple (C, f, σ) where

1. $C \to U$ is a flat, proper family of connected curves with only nodes over a k-scheme U,
2. $\sigma : P \times_k U \to C$ is an embedding, and
3. $f : C \to X \times_k U$ is a U-morphism such that (C_u, f_u, σ_u) is a stable curve (defined over $k(u)$) with base point $t_u(P_u)$ for every $u \in U$.

A basic theorem, due to Kontsevich and Manin [19] (in the context of algebraic stacks) and Alexeev [1] (in the present setting of projective moduli space), asserts that one can parametrize all stable curves with a single scheme. The relevant notion is the concept of coarse moduli spaces, introduced by Mumford. A more general version is proved in Section 10.

Theorem 40. *Let X be a projective scheme over a field k with an ample divisor H, P a smooth zero dimensional k-scheme and $t : P \to X$ a k-morphism. Fix integers g and d. Then there is a projective k-scheme $\overline{M}_g(X, d, t)$ which is a coarse moduli space for all genus g stable curves (C, f, σ) with base point $t(P)$ such that $f_*[C] \cdot H = d$.* ∎

41 (What does this mean?). We want to say that there is a "natural" correspondence between the set of all stable curves and the points of $\overline{M}_g(X, d, t)$. For our purposes this means the following two requirements.

1. There is a bijection of sets

$$\Phi : \left\{ \begin{array}{c} \text{genus } g \text{ stable curves over } \bar{k} \\ (C, f, \sigma) \\ \text{with base point } t(P) \\ \text{such that } f_*[C] \cdot H = d \end{array} \right\} \xrightarrow{\cong} \left\{ \bar{k}\text{-points of } \overline{M}_g(X, d, t) \right\}.$$

2. If $(\mathcal{C}/U, f, \sigma)$ is a family of degree d stable curves with base point $t(P)$, then there is a unique morphism $M_U : U \to \overline{M}_g(X, d, t)$ such that

$$M_U(u) = \Phi(C_u, f_u, \sigma_u) \quad \text{for every } u \in U(\bar{k}).$$

M_U is called the *moduli map.*

Given a stable curve (C, f, σ) with base point $t(P)$, we denote by $[(C, f, \sigma)]$ the corresponding point $\Phi((C, f, \sigma))$ in $\overline{M}_g(X, d, t)$.

We usually do not care about the degree, and use the scheme $\overline{M}_g(X, t)$ which is the union of all the $\overline{M}_g(X, d, t)$. The larger scheme $\overline{M}_g(X, t)$ is not projective since it has infinitely many connected components.

The requirements above seem somewhat awkward, but one cannot achieve more in general. For instance, (1) usually fails over nonclosed fields. The genus zero curves $(x_0^2 + x_1^2 + x_2^2 = 0) \subset \mathbb{P}^2_{\mathbb{R}}$ and $\mathbb{P}^1_{\mathbb{R}}$ are isomorphic over $\mathbb{C}$ but not over $\mathbb{R}$. Over $\mathbb{Q}$ there are infinitely many nonisomorphic genus zero curves (all isomorphic over $\mathbb{C}$). These examples illustrate that Φ is not injective over nonclosed fields. In general, Φ is also not surjective, but such examples are harder to get.

We need to know a little about the infinitesimal description of $\overline{M}_g(X, t)$ near a stable curve (C, f, σ). As usual, if X is smooth, this is governed by the cohomology groups of the twisted pull back of the tangent bundle, $f^*T_X(-\sigma(P))$. The moduli space is usually very complicated, but its local structure is nice in some cases.

Trouble comes from two sources. First, if $H^1(C, f^*T_X(-\sigma(P)))$ is large then one can say little beyond a dimension estimate, as in Theorem 5.3. Second, if C has automorphisms commuting with f and σ, then $\overline{M}_g(X, t)$ looks like the quotient of a smooth variety by a finite group. This can be quite complicated, especially in positive characteristic. This accounts for the somewhat messy formulation of Theorem 42.5. Fortunately, this seemingly cumbersome form is more useful for many purposes. A proof is given in Section 10.

Theorem 42. *Let X be a smooth projective variety over a field k. Let P be a smooth zero dimensional k-scheme and $t : P \to X$ a k-morphism. Set $|P| := \dim_k \mathcal{O}_P$. Let (C, f, σ) be a stable curve of genus g with base point $t(P)$. Assume $H^1\big(C, f^*T_X(-\sigma(P))\big) = 0$. Then*

1. *There is a unique irreducible component (defined over k)*
$$\operatorname{Comp}(C, f, \sigma) \subset \overline{M}_g(X, t)$$
passing through $\big[(C, f, \sigma)\big]$.
2. *The component $\operatorname{Comp}(C, f, \sigma)$ is geometrically irreducible.*
3. *There is a dense open subset of $\operatorname{Comp}(C, f, \sigma)$ that corresponds to irreducible curves.*
4. *The dimension of $\operatorname{Comp}(C, f, \sigma)$ is*
$$-\big(f_*[C] \cdot K_X\big) + (\dim X - 3)(1-g) - |P|(\dim X - 1).$$
5. *There is a smooth k-scheme U with a rational point $u \in U(k)$ and a family of stable curves*
$$(\mathcal{C}/U, F_U : \mathcal{C} \to X \times U,\ \sigma_U : P \times U \to \mathcal{C})$$
such that $(\mathcal{C}_u, F_u, \sigma_u) \cong (C, f, \sigma)$ and the corresponding moduli map
$$M_U : (U, u) \to \big(\operatorname{Comp}(C, f, \sigma), \big[(C, f, \sigma)\big]\big)$$
is quasifinite and open.

In fact, if C has no automorphisms commuting with f and σ, we can choose M_U to be an open embedding.

9. Finding Rational Curves over Nonclosed Fields

Let X be a smooth, projective, separably rationally connected variety over a field k. By definition, X has plenty of rational curves over $\bar{k}$, but we know very little about rational curves over k.

It is possible that $X(k) = \varnothing$, and then we certainly do not have any morphism $\mathbb{P}^1 \to X$ defined over k.

Thus assume there exists a point $x \in X(k)$. Can we find a rational curve $f : \mathbb{P}^1 \to X$ defined over k and passing through x? It is quite unlikely that such a curve always exists, but no counterexample is known. Here we discuss a weaker version of the problem, which allows us to settle the above question for certain fields.

Consider the space $\overline{M}_0(X, \operatorname{Spec} k \to x)$ of genus zero stable curves through x and let Z_i be its irreducible components over $\bar{k}$. The Galois group $\operatorname{Gal}(\bar{k}/k)$ acts on the set of all components. If there is a free rational curve (defined over $\bar{k}$) $f : \mathbb{P}^1_{\bar{k}} \to X_{\bar{k}}$, $f(0) = x$, then $[(\mathbb{P}^1, f, \operatorname{Spec} k \to 0)]$ lies in a unique $Z_i = \operatorname{Comp}(f)$ by Theorem 42. If, in addition, f is defined over k, then $\operatorname{Comp}(f)$ is a geometrically irreducible component of $\overline{M}_0(X, \operatorname{Spec} k \to x)$ defined over k.

In order to answer the above question, we do the following. First we construct geometrically irreducible components Z_i of $\overline{M}_0(X, \operatorname{Spec} k \to x)$ that are defined over k. This is done over any field. For local fields we can then find k-points in Z_i corresponding to irreducible curves through x. For arbitrary fields this seems rather difficult.

The main method is to assemble combs from given morphisms. This is what we define next.

Definition 43. Let X be a k-variety and $x \in X(k)$. Assume that we have $\bar{k}$-morphisms (not necessarily defined over k) $f_i : \mathbb{P}^1_{\bar{k}} \to X_{\bar{k}}$ such that $f_i(0) = x$ for $i = 1, \ldots, n$.

Let $C := C_0 \cup \cdots \cup C_n$ be a rational comb (over $\bar{k}$) with handle C_0. Choose identifications $C_i \cong \mathbb{P}^1$ such that the node is identified with 0. Define a morphism

$$F : C \to X \quad \text{by} \quad F|_{C_i} = f_i \quad \text{and} \quad F(C_0) = \{x\}.$$

The pair (C, F) is called a *comb assembled from the f_i*. If $n \geq 3$, (C, F) is a stable curve over $X_{\bar{k}}$, provided none of the f_i are constant.

A comb assembled from the f_i is not unique, but all the combs assembled from the f_i form an irreducible family (parametrized by the space of n points on $C_0 \cong \mathbb{P}^1$).

It is natural to think of the f_i as $\bar{k}$-points of $\overline{M}_0(X, \operatorname{Spec} k \to x)$. In this context, we pick an extra point $p_0 \in C_0(\bar{k})$ distinct from the nodes of C and set $\sigma : \operatorname{Spec} \bar{k} \to p_0$. This makes (C, F, σ) into a stable curve (defined over $\bar{k}$) with base point x, provided that $n \geq 2$ and none of the f_i are constant. We call this triple a *base pointed comb assembled from the f_i*.

Assume next that the

$$f_i : \mathbb{P}^1 \to X, \; f_i(0) = x, \quad i = 1, \ldots, n,$$

form a conjugation invariant set over k. We would like to assemble all the f_i into an element of $\overline{M}_0(X, \operatorname{Spec} k \to x)$ that is defined over k. The construction above has to be carried out carefully: even if we choose C to be a comb over k, the morphism F is not defined over k in general. In order to achieve this, we need that the induced scheme structure on the subset of nodes $\{P_i = C_0 \cap C_i\} \subset C_0$ matches the one on $\left\{[(C_i, f_i)]_{i=1,\dots,n}\right\} \subset \overline{M}_0(X, \operatorname{Spec} k \to x)$. This is the role of the scheme T below.

View each f_i as a $\bar{k}$-point of $\overline{M}_0(X, \operatorname{Spec} k \to x)$. This identifies the collection $\{f_i\}$ with a zero dimensional smooth k-subscheme $T \subset \overline{M}_0(X, \operatorname{Spec} k \to x)$. (See Remark 45 for some further clarification in positive characteristic.) The f_i correspond to a k-morphism $f_T : \mathbb{P}^1_T \to X$ which maps $\{0\} \times T$ to x. This construction is interesting only when none of the f_i corresponds to a k-point of $\overline{M}_0(X, \operatorname{Spec} k \to x)$, so we assume that T has no k-points.

Let $C_0 = \mathbb{P}^1$ and fix an embedding $j : T \hookrightarrow C_0$. This determines a rational comb C over k with teeth $\mathbb{P}^1_T$ whose nodes coincide with $j(T)$. Now define a morphism

$$F : C \to X \quad \text{by} \quad F|_{\mathbb{P}^1_T} = \{f_i\} \quad \text{and} \quad F(C_0) = \{x\}.$$

Notice that C and F are both defined over k. Since T has no k-points, we can pick a point $p_0 \in C_0(k)$ distinct from the nodes of C and set $\sigma : \operatorname{Spec} k \to p_0$. The triple (C, F, σ) is called a *pointed k-comb assembled from the f_i*. If $n \geq 2$, (C, F, σ) is a stable curve over X with base point x, provided none of the f_i are constant.

As before, a k-comb assembled from the f_i is not unique, but all $\bar{k}$-combs assembled from the f_i form an irreducible family defined over k (parametrized by the space of all $\bar{k}$-embeddings $T \cup \operatorname{Spec} k \hookrightarrow \mathbb{P}^1$).

Proposition 44. *Let k'/k be a finite, separable field extension. Let X be a smooth projective variety over k and let $x \in X(k)$ be a point. Let*

$$f_i : \mathbb{P}^1 \to X, \; f_i(0) = x, \quad i = 1, \dots, n,$$

be a set of nonconstant free k'-morphisms together defined over k. Let (C, F, σ) be a pointed k-comb assembled from the f_i, so that (C, F, σ) is a stable curve defined over k with base point x.

Then there is a smooth k-scheme U with a k-point $u \in U(k)$ and a family of stable curves

$$(\mathcal{C}/U, F_U : \mathcal{C} \to X \times U, \; \sigma_U : P \times U \to \mathcal{C})$$

such that $(C_u, F_u, \sigma_u) \cong (C, F, \sigma)$ *and the corresponding moduli map*

$$M_U : (U, u) \to \big(\operatorname{Comp}(C, F, \sigma), [(C, F, \sigma)]\big)$$

is quasifinite and open.

Moreover, the irreducible component $\operatorname{Comp}(C, F, \sigma) \subset \overline{M}_0(X, \operatorname{Spec} k \to x)$ *is geometrically irreducible and independent of the choice of the comb.*

Proof. Denote by C_0 the handle of C and by $C_1, \ldots, C_n$ its teeth. Let $p_0 = \sigma(\operatorname{Spec} k) \in C_0$.

Set $E = F^*T_X(-p_0)$. Since F maps C_0 to a point, $f^*T_X|_{C_0} = \mathcal{O}_{C_0}^{\oplus n}$. Thus $H^1(C_0, E|_{C_0}) = 0$. For any other C_i, $E|_{C_i}(-1) \cong f_i^*T_X(-1)$, and this has vanishing H^1 since f_i is free. Thus $H^1(C, F^*T_X(-p_0)) = 0$ by Exercise 18. The first assertion then follows from Theorem 42, and so does geometric irreducibility of $\operatorname{Comp}(C, F, \sigma)$.

Theorem 42 also implies that each $[(C, F, \sigma)]$ lies in a unique irreducible component $\operatorname{Comp}(C, F, \sigma) \subset \overline{M}_0(X, \operatorname{Spec} k \to x)$. All $\bar{k}$-combs assembled from the f_i form an irreducible family defined over k. Hence they all lie in the same irreducible component of $\overline{M}_0(X, \operatorname{Spec} \bar{k} \to x)$, and $\operatorname{Comp}(C, F, \sigma)$ does not depend on the choice of the comb. ■

Remark 45. In positive characteristic we have to deal with the possibility that $f_i : \mathbb{P}^1 \to X$ is not defined over a separable field extension. Since f_i is free, its irreducible component in $\operatorname{Hom}(\mathbb{P}^1, X, 0 \mapsto x)$ is generically smooth, thus it contains a dense set of points defined over separable field extensions.

As an application, we prove a result about separably rationally connected varieties over local fields.

Definition 46. For our purposes *local fields* are:

1. Finite degree extensions of the p-adic fields $\mathbb{Q}_p$;
2. Power series fields over finite fields, $\mathbb{F}_q((t))$;
3. $\mathbb{R}$ and $\mathbb{C}$.

For all of these fields we have an analytic implicit function theorem, thus if X is a variety of dimension n over a local field k, and $x \in X(k)$ is a smooth point, then a neighborhood of x in the analytic variety $X(k)$ is biholomorphic to a neighborhood of 0 in k^n.

In particular, if X has *one* smooth k-point then k-points are Zariski dense.

Theorem 47 [14]. *Let k be a local field and X a smooth, projective, separably rationally connected variety defined over k. Let $x \in X(k)$ be a point.*

Then there is a nonconstant free morphism defined over k

$$f : \mathbb{P}^1 \to X \quad \text{such that} \quad f(0) = x.$$

Proof. By assumption and by Remark 45, there is such a morphism g over a separable extension of finite degree k'/k. Use g and all of its conjugates to assemble a k-comb $(C, F, \operatorname{Spec} k \to p_0)$ that is a stable curve with base point x.

By Proposition 44, there is a smooth pointed k-scheme (U, u) and a dominant, quasifinite morphism

$$(U, u) \to \big(\operatorname{Comp}(C, F, \operatorname{Spec} k \to p_0), \big[(C, F, \operatorname{Spec} k \to p_0)\big]\big).$$

Since k is a local field, $U(k)$ is Zariski dense in U, thus k-points are Zariski dense in $\operatorname{Comp}(C, F, \operatorname{Spec} k \to p_0)$. By Theorem 42.3, there exists a k-point $\big[(C', F', \sigma')\big]$ that corresponds to an irreducible curve $F' : C' \to X$ through x. Since C' comes with a k-point $\sigma'(\operatorname{Spec} k)$, there is an isomorphism $\mathbb{P}^1 \to C'$ defined over k. So we obtain a nonconstant k-morphism $\mathbb{P}^1 \to X$ whose image passes through x. ■

The construction of Proposition 44 can be made simultaneously for all points. This is the content of the next proposition.

Proposition 48. *Let X be a smooth, projective, separably rationally connected variety over a field k.*

There is an X-scheme U and a family of stable curves

$$(\mathcal{C}/U, F_U : \mathcal{C} \to X \times U,\ \sigma_U : \operatorname{Spec} k \times U \to \mathcal{C})$$

such that for each $x \in X$ the corresponding moduli map

$$M_{U_x} : U_x \to \overline{M}_0\big(X_{k(x)}, \operatorname{Spec} k(x) \to x\big)$$

is quasifinite and open and maps U_x onto an open subset of

$$\overline{M}_0\big(X_{k(x)}, \operatorname{Spec} k(x) \to x\big)$$

parametrizing free stable curves.

Proof. Let $x_g \in X$ be the generic point. As in Proposition 44, we obtain a $k(X)$-scheme U and a family of stable curves

$$\big(\mathcal{C}/U, F_U : \mathcal{C} \to X_{x_g} \times U,\ \sigma_U : \operatorname{Spec} k(X) \times U \to \mathcal{C}\big)$$

such that for each $x \in X$ the corresponding moduli map is quasifinite and open and maps U onto an open subset of $\overline{M}_0\big(X_{k(X)}, \operatorname{Spec} k(X) \to x_g\big)$ parametrizing free stable curves.

By specialization, for x in an open set $X^0 \subset X$ we obtain a family of stable curves

$$\big(\mathcal{C}/U_x, F_{U_x} : \mathcal{C}_{U_x} \to X_{k(x)} \times U_x, \operatorname{Spec} k(x) \times U_x \to \mathcal{C}_{U_x}\big)$$

such that the corresponding moduli map is quasifinite and open and maps U_x onto an open subset of $\overline{M}_0\big(X_{k(x)}, \operatorname{Spec} k(x) \to x\big)$ parametrizing free stable curves.

Repeat the argument with x_g replaced with the generic points of $X \setminus X^0$. Continuing in this way we eventually take care of all points of X.

It is quite likely that one can choose U to be irreducible and $U \to X$ smooth but the above construction gives neither. ∎

10. Moduli Spaces of Stable Curves

The aim of this rather technical section is to prove Theorems 50 and 51, which are generalizations of Theorems 40 and 42. The main outlines of the proofs follow the general philosophy of constructing moduli spaces. The closest example is the special case explained in [8]. There are two key differences in the setup.

First, in [8] the marking of curves over $\mathbb{C}$ was done by choosing n smooth points of a curve. In the general setting it is more natural to let the marking be an embedding of a zero dimensional scheme Q into the smooth locus of a curve. The theory is interesting only when Q has embedding dimension at most 1, and the global theory works best only if it is smooth.

Second, we also want to account for the possibility that some of the marked points have a fixed image. For existence questions this is not much different, but the local structure has to be worked out by a somewhat different construction.

We also drop the "convexity" condition of [8], and replace it with the assumption $H^1\big(C, f^*T_{X_k}(-\sigma(P_k))\big) = 0$, which is essentially the local version of it.

First we generalize the notion of stable curves, defined in Section 8, to the relative setting. This means that instead of working with schemes over a field k, we fix a Noetherian scheme S, and work with schemes over S.

Definition 49. Fix a Noetherian scheme S. Let X be a scheme over S, let $P \to S$ and $Q \to S$ be finite, flat morphisms and let $t : P \to X$ be a morphism. A *stable curve over* X *with base point* $t(P)$ *and marking* Q is a quadruple (C, f, σ, ρ) where

1. $C \to U$ is a flat, proper family of connected curves with only nodes over an S-scheme U,
2. $f : C \to X \times_S U$ is a U-morphism,
3. $\sigma : P \times_S U \to C$ and $\rho : Q \times_S U \to C$ are embeddings into the smooth locus of $C \to U$ with disjoint images,
4. $f \circ \sigma$ is the morphism $P \times_S U \to X \times_S U$ induced by t, and
5. for every point $\operatorname{Spec} k \to U$, C_k has only finitely many automorphisms commuting with f_k and fixing the closed points of $\sigma_k(P_k)$ and $\rho_k(Q_k)$.

Note: Condition 5 is equivalent to saying that every rational component of $C_{\bar{k}}$ that is mapped to a point by $f_{\bar{k}}$ has at least 3 nodes or marked points on it. We allow marking by a nonreduced P_k, but a node plus a single point in P_k with a length 3 nilpotent structure is not enough for stability, although it is enough to kill all commuting automorphisms. The reason for imposing this stronger condition is that without it we get a nonseparated moduli problem.

As a simple example, consider a smooth curve B of genus $g \geq 2$ defined over a field k. Set $X = B$ and $P = \operatorname{Spec} k[t]/(t^3)$. Take $U = \mathbb{A}^1$ and consider the family $C = B \times U \to U$ with a section $S = \{b\} \times U$. Denote by $\mathcal{I}_S$ the ideal sheaf of S in C. Let σ map $P \times U$ onto the subscheme of C with ideal sheaf $\mathcal{I}_S^3$. Now let $f' : C' \to X \times U$ be the blow up of $(b, 0)$, let $\sigma' : P \times U \to C'$ be the map induced by σ and consider the triple (C', f', σ'). This shows that the identity map $B \to X$ with $P \hookrightarrow B$ degenerates to $B' \to X$ where $B' = B \cup \mathbb{P}^1$ and P is embedded into $\mathbb{P}^1$.

In the sequel we assume that $\deg(P/S)$ and $\deg(Q/S)$ are actually constants. This is the case if S is connected, so this is not a significant restriction. We use the shorthand $|P| = \deg(P/S)$ and $|Q| = \deg(Q/S)$.

The next theorem is a generalization of Theorem 40.

Theorem 50. *Fix a Noetherian base scheme S. Let X be a projective scheme over S with an ample divisor H. Let $P \to S$ and $Q \to S$ be finite and flat morphisms over S of relative embedding dimension at most 1 and let $t : P \to X$ be a morphism. Fix integers g and d. Then there is a separated algebraic space $\overline{M}_{g,Q}(X,d,t)$ of finite type over S which is a coarse moduli space for all genus g stable curves (C,f,σ,ρ) with base point $t(P)$ and marking Q such that $\deg_C f^*H = d$.*

Moreover, the space $\overline{M}_{g,Q}(X,d,t)$ is projective over S if the maps $P \to S$ and $Q \to S$ are smooth.

It is quite likely that $\overline{M}_{g,Q}(X,d,t)$ is always quasi-projective over S, but the proof does not give this. The reason is that the projectivity criterion of [11] used in [8] does not have a good quasi-projective analog.

As explained in [8, 5.1], one can use mH to embed X into $\mathbb{P}^M_S$ and deduce the existence of $\overline{M}_{g,Q}(X,d,t)$ from the existence of $\overline{M}_{g,Q}(\mathbb{P}^M_S, md, t)$. Thus we can restrict ourselves to the case when X is smooth over S. This is the only case in which the local structure of $\overline{M}_{g,Q}(X,d,t)$ is easy to describe. Such a description is given in the next theorem, which is a generalization of Theorem 42. As before, we work with the space $\overline{M}_{g,Q}(X,t)$, which is the union of all the $\overline{M}_{g,Q}(X,d,t)$, and we denote by $\left[(C,f,\sigma,\rho)\right]$ the point in $\overline{M}_{g,Q}(X,t)$ corresponding to a stable curve (C,f,σ,ρ).

Theorem 51. *Let S be a normal scheme and X a smooth projective scheme over S. Let $P \to S$ and $Q \to S$ be finite and flat S-schemes of relative embedding dimension at most 1 and let $t : P \to X$ be a morphism. Let k be a field, $\operatorname{Spec} k \to s \in S$ a point and*

$$(C \to \operatorname{Spec} k, f : C \to X_k,\ \sigma : P_k \to C,\ \rho : Q_k \to C)$$

a stable curve of genus g and degree d with base point $t_k(P_k)$ and marking Q_k over k. Assume $H^1\big(C, f^*T_{X_k}(-\sigma(P_k))\big) = 0$. Then

1. *There is a unique irreducible component $\operatorname{Comp}(C,f,\sigma,\rho) \subset \overline{M}_{g,Q}(X,t)$ passing through $\left[(C,f,\sigma,\rho)\right]$.*
2. *If k is the residue field of s in S, then the unique irreducible component of the fiber of $\overline{M}_{g,Q}(X,d,t) \to S$ over s containing $\left[(C,f,\sigma,\rho)\right]$ is geometrically irreducible.*
3. *There is a dense open subset of $\operatorname{Comp}(C,f,\sigma,\rho)$ that corresponds to irreducible stable curves.*

4. *The dimension of* Comp (C, f, σ, ρ) *is*

$$-\big(f_*[C]\cdot K_X\big) + (\dim X - 3)(1-g) + \dim S + |Q| - |P|(\dim X - 1).$$

5. *Assume $k/k(s)$ is a finite field extension, where $k(s)$ denotes the residue field of s in S. Then there is an S-scheme U, smooth over S, with a point* Spec $k \stackrel{\cong}{\to} u \in U$ *and a family of stable curves*

$$(C/U, F_U : C \to X_U,\ \sigma_U : P_U \to C,\ \rho_U : Q_U \to C)$$

such that $(C_u, F_u, \sigma_u, \rho_u) \cong (C, f, \sigma, \rho)$ and the corresponding moduli map

$$M_U : (U, u) \to \big(\operatorname{Comp}(C, f, \sigma, \rho), [(C, f, \sigma, \rho)]\big)$$

is quasifinite and open.

If C has no automorphism commuting with f, σ and ρ, we can choose M_U to be an open embedding.

52 (Outline of the proof). Let (C, f, σ, ρ) be a stable curve with base point $t(P)$, where $h : C \to U$ is a U-scheme and U is a connected S-scheme. The starting point, as in [8, 2.3], is the observation that $\omega_{C/U}\big(\sigma(P) + \rho(Q)\big) \otimes f^*\mathcal{O}_X(3H)$ is h-ample on C/U. Let L be the above line bundle raised to the $\big(2+|P|\big)$-th power. Then $L\big(-\sigma(P)\big)$ is h-very ample and $R^1h_*L = R^1h_*L\big(-\sigma(P)\big) = 0$. Let $N+1$ be the rank of the vector bundle h_*L (where N depends on the genus of C, $|P|$, $|Q|$ and d). After replacing U with an open affine cover, we may assume that $h_*L \cong \mathcal{O}_U^{N+1}$. Thus we have an embedding $\delta : C/U \hookrightarrow \mathbb{P}^N_U$. This is of course not unique, but defined only up to automorphisms of $\mathbb{P}^N_U$.

Let us now look at the diagonal map

$$(f, \delta) : C/U \hookrightarrow X \times_U \mathbb{P}^N.$$

Our strategy has 3 main parts:

1. Identify all possible lifts (f, δ) with some subset of a Hilbert scheme. This is straightforward if $P = \varnothing = Q$, but needs an extra twist otherwise.
2. Construct $\overline{M}_{g,Q}(X, d, t)$ as the quotient of the above subset of a Hilbert scheme by $PGL(N+1)$.
3. Read off properties of $\overline{M}_{g,Q}(X, d, t)$ through the properties of the Hilbert scheme.

53 (Construction of a related Hilbert scheme).

In order to deal with the base point $t(P)$, let us fix an embedding $t' : P_U \hookrightarrow \mathbb{P}^N_U$ that is nondegenerate (that is, over any algebraically closed field K, the image $t'(P_K)$ spans a subspace of dimension $|P| - 1$). First we consider only those liftings δ such that $t' = \delta \circ \sigma$. In this case the image of (f, δ) passes through the image of (t, t'), so it is more natural to view it as an embedded curve in the scheme $B_{t'} = B_{t'}(X \times_U \mathbb{P}^N)$ obtained by blowing up the image of (t, t').

We need to check a few things since the scheme $B_{t'}(X \times_U \mathbb{P}^N)$ is singular if P is not smooth. Let us consider a single fiber of $X \times_U \mathbb{P}^N \to U$. We assume that P has embedding dimension at most 1 and (t, t') is an embedding, thus in suitable local analytic coordinates $z_1, \dots, z_k$ (where $k = \dim X/U + N$) we can write

$$C = (z_1 = \cdots = z_{k-1} = 0) \quad \text{and} \quad P = (z_1 = \cdots = z_{k-1} = z_k^d = 0).$$

The key chart of the blow up is given by coordinates $z'_1, \dots, z'_{k-1}, z_1, \dots, z_k$ and relations

$$z_i = z'_i z_k^d \quad \text{for } i = 1, \dots, k-1.$$

This chart is smooth with coordinates $z'_1, \dots, z'_{k-1}, z_k$ and the birational transform of C is $C' = (z'_1 = \cdots = z'_{k-1} = 0)$. Thus $B_{t'}$ is smooth along C'. (All the other local charts of the blow up are singular.)

Inside $\operatorname{Hilb}\left(B_{t'}(X \times_U \mathbb{P}^N)\right)$ consider the open subset

$$W = W_{U,t',g}(X, d, t)$$

parametrizing 1-dimensional subschemes D such that

1. D is a connected curve of genus g contained in the smooth locus of $B_{t'}$ and with at worst nodes,
2. the natural projection $D \to \mathbb{P}^N$ is a nondegenerate embedding,
3. D is smooth at the points of E, and $E \cap D$ projects isomorphically onto the image of (t, t').

We still have to add in the marking by Q. To do this consider the universal family $\mathbf{D} \to W_{U,t',g}(X, d, t)$ and let Q_W denote the pull back of Q to $W_{U,t',g}(X, d, t)$. Consider the open subset

$$J_{U,t',g,Q}(X, d, t) \subset \operatorname{Hom}_W(Q_W, \mathbf{D}),$$

parametrizing curves D and homomorphisms $\rho : Q \to D$ such that

4. the map ρ is an embedding into the smooth locus and its image is disjoint from $E \cap D$,
5. the natural projection $D \to X$ is a stable curve of degree d with base point $t(P)$ and marking Q,
6. on every irreducible component of every geometric fiber of $D \to U$, the degree of $\mathcal{O}_{\mathbb{P}^N}(2 + |P|)$ is the same as the degree of

$$\omega_{D/U} \otimes \mathcal{O}_X(3H) \otimes \mathcal{O}_{B_{t'}}(E) \otimes \mathcal{O}_D(\rho(Q)).$$

(The various pull backs to D have been omitted from the notation.)

Since we are looking only at those morphisms in $\mathrm{Hom}_W(Q_W, \mathbf{D})$ that land in the smooth locus of $\mathbf{D} \to W$, we conclude from the general case of Theorem 5 given in [12, I.2.17] that

$$J_{U,t',g,Q}(X, d, t) \to \mathrm{Hilb}\left(B_{t'}(X \times_U \mathbb{P}^N)\right)$$

is smooth of relative dimension $|Q|$.

If the genus of C is zero, this is the subscheme we want. In general, based on our construction we also want to achieve that

7. the pull back of $\mathcal{O}_{\mathbb{P}^N}(2 + |P|)$ to D is isomorphic to

$$\omega_{D/U} \otimes \mathcal{O}_X(3H) \otimes \mathcal{O}_{B_{t'}}(E) \otimes \mathcal{O}_D(\rho(Q)).$$

As explained in [8, 2.2], the latter is a locally closed condition and defines $J^*_{U,t',g,Q}(X, d, t) \subset J_{U,t',g,Q}(X, d, t)$, which is open if $g = 0$ and has codimension g in general.

Ultimately we do not want to fix the embedding $t' : P \to \mathbb{P}^N$. To achieve this, let $\mathrm{Hom}^1(P, \mathbb{P}^N)$ denote the open subset of all nondegenerate embeddings. We have a universal embedding

$$T : P \times_S \mathrm{Hom}^1(P, \mathbb{P}^N) \to \mathbb{P}^N \times_S \mathrm{Hom}^1(P, \mathbb{P}^N).$$

Out of T and t we get a diagonal embedding

$$(t, T) : P \times_S \mathrm{Hom}^1(P, \mathbb{P}^N) \to X \times_S \mathbb{P}^N \times_S \mathrm{Hom}^1(P, \mathbb{P}^N).$$

Observe that the automorphism group $\mathrm{Aut}(\mathbb{P}^N)$ acts on both sides and (t, T) is equivariant. Let $B \to X \times_S \mathbb{P}^N \times_S \mathrm{Hom}^1(P, \mathbb{P}^N)$ denote the blow up of the image. We can now construct

$$\begin{aligned} W_g(X, d, t) &:= W_{\mathrm{Hom}^1(P,\mathbb{P}^N),(t,T),g}(X, d, t) \\ &\subset \mathrm{Hilb}\left(X \times_S \mathbb{P}^N \times_S \mathrm{Hom}^1(P, \mathbb{P}^N)\right) \end{aligned}$$

and

$$J_{g,Q}(X,d,t) := J_{\mathrm{Hom}^1(P,\mathbb{P}^N),(t,T),g,Q}(X,d,t) \to W_g(X,d,t).$$

The pullback of $J_{g,Q}(X,d,t)$ by the morphism $U \to S$ and the embedding t' is precisely our earlier $J_{U,t',g,Q}(X,d,t)$. As before we get that:

Claim 54. With the above notation,

$$J_{g,Q}(X,d,t) \to W_g(X,d,t)$$

is smooth of relative dimension $|Q|$. ■

The automorphism group $\mathrm{Aut}\,(\mathbb{P}^N)$ acts on B, hence it also acts on $\mathrm{Hilb}\,(B)$ and on $J_{g,Q}(X,d,t)$. From our construction we have the key observation:

Claim 55. Let K be any algebraically closed field and $\mathrm{Spec}\,K \to S$ a morphism.

1. If $g = 0$, there is a one-to-one correspondence between the $PGL(N+1,K)$ orbits on $J_{g,Q}(X_K,d,t_K)$ and genus zero stable curves over K of degree d with base point $t(P_K)$ and marking Q_K.
2. If $g \geq 0$, there is a one-to-one correspondence between the $PGL(N+1,K)$ orbits on $J_{g,Q}(X_K,d,t_K)$ and genus zero stable curves over K of degree d with base point $t(P_K)$ and marking Q_K plus the choice of a line bundle which has degree zero on every irreducible component of the curve. ■

Remark 56. The slight mystery is why we need K algebraically closed. This is already visible in the much simpler example of the hyperelliptic plane curves $ay^2 = f(x)$, where we can even fix f for simplicity. The multiplicative group K^* acts on the family by $(x,y) \mapsto (x,ty)$ and over an algebraically closed field we have a single orbit. On the other hand, if $K = \mathbb{Q}$, the $\mathbb{Q}^*$-action on the $\mathbb{Q}$-points has infinitely many orbits, parametrized by the square free integers. Upon deeper analysis one sees that all of this is caused by the automorphism $(x,y) \mapsto (x,-y)$ of the hyperelliptic plane curves.

This completes the first step of the plan, and next we need to take the quotient $J(X,d,t)/\,\mathrm{Aut}\,(\mathbb{P}^N)$.

57 (Construction of $\overline{M}_g(X,d,t)$ as a quotient space).

As we noted above, we would like to find a morphism

$$q : J_{g,Q}(X,d,t) \to \overline{M}_{g,Q}(X,d,t)$$

with the property that for any algebraically closed field K and for any morphism $\operatorname{Spec} K \to S$ the fibers of

$$q_K : J_{g,Q}(X_K,d,t_K) \to \overline{M}_{g,Q}(X_K,d,t_K)$$

are precisely the $PGL(N+1,K)$ orbits on $J_{g,Q}(X_K,d,t_K)$.

At this point it is very convenient to use the general quotient theorems about algebraic group actions established in [13] and [10]. First we need to check a technical condition about the $\operatorname{Aut}(\mathbb{P}^N)$-action on $J_{g,Q}(X,d,t)$, namely that it is *proper.* Properness of the group action is equivalent to the separatedness of the family considered in [8, 4.2]. We need the following:

Proposition 58. *In the setting of Theorem* 50 *assume that S is the spectrum of a complete DVR with quotient field K and residue field k. Let $(C_K, f_K, \sigma_K, \rho_K)$ be a stable curve over K.*

1. *There is at most one way to extend it to a stable curve $(C_S, f_S, \sigma_S, \rho_S)$ over S.*
2. *If $P \to S$ and $Q \to S$ are smooth, then there is a scheme S', which is also the spectrum of a DVR, and a local map $S' \to S$ such that the pull back of $(C_K, f_K, \sigma_K, \rho_K)$ can be extended to $(C_{S'}, f_{S'}, \sigma_{S'}, \rho_{S'})$.*

Proof. For the first part, we plan to follow the proof of the uniqueness of canonical models (see for instance [18, 3.52]).

If we have two extensions, they are dominated by a common resolution. So it is enough to show how to recover C_S from a resolution of it. The notion of resolution here is slightly different from the usual one since the general fiber may also be singular. Thus we allow 3 types of operations:

1. blowing up a smooth point of the special fiber, introducing a (-1)-curve,
2. blowing up an isolated singular point of the special fiber, introducing one or two (-2)-curves, and
3. blowing up a nonisolated singular point of the special fiber, introducing two (-1)-curves in the normalization of C_S.

We would like to consider the pair $(C_S, \sigma_S(P_S) + \rho_S(Q_S))$, but it is not canonical. Instead we do the following. Let D_i be an irreducible component of $\sigma_S(P_S) + \rho_S(Q_S)$ with central fiber $(D_i)_k = \operatorname{Spec} k'[t]/(t^{d_i})$ for some $k' \supset k$ and $d_i \geq 1$. (Here we use that S is complete.) Set $D = \sum \frac{1}{d_i} D_i$ as a $\mathbb{Q}$-divisor on C_S. The reason for introducing D is to assure that every irreducible component of D has local intersection number 1 with C_k at every geometric intersection point. Our stability assumption 49.5 was made to assure that $\omega_{C_S/S}(D)$ is relatively ample.

We also see that under a blowup $C'_S \to C_S$, the exceptional curves appear in $\omega_{C'_S/S}(D)$ with coefficient 0, unless we blow up a smooth point not on D, when the coefficient is 1. Thus if $g : C_S^m \to C_S$ is any resolution, the global sections of $\left(\omega_{C_S^m/S}(D)\right)^{\otimes n}$ all pull back from $\left(\omega_{C_S/S}(D)\right)^{\otimes n}$ for $n \geq 1$. Thus C_S is the image of C_S^m under the morphism given by the global sections of $\left(\omega_{C_S^m/S}(D)\right)^{\otimes n}$ for some $n \gg 1$.

Assume next that the maps $P \to S$ and $Q \to S$ are smooth. In this case the existence proof of [8, 4.2] works, except that one should use the semi stable reduction theorem in arbitrary characteristic as proved in [2]. ∎

Thus we conclude that $\overline{M}_{g,Q}(X,d,t)$ exists as an algebraic space that is separated and of finite type over S. If $P \to S$ and $Q \to S$ are smooth, the properness of $\overline{M}_{g,Q}(X,d,t)$ over S follows from Proposition 58. The projectivity over S can be checked as in [8, 3–4], but it is actually not important for any of what follows.

59 (Local structure of $\overline{M}_{g,Q}(X,d,t)$).

We have constructed $\overline{M}_{g,Q}(X,d,t)$ as a quotient of $J_{g,Q}(X,d,t)$, and so we expect to read off the local properties of $\overline{M}_{g,Q}(X,d,t)$ from the local properties of $J_{g,Q}(X,d,t)$. The latter is actually much nicer, and parts 1,2,3 and 5 of Theorem 51 follow from the corresponding properties of $J_{g,Q}(X,d,t)$.

Proposition 60. *Let S be a normal scheme and X a smooth projective scheme over S. Let P, Q be finite and flat S-schemes of relative embedding dimension at most 1 and let $t : P \to X$ be a morphism. Let k be a field, $\operatorname{Spec} k \to s \in S$ a point in S and*

$$(C, f : C \to X_k,\ \sigma : P_k \to C,\ \rho : Q_k \to C)$$

a stable curve of genus g and degree d with base point $t_k(P_k)$ and marking Q_k over k. Let L be the line bundle on C defined as in Paragraph 52. Let $\delta : C \hookrightarrow \mathbb{P}_k^N$ be a nondegenerate embedding by all sections of L

and let $[(C, f, \sigma, \delta, \rho)]$ *be the corresponding point in* $J_{g,Q}(X, d, t)$. *Assume* $H^1(C, L(-\sigma(P_k))) = 0$ *and* $H^1(C, f^*T_{X_k}(-\sigma(P_k))) = 0$. *Then*

1. *The map* $J_{g,Q}(X, d, t) \to \mathrm{Hom}^1(P, \mathbb{P}^N)$ *is smooth at* $[(C, f, \sigma, \delta, \rho)]$.
2. *There is a unique irreducible component*
$$\mathrm{Comp}\,(C, f, \sigma, \delta, \rho) \quad \text{of} \quad J_{g,Q}(X, d, t)$$
passing through $[(C, f, \sigma, \delta, \rho)]$.
3. *The dimension of* $[(C, f, \sigma, \delta, \rho)]$ *is*
$$-(f_*[C] \cdot K_X) + (\dim X - 3)(1 - g)$$
$$+ |Q| - |P|(\dim X - 1) + \dim S + g + (N + 1)^2 - 1.$$
4. *A general point of* $\mathrm{Comp}\,(C, f, \sigma, \delta, \rho)$ *corresponds to a smooth stable curve.*

Proof. Recall from Claim 54 that the map $J_{g,Q}(X, d, t) \to W_g(X, d, t)$ is smooth of relative dimension $|Q|$, thus the main point is to prove the analogous statements for the map $W_g(X, d, t) \to \mathrm{Hom}^1(P, \mathbb{P}^N)$.

The local structure of the Hilbert scheme is governed by the normal bundle (see Paragraph 25). Thus the key to prove Proposition 60 is to understand the normal bundle of the image of

$$C \xrightarrow{(f,\delta)} X \times_k \mathbb{P}^N \dashrightarrow B_{t'}(X \times_k \mathbb{P}^N).$$

Going from the embedding into $X \times_k \mathbb{P}^N$ to the blowup is straightforward, and the important part is to describe the normal bundle of the image of C in $X \times_k \mathbb{P}^N$.

The argument has three main ingredients.

5. General remarks about the "second exact sequence".
6. Computing the normal bundle of a curve in $\mathbb{P}^N$.
7. Relating the normal bundle of $C \subset X \times \mathbb{P}^N$ to f^*T_X.

61 (The second exact sequence).

Let Z be a closed subscheme of Y with ideal sheaf $\mathcal{I}_Z$. The so-called "second exact sequence" (see [9, II.8.4.A]) is

$$\mathcal{I}_Z/\mathcal{I}_Z^2 \xrightarrow{\delta} \Omega^1_Y|_Z \to \Omega^1_Z \to 0.$$

In general it is not left exact, but it is easily seen to be left exact at points where Z and Y are smooth. If such points are dense in Z, Z is reduced and $\mathcal{I}_Z/\mathcal{I}_Z^2$ is locally free, then δ can have no kernel at all, so we also have left exactness. This holds if, for instance, Y is smooth and Z is a reduced local complete intersection subscheme.

If $C \subset \mathbb{P}^N$ is a reduced local complete intersection curve, we can compute the degree of its normal bundle from the second exact sequence as

$$\deg N_{C,\mathbb{P}^N} = (N+1)\deg C + 2p_a(C) - 2.$$

62 (Normal bundles in $\mathbb{P}^N$).

Let $Z \subset \mathbb{P}^N$ be a reduced local complete intersection subscheme. By the previous remarks, the natural map

$$T_{\mathbb{P}^N}|_Z \to N_{Z,\mathbb{P}^N}$$

is a surjection at smooth points of Z. The tangent bundle of $\mathbb{P}^N$ is the quotient of a direct sum of $N+1$ copies of $\mathcal{O}_{\mathbb{P}^N}(1)$ (see [9, II.8.13]), hence we get

$$\mathcal{O}_{\mathbb{P}^N}(1)^{N+1}|_Z \to N_{Z,\mathbb{P}^N}.$$

If $Z = C$ is a curve, we get a surjection of the first cohomologies

$$H^1\big(C, \mathcal{O}_{\mathbb{P}^N}(1)|_C\big)^{N+1} \twoheadrightarrow H^1(C, N_{C,\mathbb{P}^N}).$$

Thus if C is embedded by the sections of a line bundle L, this allows us to get vanishing results about the normal bundle of C in terms of $H^1(C,L)$.

The normal bundle of a diagonal embedding is described by the following lemma. We are interested in the case $Y = \mathbb{P}^N$.

Lemma 63. *Let X, Y be smooth varieties over a field k. Let C be a proper, reduced, local complete intersection curve. Let $f : C \to X$ be a morphism and $g : C \hookrightarrow Y$ an embedding. Let $C' \subset X \times Y$ be the image of the diagonal embedding. Then there is an exact sequence*

$$0 \to f^*T_X \to N_{C',X\times Y} \to N_{C,Y} \to 0.$$

Proof. The middle term of the second exact sequence

$$0 \to \mathcal{I}_{C'}/\mathcal{I}_{C'}^2 \to \Omega^1_{X\times Y}|_{C'} \to \Omega^1_{C'} \to 0$$

can be identified as

$$\Omega^1_{X\times Y}|_{C'} \cong f^*\Omega^1_X + \Omega^1_Y|_{C'}.$$

Since g is an embedding, the restriction $\Omega^1_Y|_C \to \Omega^1_C$ is surjective, which implies that the composition

$$\mathcal{I}_{C'}/\mathcal{I}^2_{C'} \to \Omega^1_{X\times Y}|_{C'} \to f^*\Omega^1_X$$

is surjective and its kernel is the same as the kernel of $\Omega^1_Y|_C \to \Omega^1_C$, which is just $\mathcal{I}_C/\mathcal{I}^2_C$ by the second exact sequence applied to $C \subset Y$. Putting all these together we get the exact sequence

$$0 \to \mathcal{I}_C/\mathcal{I}^2_C \to \mathcal{I}_{C'}/\mathcal{I}^2_{C'} \to f^*\Omega^1_X \to 0,$$

which is the dual of the claimed sequence. ■

64 (The local structure of $J(X, d, t)$).

We can now put things together to prove parts 1,2 and 3 of Proposition 60.

Let N_C be the normal bundle of C in $X_k \times \mathbb{P}^N$, embedded by (f, δ), and let N'_C be the normal bundle of the birational transform of C in $B_{t'}(X_k \times \mathbb{P}^N)$. Then $N'_C \cong N_C(-\sigma_k(P_k))$. Twisting the exact sequence of Lemma 63 we obtain an exact sequence

$$0 \to f^*T_X(-\sigma(P_k)) \to N'_C \to N_{C,\mathbb{P}^N}(-\sigma(P_k)) \to 0.$$

We have proved that $N_{C,\mathbb{P}^N}(-\sigma(P_k))$ is generically a quotient of $N+1$ copies of $L(-\sigma(P_k))$. This implies that $H^1(C, N'_C) = 0$. By Theorem 5.2, this implies Proposition 60.1.

The scheme $\mathrm{Hom}^1(P, \mathbb{P}^N)$ is smooth over S. Therefore, in a neighborhood of $\mathrm{Comp}(C, f, \sigma, \delta, \rho)$ the composition $W_g(X, d, t) \to \mathrm{Hom}^1(P, \mathbb{P}^N) \to S$ is smooth over the normal scheme S, hence $W_g(X, d, t)$ is normal. A normal scheme is locally irreducible, proving Proposition 60.2.

The dimension of the fiber of

$$W_g(X, d, t) \to \mathrm{Hom}^1(P, \mathbb{P}^N)$$

at $\left[(C, f, \sigma, \delta, \rho)\right]$ is $H^0(C, N'_C)$. This is computed as

$$\begin{aligned}
&h^0(C, N'_C) = \chi(C, N'_C) \\
&= \deg N'_C + (\dim X + N - 1)(1-g) \\
&= \deg f^*T_X\big(-\sigma(P_k)\big) + \deg N_{C,\mathbb{P}^N}\big(-\sigma(P_k)\big) + (\dim X + N - 1)(1-g) \\
&= -\big(f_*[C]\cdot K_X\big) - |P|(\dim X + N - 1) + (N+1)\deg L + 2g - 2 \\
&\quad +(\dim X + N - 1)(1-g) \\
&= -\big(f_*[C]\cdot K_X\big) + (\dim X - 3)(1-g) - |P|(\dim X - 1) \\
&\quad +(N+1)\deg L + N(1-g) - |P|\cdot N.
\end{aligned}$$

In order to get the dimension of $W_g(X, d, t)$, we have to add the dimension of $\mathrm{Hom}^1(P, \mathbb{P}^N)$, which is $|P|\cdot N + \dim S$. Finally using that $N = \deg L - g$ we get Proposition 60.3.

Proposition 60.4 is a special case of the next Lemma. ■

Lemma 65. *Let Y be a smooth variety over k and $C \subset Y$ a reduced curve with at worst nodes. Assume that $N_{C,Y}$ is generated by global sections and $H^1(C, N_{C,Y}) = 0$.*

Then Hilb (Y) *has a unique irreducible component containing $[C]$. This component is smooth at $[C]$ and a nonempty open subset of it parametrizes irreducible and smooth curves in Y.*

Proof. It is enough to prove that a general 1-parameter deformation has no nodes near a given node of C.

This is a local question, so choose local (analytic or formal) coordinates so that C is given by

$$y_1y_2 = y_3 = \cdots = y_n = 0.$$

The coordinates $y_3, \ldots, y_n$ generate a well-defined subspace of $N_{C,Y}\big|_P$. Assume that our 1-parameter deformation corresponds to a section of $N_{C,Y}$ that is not in that subspace at P. Then we have equations

$$y_1y_2 + tg(t, \mathbf{y}) = y_3 + tg_3(t, \mathbf{y}) = \cdots = y_n + tg_n(t, \mathbf{y}) = 0,$$

where $g(0,0) \neq 0$ by our assumption. We can choose new coordinates $y_i + tg_i(t, \mathbf{y})$ for $i = 3, \ldots, n$, and call them again y_i, to get a simpler system

$$y_1y_2 + t\big(a + G(t, \mathbf{y})\big) = y_3 = \cdots = y_n = 0,$$

where $a \neq 0$ and $G(0,0) = 0$. The singular points are given by equations $y_3 = \cdots = y_n = 0$ and

$$y_2 + t\frac{\partial G}{\partial y_1} = y_1 + t\frac{\partial G}{\partial y_2} = 0.$$

We can view these as equations that can be substituted back into $y_1 y_2 + t(a + G(t, \mathbf{y})) = 0$ to get a new equation for the supposed singular point:

$$ta = -tG(t, \mathbf{y}) - t^2 \frac{\partial G}{\partial y_1}\frac{\partial G}{\partial y_2}.$$

The latter has no solutions for $t \neq 0$ and $t, y_1, \ldots, y_n$ small since $a \neq 0$ and $G(0,0) = 0$. ∎

66 (The case $g \geq 1$). The only remaining problem is that we have worked with $J_{g,Q}(X, d, t)$ but, as noted in 53.7, we would like to conclude similar results about $J^*_{g,Q}(X, d, t)$. The latter is a subscheme of $J_{g,Q}(X, d, t)$, but it is better to view $J_{g,Q}(X, d, t)$ as Pic^0 of the universal curve over $J^*_{g,Q}(X, d, t)$. By [7], Pic^0 of a flat family of curves is smooth, hence we conclude that $J_{g,Q}(X, d, t) \to J^*_{g,Q}(X, d, t)$ is smooth of relative dimension g. Thus the smoothness results about J translate into smoothness results about J^*.

Acknowledgments. We thank J.-L. Colliot-Thélène and the referee for a long list of corrections and suggestions. Partial financial support was provided by the NSF under grant number DMS-9970855 and DMS02-00883. The first author was partially supported by CNPq.

References

[1] V. Alexeev, Moduli spaces $M_{g,n}(W)$ for surfaces, *Higher-dimensional complex varieties* (Trento, 1994), de Gruyter (Berlin, 1996), pp. 1–22.

[2] M. Artin and G. Winters, Degenerate fibres and stable reduction of curves, *Topology,* **10** (1971), 373–383.

[3] C. H. Clemens and P. Griffiths, The Intermediate Jacobian of the cubic three-fold, *Ann. of Math.,* **95** (1972), 281–356.

[4] P. Deligne and D. Mumford, The irreducibility of the space of curves of given genus, *Publ. Math. IHES,* **36** (1969), 75–109.

[5] T. Graber, J. Harris and J. Starr, Families of Rationally Connected Varieties, *J. Amer. Math. Soc.,* **16** (2003), 29–55.

[6] A. Grothendieck, *Technique de construction et théorèmes d'existence en géométrie algébrique IV: Les schémas de Hilbert,* Séminaire Bourbaki, 1960–1961, exposé 221, Soc. Math. France (Paris, 1962).

[7] A. Grothendieck, *Technique de descente et théorèmes d'existence en géométrie algébrique V. Les schémas de Picard: théorèmes d'existence; VI. Les schémas de Picard: propriétés générales,* Séminaire Bourbaki, Vol. 7 exposés 232, 236, Soc. Math. France (Paris, 1962).

[8] W. Fulton and R. Pandharipande, Notes on stable maps and quantum cohomology, in: *Algebraic geometry—Santa Cruz 1995* (S. Bloch et al., eds.) Proc. Sympos. Pure Math., vol. 62, Part 2, Amer. Math. Soc. (Providence, RI, 1997), pp. 45–96.

[9] R. Hartshorne, *Algebraic Geometry,* Graduate Texts in Mathematics, vol. 52, Springer (1977).

[10] S. Keel and S. Mori, Quotients by groupoids, *Ann. of Math.,* **145** (1997), 193–213.

[11] J. Kollár, Projectivity of complete moduli, *J. Differential Geom.,* **32** (1990), 235–268.

[12] J. Kollár, *Rational curves on algebraic varieties,* Ergebnisse der Mathematik und ihrer Grenzgebiete (3), vol. 32, Springer (1996).

[13] J. Kollár, Quotient spaces modulo algebraic groups, *Ann. of Math.,* **145** (1997), 33–79.

[14] J. Kollár, Rationally connected varieties over local fields, *Ann. of Math.,* **150** (1999), 357–367.

[15] J. Kollár, Which are the simplest Algebraic Varieties?, *Bull. Amer. Math. Soc.,* **38** (2001), 409–433.

[16] J. Kollár, *Specialization of zero cycles,* preprint `math.AG/0205148`.

[17] J. Kollár, Y. Miyaoka and S. Mori, Rationally Connected Varieties, *J. Algebraic Geom.,* **1** (1992), 429–448.

[18] J. Kollár and S. Mori, *Birational geometry of algebraic varieties,* with the collaboration of C. H. Clemens and A. Corti. Cambridge Tracts in Mathematics, vol. 134. Cambridge University Press (1998).

[19] M. Kontsevich and Y. I. Manin, Gromov–Witten classes, quantum cohomology, and enumerative geometry, *Comm. Math. Phys.*, **164** (1994), 525–562.

[20] A. F. Lopez, Noether–Lefschetz Theory and the Picard Group of Projective Surfaces, *Mem. Amer. Math. Soc.*, vol. 89 (1991).

[21] G. Xu, Subvarieties of General Hypersurfaces in Projective Space, *J. Differential Geom.*, **39** (1994), 139–172.

Carolina Araujo
Mathematics Department
Princeton University
Princeton
NJ 08544
carauj o@math.princeton.edu

János Kollár
Mathematics Department
Princeton University
Princeton
NJ 08544
kollar@math.princeton.edu

BOLYAI SOCIETY
MATHEMATICAL STUDIES, 12

Higher Dimensional Varieties and Rational Points, pp. 69–92.

Rationally Connected Varieties and Fundamental Groups

JÁNOS KOLLÁR

The Lefschetz hyperplane theorem says that if $X \subset \mathbb{P}^N$ is a smooth projective variety over $\mathbb{C}$ and $C \subset X$ is a smooth curve which is a complete intersection of hyperplanes with X then

$$\pi_1^{\mathrm{top}}(C) \to \pi_1^{\mathrm{top}}(X) \quad \text{is surjective,}$$

where π_1^{top} is the topological fundamental group. Later a quasiprojective version of this result was also established (see, for instance, [13] for a discussion and further generalizations). This says that if $X^0 \subset X$ is open and $C \subset X$ as above is *sufficiently general*, then

$$\pi_1^{\mathrm{top}}(C \cap X^0) \to \pi_1^{\mathrm{top}}(X^0) \quad \text{is surjective.}$$

On a rationally connected variety one would like to use rational curves to obtain a similar result. Complete intersection curves are essentially never rational. (For instance, if $X \subset \mathbb{P}^n$ is a hypersurface then a general complete intersection with hyperplanes is a rational curve iff X is a hyperplane or a quadric.) Therefore we have to proceed in a quite different way.

Let X be a smooth, projective, rationally connected variety over $\mathbb{C}$ and $X^0 \subset X$ an open set whose complement is a normal crossing divisor $\sum D_i$. Let $C \subset X$ be a smooth rational curve which intersects $\sum D_i$ transversally everywhere. It is then easy to see that the *normal* subgroup of $\pi_1^{\mathrm{top}}(X^0)$ generated by the image of $\pi_1^{\mathrm{top}}(C \cap X^0)$ is $\pi_1^{\mathrm{top}}(X^0)$ itself. One can also achieve that the image of $\pi_1^{\mathrm{top}}(C \cap X^0)$ has finite index in $\pi_1^{\mathrm{top}}(X^0)$. The map $\pi_1^{\mathrm{top}}(C \cap X^0) \to \pi_1^{\mathrm{top}}(X^0)$ is, however, not always onto. Some examples are given in [19, 4.5].

The group $\pi_1^{\mathrm{top}}(X^0)$ is typically an infinite group and in many questions the difference between it and its finite index subgroups is minor. However, in the arithmetic applications, for instance in (34) and in [22] it is crucial to prove surjectivity.

In order to work in arbitrary characteristic, we should use the algebraic fundamental group, which we denote by π_1. (A very good introduction to algebraic fundamental groups is [24]. A more thorough treatment is given in [27] and the ultimate reference is [14].) All the crucial points of the arguments in this paper can be seen by concentrating on the case of complex varieties and the topological fundamental group.

We care only about finite quotients of the algebraic fundamental group. Classically, quotients of the fundamental group correspond to covering spaces, and similarly, finite étale covers correspond to finite index open subgroups of the algebraic fundamental group. (The relevant topology plays no essential role in what follows.) A consequence of this is that a surjectivity between fundamental groups can be checked on finite quotients. We adopt this as a definition:

Definition-Theorem 1 (cf. [27, p. 94]). Let k be a field and $p : Y \to X$ a morphism of normal, geometrically connected k-schemes. Then $\pi_1(Y) \to \pi_1(X)$ is *surjective* if for every finite étale cover $X' \to X$ with X' connected, the pull back $Y \times_X X'$ is also connected.

Similarly, the image of $\pi_1(Y) \to \pi_1(X)$ is said to have index $\leq m$ if for every finite étale cover $X' \to X$ with X' connected, the pull back $Y \times_X X'$ has $\leq m$ connected components.

There are infinite groups without finite quotients. Thus there are compact manifolds which are not simply connected but have no finite sheeted covering spaces. It is not known if this also occurs for smooth projective varieties over $\mathbb{C}$. In any case, the algebraic fundamental group detects only the finite quotients of the topological fundamental group in characteristic zero.

Remark 2 (Base points). As in topology, the fundamental group can be defined only with a fixed base point. Fundamental groups of the same scheme with different base points are isomorphic, but the isomorphism is well defined only up to inner automorphisms. For some questions, for instance for surjectivity of maps, this does not matter. In such cases we occasionally omit the base point from the notation.

It is, however, important to keep in mind that ignoring the base point can easily lead to wrong conclusions.

In algebraic geometry the base point is a *geometric point* of X. That is, a morphism $\bar{x} \to X$ where $\bar{x}$ is the spectrum of an algebraically closed field.

If $x \in X$ is a point and K is any algebraically closed field containing the residue field $k(x)$, then we get a geometric point $\bar{x} := \operatorname{Spec} K$ with a map $\bar{x} \to X$. The resulting fundamental group $\pi_1(X, \bar{x})$ does not depend on the choice of K and we usually denote it by $\pi_1(X, x)$.

As in topology, we have the equivalence

$$\left\{ \begin{matrix} \text{finite index open} \\ \text{subgroups of } \pi_1(X,x) \end{matrix} \right\} \Leftrightarrow \left\{ \begin{matrix} \text{connected finite étale} \\ \text{covers of } X \text{ plus a} \\ \text{lifting of the base point} \end{matrix} \right\}$$

1. Statements of the Main Results

Separably rationally connected (or SRC) varieties were introduced in [21]. See [1, Sec. 34] in this volume for the definition and basic properties. (Note in particular that an SRC variety is always assumed to be geometrically connected.) A weaker version of our main Theorem (3) was proved in [19]. The key improvement is that the present proof, besides being simpler, does not need the resolution of singularities, thereby extending the range of applications to positive characteristic. It is also nicer that the choice of the curve C is largely independent of the open set X^0.

We formulate the main result in two versions, with fixed base point and with variable base point. Both are quite useful and it seems very cumbersome to have a unified statement.

Theorem 3. *Let X be a smooth, projective, SRC variety over a field k and $x \in X(k)$. Then there is a dominant family of rational curves through x (defined over k)*

$$F : W \times \mathbb{P}^1 \to X, \quad F\big(W \times (0{:}1)\big) = \{x\}$$

with the following properties:

1. *W is geometrically irreducible, smooth and open in* $\operatorname{Hom}\big(\mathbb{P}^1, X, (0{:}1) \mapsto x\big)$ *and $F : W \times (1{:}0) \to X$ is smooth.*
2. *There is a k-compactification $\bar{W} \supset W$ such that $\bar{W}$ has a smooth k-point.*
3. *The pullback $F_w^* T_X$ is ample for every $w \in W$.*

4. *Let $K \supset k$ be an algebraic closure and $X_K^0 \subset X_K$ any open K-subvariety containing x. Then for every $w \in W(K)$ the induced map on the geometric fundamental groups*

$$\pi_1\big(F^{-1}(X_K^0), (w, (0{:}1))\big) \twoheadrightarrow \pi_1(X_K^0, x) \quad \textit{is surjective.}$$

Remark 4. From the proof we see that we have considerable freedom in constructing W and F. Thus we can construct them with additional properties that are useful in some applications.

5. If $\dim X \geq 3$ then we can also assume that F is an embedding on every $\mathbb{P}^1_w$. If $\dim X = 2$ then we can also assume that F is an immersion on every $\mathbb{P}^1_w$.
6. Given an integer d, we can also assume that $F_w^* T_X(-d)$ is ample for every $w \in W$.
7. Fix an embedding $X \subset \mathbb{P}^N$. Then the degrees of the curves $F(\mathbb{P}^1_w)$ can be bounded from above in terms of the degree and dimension of X (and the above integer d).

We point out at the end of the proof (26) how to achieve these additional properties.

These comments also apply to (6).

Based on topology, one would expect that for $w \in W(K)$ general, the induced map

$$\pi_1\big(\mathbb{P}^1_w \cap F^{-1}(X_K^0), (0{:}1)\big) \to \pi_1(X_K^0, x)$$

is surjective. This is indeed so in characteristic zero, but in characteristic p the étale covers of p-power degree behave unexpectedly (see, for instance, example (15). The following weaker result, however, still holds, and this is enough for most applications.

Corollary 5. *Notation and assumptions as in (3). Let $\pi_1(X_K^0, x) \twoheadrightarrow G$ be a finite quotient (whose kernel is open). Then there is a dense open subset $W_G^0 \subset W$ such that for every $w \in W_G^0(K)$ the composed map*

$$\pi_1\big(\mathbb{P}^1_w \cap F^{-1}(X_K^0), (0{:}1)\big) \to \pi_1(X_K^0, x) \to G \quad \textit{is surjective.}$$

The base point free version of (3) is the following:

Theorem 6. *Let X be a smooth, projective SRC variety over a field k. Then there is a dominant family of rational curves (defined over k)*

$$F : W \times \mathbb{P}^1 \to X$$

with the following properties:

1. *W is geometrically irreducible and smooth.*
2. *F is smooth and $F_w^* T_X$ is ample for every $w \in W$.*
3. *Let $K \supset k$ be an algebraic closure and $X_K^0 \subset X_K$ any open K-subvariety. Then for every $z \in F^{-1}X_K^0(K)$ the induced map on the geometric fundamental groups*

$$\pi_1(F^{-1}X_K^0, z) \twoheadrightarrow \pi_1\big(X_K^0, F(z)\big) \quad \text{is surjective.}$$

Corollary 7. *Notation and assumptions as in (6). Let Y_K be an irreducible K-variety and $h : Y_K \to X_K$ a dominant morphism.*

Then there is a dense open set $W_Y^0 \subset W_K$ such that for every $w \in W_Y^0(K)$ the fiber product $Y_K \times_{X_K} \mathbb{P}^1_w$ is irreducible.

2. Fundamental Groups in Families

In this section we gather some fundamental group lemmas. The main result is (17) comparing the fundamental groups of different fibers of a morphism.

Lemma 8. *Let X be a normal variety and $X^0 \subset X$ an open subvariety. Then for every geometric point $x \to X^0$ the induced map $\pi_1(X^0, x) \twoheadrightarrow \pi_1(X, x)$ is surjective.*

Proof. Let $p : Y \to X$ be a finite étale cover with Y connected. Since Y is normal, it is also irreducible. Then $p^{-1}(X^0) \subset Y$ is a dense open subscheme, hence also irreducible and connected. ■

Lemma 9. *Let $p : Y \to X$ be a dominant morphism of normal varieties. Then there is an open subset $X^0 \subset X$ and a factorization*

$$p^{-1}(X^0) \xrightarrow{q} Z^0 \xrightarrow{r} X^0$$

such that q has only geometrically irreducible fibers and r is finite and étale. Z^0 is also normal and if Y is irreducible then so is Z^0.

Proof. Let $\bar{Y} \supset Y$ be a normal variety such that p extends to a proper morphism $\bar{p} : \bar{Y} \to X$. Let $\bar{Y} \to W \to X$ be the Stein factorization of $\bar{p}$. $W \to X$ is finite and dominant, hence W is the normalization of X in the function field $k(W)$. Let $F \subset k(W)$ denote the separable closure of $k(X)$ in $k(W)$. Let Z be the normalization of X in the function field F. Then $k(W)/k(Z)$ is purely inseparable and $k(Z)/k(X)$ is separable. Thus $Z \to X$ is étale over an open set $X^1 \subset X$.

Next we claim that there is an open subset $W^1 \subset W$ such that the fibers of $\bar{Y} \to W$ are geometrically irreducible over W^1. This is a little tricky in positive characteristic. Set $K = k(W)$.

The generic fiber $\bar{Y}_K$ is normal and $H^0(\bar{Y}_K, \mathcal{O}) = K$ by the definition of Stein factorization. Unfortunately this does not imply that the geometric generic fiber $\bar{Y}_{\bar{k}}$ is also normal. Nonetheless, by (10) the geometric generic fiber $\bar{Y}_{\bar{K}}$ is irreducible.

Thus there is an open subset $W^1 \subset W$ such that the fibers of $\bar{Y} \to W$ are geometrically irreducible over W^1 (cf. [8, IV.12]). Let $X^2 \subset X$ be an open subset whose preimage in W is contained in W^1. Taking $X^0 := X^1 \cap X^2$ we are done. ∎

Lemma 10. *Let Z_K be a normal proper variety over a field K and assume that $H^0(Z, \mathcal{O}_Z) = K$.*

Then Z is geometrically irreducible, that is, $Z_{\bar{K}}$ is irreducible.

Proof. If L/K is a separable field extension then $Z_L := Z_K \times_K \operatorname{Spec} L$ is also normal (see, for instance [23, 21.E]). If we are in characteristic zero then $\bar{K}/K$ is separable, thus $Z_{\bar{K}}$ is also normal. Cohomologies of sheaves change by tensoring under field extensions, thus $H^0(Z_{\bar{K}}, \mathcal{O}_{Z_{\bar{K}}}) = \bar{K}$, and so $Z_{\bar{K}}$ is connected. A normal and connected scheme is irreducible.

In positive characteristic we have a problem with inseparable extensions of K.

Let L/K be a finite degree purely inseparable extension. Let Z_L^n denote the normalization of Z_L. Then Z_L^n is also the normalization of Z_K in the function field $L(Z)$. $L(Z)/K(Z)$ is purely inseparable, thus there is a p-power q such that the q-power map $F_q : g \mapsto g^q$ maps $L(Z)$ into $K(Z)$. This implies that, up to a power of the Frobenius map, $Z_L^n \to Z_K$ can be inverted. In particular, $Z_L^n \to Z_K$ is a *homeomorphism.*

Every finite extension of K can be written as a separable extension followed by an inseparable one. Putting these two steps together we get that if Y_K is normal and geometrically connected then it is geometrically irreducible. ∎

Lemma 11. *Let $p : Y \to X$ be a dominant morphism of geometrically irreducible normal varieties over an algebraically closed field k. Let $\bar{y} \to Y$ be a geometric point of Y and $\bar{x} = p(\bar{y})$. Then* $\operatorname{im}\left[\pi_1(Y, \bar{y}) \to \pi_1(X, \bar{x})\right]$ *is a finite index closed subgroup of $\pi_1(X, \bar{x})$.*

Proof. Let $Y^0 := p^{-1}(X^0) \xrightarrow{q} Z^0 \xrightarrow{r} X^0$ be as in (9). The group $\pi_1(Z^0)$ has finite index closed image in $\pi_1(X^0)$, which surjects onto $\pi_1(X)$. Thus it is enough to prove that $\pi_1(Y^0)$ surjects onto $\pi_1(Z^0)$.

Let $Z^1 \to Z^0$ be a connected étale cover. Then $Y^0 \times_{Z^0} Z^1 \to Z^1$ has geometrically irreducible fibers, hence the fiber product is also connected. Thus $\pi_1(Y^0)$ surjects onto $\pi_1(Z^0)$ according to the definition (1). ∎

The following result was stated by [4] over $\mathbb{C}$ but the proof applies in any characteristic:

Corollary 12. *Let X be a normal, proper rationally connected variety over an algebraically closed field k. Then $\pi_1(X)$ is finite.*

Proof. Let $x \in X$ be any point. Since X is rationally connected, there is a dominant morphism

$$F : W \times \mathbb{P}^1 \to X \quad \text{such that} \quad F\big(W \times (0{:}1)\big) = \{x\}.$$

By (11) the image of $\pi_1(W \times \mathbb{P}^1)$ has finite index in $\pi_1(X)$. On the other hand, $\pi_1(W \times \mathbb{P}^1) = \pi_1\big(W \times (0{:}1)\big)$ and the latter is mapped to the identity. Thus $\pi_1(X)$ is finite. ∎

In the smooth case, we even have simple connectedness. In characteristic zero this was proved by [4], it is also explained in [7]. The positive characteristic case follows from a recent result of [16], a proof is written up in [6, 3.6].

Theorem 13 [20]. *Let X be a smooth projective SRC variety over an algebraically closed field k of arbitrary characteristic. Then $\pi_1(X) = \{1\}$.* ∎

14 (Fundamental groups in families). We also need to know how the fundamental group varies in families. In characteristic zero the question is topological and easy. There are some new twists in positive characteristic. Several of these are illustrated by the following example:

Example 15. Let us work over an algebraically closed field of characteristic p. Consider the surface $S := (f = 0) \subset \mathbb{A}^3$ where $f = y + z - xz^p + z^{2p}$. The derivative $\partial f/\partial z = 1$, so the projection onto the $z = 0$ plane $S \to \mathbb{A}^2$ is finite and étale. (This already shows that $\pi_1(\mathbb{A}^2)$ is nontrivial.)

Consider the line $L_c := (y - cx = 0)$. The preimage of L_c in S is isomorphic to the curve $cx + z - xz^p + z^{2p} = 0$. If $c = 0$ then this curve is reducible but if $c \neq 0$ then it is irreducible. This shows that the image of

$$\pi_1(\mathbb{A}^1) \cong \pi_1(L_c) \to \pi_1(\mathbb{A}^2)$$

depends on c, and in fact the map is not surjective for $c = 0$. Since any two lines are equivalent under automorphisms of $\mathbb{A}^2$, we get that for any line $L \subset \mathbb{A}^2$, the natural map

$$\pi_1(\mathbb{A}^1) \cong \pi_1(L) \to \pi_1(\mathbb{A}^2)$$

is not surjective. It is not hard to see, however, that the following slightly weaker version holds:

Let $\pi_1(\mathbb{A}^2) \to G$ be a finite quotient (with open kernel). Then the composite

$$\pi_1(\mathbb{A}^1) \cong \pi_1(L) \to \pi_1(\mathbb{A}^2) \to G$$

is surjective for a general line L, the notion of general depending on G.

The latter observation holds very generally:

Proposition 16. *Let k be an algebraically closed field. Let $f : Y \to X$ be a morphism of irreducible and normal k-varieties. Let $S \subset Y$ be an irreducible sub-variety such that $S \to X$ is dominant. For $s \in S(k)$ let Y_s denote the fiber of f over $f(s)$. Let Z be a variety, $z \in Z(k)$ a point and $\pi_1(Z, z) \to G$ a finite quotient (with open kernel). Let $h : Y \to Z$ be a morphism such that $h(S) = \{z\} \in Z$.*

Then there is a dense open subset $S^0 \subset S$ such that for every $s \in S^0(k)$ the two maps

$$\pi_1(Y_s, s) \to \pi_1(Z, z) \to G \quad \text{and} \quad \pi_1(Y, s) \to \pi_1(Z, z) \to G$$

have the same image.

Proof. We are allowed to change base from X to S and then replace S by an open subset. Thus we can assume that $S \cong X$ is a section.

Let $Z' \to Z$ be the cover corresponding to the image of $\pi_1(Y, s) \to \pi_1(Z, z) \to G$. We need to prove that the number of irreducible components of $Z' \times_Z Y$ is the same as the number of irreducible components of $Z' \times_Z Y_s$ for $s \in S^0$. Let $W \subset Z' \times_Z Y$ be an irreducible component. Then W is an irreducible étale cover of Y which is trivial over S. Let $g : W \to X$ be the composite.

As in (9) there are open sets $X^0 \subset X$, $W^0 \subset W$ and a factorization $W^0 \to V^0 \to X^0$ where $W^0 \to V^0$ has geometrically irreducible fibers and $V^0 \to X^0$ is finite and étale. Since $S \subset W$ is a section of g, we have $X^0 \cong S \cap W^0 \to V^0 \to X^0$, thus $V^0 \cong X^0$. Therefore $W^0 \to X^0$ has geometrically irreducible fibers. ∎

As a corollary we obtain a lower semicontinuity statement for fundamental groups of fibers. We do not compare the groups themselves, just their images in finite groups whenever this can be done sensibly.

Corollary 17. *Let k be an algebraically closed field. Let $f : Y \to X$ be a morphism of irreducible and normal k-varieties with a section $S \subset Y$. For $s \in S$ let Y_s denote the fiber of f over $f(s)$. Let Z be a k-variety, $z \in Z(k)$ a point and $h : Y \to Z$ a morphism such that $h(S) = \{z\} \in Z$.*

1. *For any $s, s' \in S$,*

$$\mathrm{im}\left[\pi_1(Y_s, s) \to \pi_1(Z, z)\right] \subset \mathrm{im}\left[\pi_1(Y, s') \to \pi_1(Z, z)\right].$$

2. *Let $S^0 \subset S$ be any dense open subset. Then*

$$\mathrm{im}\left[\pi_1(Y, s') \to \pi_1(Z, z)\right] = \left\langle \mathrm{im}\left[\pi_1(Y_s, s) \to \pi_1(Z, z)\right] \ : \forall s \in S^0 \right\rangle.$$

Proof. The maps $Y_s \to Y \to Z$ give the inclusion

$$\mathrm{im}\left[\pi_1(Y_s, s) \to \pi_1(Z, z)\right] \subset \mathrm{im}\left[\pi_1(Y, s) \to \pi_1(Z, z)\right].$$

Since S is irreducible and contracted to z,

$$\mathrm{im}\left[\pi_1(Y, s) \to \pi_1(Z, z)\right] = \mathrm{im}\left[\pi_1(Y, s') \to \pi_1(Z, z)\right]$$

for any $s, s' \in S$, proving (1).

This implies the containment $\supset$ in (2), and we only need the reverse inclusion. By the correspondence between the fundamental group and finite étale covers we need to prove that if $Z' \to Z$ is a finite étale cover such that $Y_s \times_Z Z'$ is reducible for every $s \in S^0$ then $Y \times_Z Z'$ is also reducible. This is so by (16). ∎

Remark 18. Several times we will apply (17) to smoothings of combs, as defined in [1, 43]. Let $C = C_0 \cup \cdots \cup C_n$ be a pointed comb with handle $C_0 \ni p$. Let $c_i \in C_i$ be the nodes and $C^* \subset C$ an open subset containing C_0. The injections $C_i \hookrightarrow C$ induce maps $\pi_1(C^* \cap C_i, c_i) \to \pi_1(C^*, c_i)$. (These are in fact injections, but this is not crucial to us.) Thus we obtain maps $\pi_1(C^* \cap C_i, c_i) \to \pi_1(C^*, p)$ which are well defined modulo conjugation by an element of $\pi_1(C_0, p)$.

In particular, if $F : C \to X$ is assembled from the maps $f_i : C_i \to X$ and $X^0 \subset X$ is an open subset containing x then

$$\operatorname{im}\left[\pi_1(F^{-1}(X^0), p) \to \pi_1(X^0, x)\right]$$

contains all the images

$$\operatorname{im}\left[\pi_1(f_i^{-1}(X^0), c_i) \to \pi_1(X^0, x)\right] \quad \text{for } i = 1, \ldots, n.$$

At least when the base field is algebraically closed, the conjugation by elements of $\pi_1(C_0, p)$ does not matter since $F(C_0) = \{x\}$ so $\pi_1(C_0, p)$ is mapped to the identity.

Note that a similar assertion would not hold without the assumption $F(C_0) = x$.

3. Proofs of the Theorems

First we reduce (3) to the a priori much weaker (19). Then in the second part we prove (19). Finally we explain that (3) implies the other statements. We repeatedly use several properties of stable curves discussed in [1, Sec.8].

Proposition 19. *Let X be a smooth, projective, SRC variety over an algebraically closed field k and $x \in X(k)$. Then there is a family of free genus zero stable maps with base point x*

$$(C/S,\ F : C \to X \times S,\ \sigma : S \to C)$$

parametrized by a scheme of finite type S such that for every open subset $X^0 \subset X$ with $x \in X^0$ the images

$$\operatorname{im}\left[\pi_1\left(F_s^{-1}(X^0),\ \sigma_s(s)\right) \to \pi_1(X^0, x)\right] \quad \textit{for all } s \in S$$

(topologically) generate $\pi_1(X^0, x)$.

20 (Proof of (19) $\Rightarrow$ (3)). Assume first that k is algebraically closed. We start with the family $(C/S,\ F : C \to X \times S,\ \sigma : S \to C)$ in (19) and using it we repeatedly construct other families until we end up with one as required for (3).

$(C/S,\ F : C \to X \times S,\ \sigma : S \to C)$ gives the moduli morphism $S \to \bar{M}_0(X, \operatorname{Spec} k \to x)$ (cf. [1, 41]); let S^* be its image (as a constructible set).

Pick any $s \in S(k)$ and let

$$\begin{array}{ccc} C_s \subset C_{U_s} & \stackrel{\Phi_{U_s}}{\to} & X \\ \downarrow \quad\ \ \downarrow & & \\ s \ \in \ U_s & & \end{array}$$

be a versal family as in [1, 42.5].

To each U_s there corresponds a natural morphism $U_s \to \bar{M}_0(X, \operatorname{Spec} k \to x)$ with open image. Finitely many of these cover S^*. From now on we work with these finitely many families

$$(C_{U_s}/U_s,\ \Phi_{U_s} : C_{U_s} \to X \times U_s,\ \sigma_{U_s} : U_s \to C_{U_s}).$$

From (17) we see that the conclusion of (19) remains true if we replace $\cup U_s$ by a dense open subset. By [1, 42.3] the points corresponding to irreducible curves are dense in each U_s, thus we may replace each U_s by the open subset of irreducible and free curves.

We now have finitely many (say N) families of stable and free maps

$$\Phi_j : U_j \times \mathbb{P}^1 \to X \quad \text{with} \quad \Phi_j(U_j \times (0{:}1)) = x,$$

where each U_j is irreducible. We are free to add one more family $\Phi_{N+1} : U_{N+1} \times \mathbb{P}^1 \to X$ consisting of very free curves through x. In order to create a single family, we use a trick introduced in [18].

Pick points $u_j \in U_j$ and, as in [1, 43], assemble a pointed comb out of the maps $\Phi_{j,u_j} : \mathbb{P}^1 \to X$. Since $N + 1 \geq 2$, these pointed combs are stable genus zero curves over X. All these combs are parametrized by an irreducible variety (the product of the U_j times the space of $N + 1$ points on $\mathbb{P}^1$). Thus all these combs are in the same irreducible component $\mathcal{U} \subset \bar{M}_0(X, \operatorname{Spec} k \to x)$.

As before, using (18) we can first replace the families $\Phi_j : U_j \times \mathbb{P}^1 \to X$ with the irreducible component $\mathcal{U}$ of the family of stable curves containing all of the above combs

$$f_C : C = C_0 \cup \cdots \cup C_{N+1} \to X,$$

and then by (17) with the open subfamily $\mathcal{U}^0 \subset \mathcal{U}$ of irreducible and free curves. The moduli map $\mathcal{U}^0 \to \operatorname{Hom}(\mathbb{P}^1, X, (0{:}1) \mapsto x)$ is dominant onto one of the irreducible components. Let $V \subset \operatorname{Hom}(\mathbb{P}^1, X, (0{:}1) \mapsto x)$ be an open set contained in the image of $\mathcal{U}^0$.

Thus we obtain a single irreducible family of stable maps

$$\Phi : V \times \mathbb{P}^1 \to X \quad \text{with} \quad \Phi(V \times (0{:}1)) = x.$$

We claim that this satisfies all the requirements of (3) for k algebraically closed.

Indeed, V is geometrically irreducible, smooth and open in

$$\operatorname{Hom}(\mathbb{P}^1, X, (0{:}1) \mapsto x)$$

by construction. By [17, II.3.4], the tangent map of $F : (V \times (1{:}0)) \to X$ at $(v, (1{:}0))$ is given by the restriction

$$H^0(\mathbb{P}^1, F_v^* T_X(-(0{:}1))) \to F_v^* T_X|_{(1:0)} \cong T_{X, F(v,(1:0))}$$

which is surjective if $H^1(\mathbb{P}^1, F_v^* T_X(-2)) = 0$. We prove the latter below, thus $F : (V \times (1{:}0)) \to X$ is smooth, at least after passing to a suitable open subset of V.

Condition (3.2) is not interesting when k is algebraically closed.

Ampleness of a vector bundle E on $\mathbb{P}^1$ is equivalent to $H^1(\mathbb{P}^1, E(-2)) = 0$. By upper semicontinuity, $H^1(\mathbb{P}^1, F_v^* T_X(-2)) = 0$ for general $v \in V$ if $H^1(C, f_C^* T_X(-2p)) = 0$ for one of the combs $f_C : C \to X$ and for some smooth point $p \in C$. Pick $p \in C_{N+1}$. By our choice of the $(N+1)$st family, $F^* T_X$ restricted to C_{N+1} is ample and all the other restrictions to C_i are semipositive. This easily implies that $H^1(C, f_C^* T_X(-2p)) = 0$, cf. [1, 18]. This proves (3.3) and also (3.1).

Applying (17) several times, we get that the closed subgroup of $\pi_1(X^0, x)$ generated by the images of $\pi_1(\mathbb{P}_v^1 \cap F^{-1}(X^0), (0{:}1)) : v \in V$ is $\pi_1(X^0, x)$ itself. On the other hand, the image of $\pi_1(F^{-1}(X_K^0), (v, (0{:}1))) \to \pi_1(X_K^0, x)$ is closed by (11), thus $\pi_1(F^{-1}(X_K^0), (v, (0{:}1))) \to \pi_1(X_K^0, x)$ is surjective, giving (3.4).

Assume next that k is not algebraically closed, and let $\bar{k}$ be its algebraic closure. The family constructed above is already defined over a finite Galois extensions k'/k with Galois group G. (This is automatic in characteristic 0. In general, note that free curves give smooth points of $\operatorname{Hom}(\mathbb{P}^1, X)$, and the smooth geometric components of a k-scheme are defined over a separable extension.) We may also assume that V has a distinguished k'-point $v \in V(k')$. By [17, II.3.14.2] we may also assume that the corresponding map $\{v\} \times \mathbb{P}^1 \to X$ is an immersion.

For every $\rho \in G$ let

$$\Phi^\rho : V^\rho \times \mathbb{P}^1 \to X$$

be the conjugate family obtained by applying ρ to all the defining equations.

We now proceed as above. Pick arbitrary points $v_\rho \in V^\rho(\bar{k})$ and assemble a pointed comb out of the maps $\Phi^\rho_{v_\rho} : \mathbb{P}^1 \to X$. All these combs are parametrized by an irreducible variety which is now defined over k. (Over k' this parameter space is the product of the V^ρ times the space of N points on $\mathbb{P}^1$. Over k, it is fibered over the Weil restriction $\mathfrak{R}_{k'/k}V$ (see, for instance [3, 7.6]) with various forms of the space of $|G|$ points on $\mathbb{P}^1$ as fibers.) As before, all these combs are in the same geometrically irreducible component of $\mathcal{W} \subset \bar{M}_0(X, \operatorname{Spec} k \to x)$.

By (17) we can replace $\mathcal{W}$ with the subfamily W of irreducible and free curves.

In order to check the required properties, the only new aspect is (3.2). The comb C assembled from the conjugates of $\{v\} \times \mathbb{P}^1$ gives a smooth k-point of $\bar{\mathcal{W}}$. Indeed, all the conjugates of $\{v\} \times \mathbb{P}^1 \to X$ are immersions, thus C has no automorphisms commuting with $C \to X$ and fixing the base point. Therefore C gives a smooth point of $\bar{\mathcal{W}}$ by [10, 2.iii]. ∎

21 (Idea of the proof of (19)). Take all possible free, stable, genus 0 curves $g : (c \in C) \to (x \in X)$ and let $H \subset \pi_1(X^0, x)$ be the subgroup (topologically) generated by all the images $\pi_1(C^0, c) \to \pi_1(X^0, x)$ where $C^0 := g^{-1}(X^0)$. We want to prove that $H = \pi_1(X^0, x)$. If not, then there is a finite index closed subgroup $H \subset H' \subsetneq \pi_1(X^0, x)$. The subgroup H' corresponds to a degree d étale cover $Y^0 \to X^0$ together with a geometric point $y \to Y^0$ lying over $x \to X^0$. Extend it to a finite morphism $p : Y \to X$.

Since X is simply connected (13), p is not étale and so it ramifies along a divisor $D \subset Y$ (using the purity of branch loci).

Pick a general point $x' \in p(D)$ and assume that $C \subset X$ is a rational curve passing through x, x' and transversal to $p(D)$ everywhere. We hope that $C \to X$ cannot be lifted to Y. The local picture we should have in mind is

$$Y = \mathbb{C}^2_{y_1,y_2} \stackrel{(y_1,y_2)\mapsto(y_1,y_2^m)}{\longrightarrow} \mathbb{C}^2_{x_1,x_2} = X.$$

Here $D = (y_2 = 0)$ and $p(D) = (x_2 = 0)$. Set $y' := (0,0) \in Y$. If C is the line $(x_1 = 0)$ through $x' := (0,0)$ then $p^{-1}(C)$ is the irreducible curve $(y_1 = 0)$ through y'. Thus the inclusion $C \hookrightarrow X$ cannot be lifted to $C \to Y$ in such a way that $x' \in C$ is mapped to $y' \in Y$.

This is, however, only the local picture. The global problem is that Y has many other points over x' besides y', and p ramifies only at some of these. The best we can say is the following.

The fiber product $C_Y := Y \times_X C$ is a smooth curve and $C_Y \to C$ definitely ramifies at $C_Y \cap D$. Thus C_Y is *not* the disjoint union of d copies of C. This means that there is at least one point $y_{i_0} \in p^{-1}(x)$ such that the inclusion $C \hookrightarrow X$ cannot be lifted to $C \to Y$ sending c to y_{i_0}. For other choices $y_i \in p^{-1}(x)$ a lifting may exist. It is very hard to tell which preimage is which.

Choosing a preimage y_i is essentially equivalent to choosing a conjugate of the subgroup H', thus we cannot move freely between the different preimages.

We circumvent this problem as follows.

First we choose a general smooth rational curve $B \subset X$ passing through a general point $z \in X$ and intersecting $p(D)$ transversally everywhere. Let $v_i \in Y$ $i = 1, \dots, d$ be all the preimages of z. As with C, there is at least one v_{i_0} such that $B \hookrightarrow X$ cannot be lifted to $B \to Y$ sending z to v_{i_0}.

Then we pick other rational curves A_k passing through x and z. We furthermore achieve (and this turns out to be easy for a *general* point) that the liftings of the curves A_k connect y with *all* the different points v_i.

This implies that at least one of the reducible curves $A_k \cup B$ cannot be lifted to Y. (See Fig. 1.)

There are only two points that need refinement in this method. First, in positive characteristic the local description of the ramification is much more complicated than over $\mathbb{C}$. Second, in order to make these choices more uniform, we should work with a whole family of curves B. This actually also takes care of the ramification problem. This is done next.

22 (Main construction). Let X be a smooth irreducible variety over an algebraically closed field k. Assume that we are given two families of curves on X

$U \stackrel{u}{\leftarrow} A \stackrel{F}{\rightarrow} X$ and $V \stackrel{v}{\leftarrow} B \stackrel{G}{\rightarrow} X$ with the following properties.

1. The k-varieties U, V are irreducible and u, v are proper, smooth morphisms with irreducible fibers.

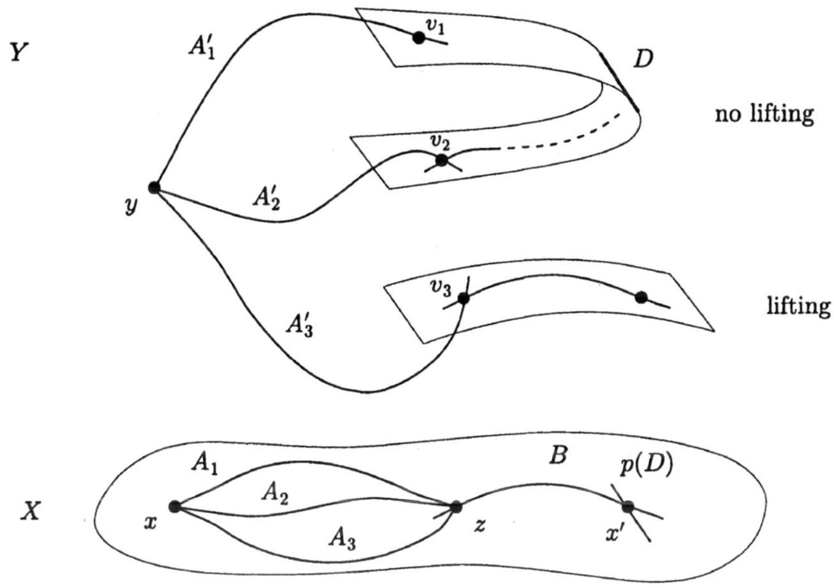

Fig. 1. A non-liftable reducible curve

2. The map F is dominant.
3. The map u has a section $s : U \to A$ such that $F \circ s$ maps U to a single point $x \in X(k)$.
4. The images of the maps $\pi_1\big(A, s(u)\big) \to \pi_1(X, x)$ for $u \in U$ topologically generate $\pi_1(X, x)$.
5. The map G is smooth and for every divisor $D \subset X$ the general $G(B_v)$ intersects D.

Every point of the fiber product $W := A \times_X B$ can be thought of as a quadruplet $(a_u \in A_u, b_v \in B_v)$ such that $F(a_u) = G(b_v)$. To this corresponds a reducible nodal curve $A_u \cup_{a_u \sim b_v} B_v$ obtained from the disjoint union of A_u and B_v by identifying the points a_u and b_v. Thus we have a flat family of reducible but connected curves

$$W \leftarrow C \xrightarrow{H} X.$$

The map $s : U \to A$ gives a section $S : W \to C$ such that $H \circ S$ maps W to $\{x\}$. (Here C can be explicitly constructed as

$$A \times_U A \times_X B \cup A \times_X B \times_V B \subset A \times_U A \times_X B \times_V B,$$

where the embedding is with $Id \times_X \Delta_B \cup \Delta_A \times_X Id$ where $\Delta_B : B \to B \times_V B$ and $\Delta_A : A \to A \times_U A$ are the diagonal embeddings. The map $C \to W$ is projection to the middle $A \times_X B$ of the above product. Finally S is given by the fiber product of s with the identity map $A \times_X B = U \times_U A \times_X B \to A \times_U A \times_X B$.)

Proposition 23. *Notation and assumptions as above. Then, for every open set $X^0 \subset X$ containing x*

$$\pi_1(X^0, x) = \Big\langle \operatorname{im}\big[\pi_1\big(C_w \cap H^{-1}(X^0), s(w)\big) \to \pi_1(X^0, x)\big] : \forall w \in W(k)\Big\rangle,$$

where $\langle \cdots \rangle$ denotes the closed subgroup generated by the images.

Proof. It is enough to show that if $(Y^0, y) \to (X^0, x)$ is a pointed finite étale cover with Y^0 irreducible and $p : Y \to X$ its extension to a (possibly ramified) finite morphism then

$$H_w : C_w = A_u \cup_{a_u \sim b_v} B_v \to X$$

cannot be lifted to Y for some w.

If $Y \to X$ is unramified then there is a $u \in U(k)$ such that $A_u \to X$ cannot be lifted by assumption 4. Thus we are left to consider the case when $Y \to X$ is ramified and every $A_u \to X$ can be lifted. We claim that in fact $F : A \to X$ lifts to $F_Y : A \to Y$ with $F_Y \circ s : U \to y$. To see this, consider the fiber product $A \times_X Y$. $s : U \to A$ and the constant map $U \to y$ provide $\sigma : U \to A \times_X Y$. Let $A_Y \subset A \times_X Y$ be the unique irreducible component containing $\sigma(U)$. By assumption $A_Y \to A$ is an isomorphism on every fiber of $A \to U$, hence an isomorphism. Thus A_Y is the graph of the required lifting F_Y.

The map F_Y is dominant, so there is an open set $X^1 \subset X^0$ such that $F_Y(A) \supset p^{-1}(X^1)$ and $X^1 \subset G(B)$.

Let $z \in X^1(k)$ be a point and $p^{-1}(z) = \{v_1, \ldots, v_m\}$. By assumption there are $u_i \in U$ and $a_i \in A_{u_i}$ such that $F_Y(a_i) = v_i$. Pick any $v \in V(k)$ and $b \in B_v$ such that $G(b) = z$.

The quadruplets $(a_i \in A_{u_i}, b \in B_v)$ give connected curves $C_i := A_{u_i} \cup_{a_i \sim b} B_v$. If $H_i : C_i \to X$ lifts to $H_{i,Y} : C_i \to Y$ then $H_{i,Y}(B_v)$ is a lifting of $G : B_v \to X$ which passes through v_i. Thus we obtain:

Claim 24. Under the above assumptions, the fiber product $B_v \times_X Y$ has m irreducible components, each isomorphic to B_v.

We show that this is impossible, which implies that our assumption is incorrect. Thus not all maps $H_i : C_i \to X$ lift to $H_{i,Y} : C_i \to Y$. This will complete the proof of (23).

It should be emphasized that at this point we only know that $B_v \times_X Y$ has m irreducible components, each isomorphic to B_v. We do not know that $B_v \times_X Y$ is a *disjoint* union of its irreducible components. The key point is precisely to establish this for general $v \in V(k)$.

Set $B_Y := B \times_X Y$. $G : B \to X$ is smooth, hence $B_Y \to Y$ is also smooth, thus B_Y is normal.

Consider the diagram

$$\begin{array}{ccccc} B_Y & \stackrel{p_B}{\to} & B & \stackrel{G}{\to} & X \\ v_Y \downarrow & & \downarrow v & & \\ V & = & V. & & \end{array}$$

In characteristic zero, B_Y normal implies that the geometric generic fiber of B_Y/V is also normal, hence it is a disjoint union of its irreducible components.

In positive characteristic, we have only a weaker result (9), which still says that the geometric generic fiber of B_Y/V is a disjoint union of its irreducible components. In particular, the generic fiber of B_Y/V is smooth over the generic fiber of B/V. Thus, after possibly shrinking V, we may assume that p_B is smooth. G is smooth by assumption, hence the composite $B_Y \to X$ is also smooth.

One can, however, factor this as $B_Y \to Y \to X$. By the purity of branch loci, $Y \to X$ ramifies over a whole divisor $D \subset X$, and so $B_Y \to X$ ramifies over every point of D. This is only possible if the image of $B_Y \to B \to X$ does not intersect D. This contradicts the assumption 5. ∎

25 (Proof of (19)). All that is left is the construction of the families $U \leftarrow A \to X$ and $V \leftarrow B \to X$ as in (22). The rest is already taken care of by (23).

For $U \leftarrow A \to X$ we take a family of rational curves through x that covers X. Property (22.4) holds by (13).

In order to get the family $B \to V$, it is enough to get a very free curve $\mathbb{P}^1 \to X$ which has positive intersection number with every divisor. A free curve has nonnegative intersection number with every divisor. Start with a family $H : U \times \mathbb{P}^1 \to X$ that shows that X is SRC. The only problem is that its image may not intersect finitely many divisors $D_1, \ldots, D_s \subset X$. By [1, 29.5] there is a free morphism $g : \mathbb{P}^1 \to X$ whose image intersects every D_i. Now take $g(\mathbb{P}^1)$ and a curve $H(\mathbb{P}^1_u)$ intersecting it. By [17, II.7] their union can be smoothed to a free curve which has positive intersection number with every divisor. The resulting morphism $G : B \to X$ is smooth by [17, II.3.5.3]. ∎

26. The additional properties listed in (4) are also easy to get.

First, by [17, II.3.14], F is automatically an embedding for general $w \in W$ if $\dim X \geq 3$, and we can replace W with an open subset at any time.

Assume now that we want more ampleness from $F_w^* T_X$. Take a family of morphisms $h : U \times \mathbb{P}^1 \to X$ such that $h_u^* T_X(-d)$ is ample and h is smooth. Let us now look at all maps of one tooth combs $g : C_0 \cup C_1 \to X$ where $g|_{C_0}$ is from the family F and $g|_{C_1}$ from the family h. Smoothing these as in (20), we get a new family $F_d : W_d \times \mathbb{P}^1 \to X$ that also satisfies (4.6)

Fix next an embedding $X \subset \mathbb{P}^N$. All smooth varieties of given degree and dimension form a bounded family (the Chow variety). Being SRC is an open property (cf. [17, IV.3.11]), hence all SRC varieties of given degree and dimension form a bounded family. This implies that we can choose bounded degree curves for the families A, B in (22). Thus we get families of bounded degree curves in (19).

There is only one problem left, namely that during the proof in (20) we have to take all conjugates of a family of curves, and their number does depend on the field. The families we have are open subsets of an irreducible component of the Hom scheme $\mathrm{Hom}(\mathbb{P}^1, X)$. We are dealing with maps of bounded degree to a bounded family of varieties. The relative Hom scheme of bounded degree maps is quasiprojective (cf. [17, I.1.10]), hence there is a bound for the number of irreducible families of bounded degree maps. The number of conjugates cannot exceed this bound. ∎

27 (Proof of (5)). From (3) we know that $\pi_1(F^{-1}(X^0)) \to G$ is surjective. Applying (16) to $F^{-1}(X^0) \to W$ we conclude that $\pi_1(\mathbb{P}^1_w \cap F^{-1}(X^0)) \to G$ is surjective for all w in some open subset $W^G \subset W$. ∎

28 (Proof of (6)). Let us start with a variety X and let $k(X)$ denote its field of rational functions. The generic point $x_g \in X$ is a $k(X)$-point of

X, thus we can apply (3) to $X_{k(X)}$ and x_g. Thus we get a geometrically irreducible $k(X)$-variety V_g and a morphism

$$F_g : V_g \times \mathbb{P}^1 \to X_{k(X)}.$$

V_g can be thought of as the generic fiber of a map $V \to X$ where V is a geometrically irreducible k-variety. F_g extends to a map which can be composed with the second projection to obtain

$$F_V : V \times \mathbb{P}^1 \dashrightarrow X \times X \to X.$$

F_V need not be everywhere defined, but it becomes a morphism after restriction to a suitable open set of the form $V' \times \mathbb{P}^1$. By choosing V' small enough we can also assume that F_V is free on $\{v\} \times \mathbb{P}^1$ for every $v \in V'$.

Let us now look at the relative Hom-scheme (cf. [17, I.1.19]) $H := \mathrm{Hom}_{V'}(V' \times \mathbb{P}^1, V' \times X)$. The restriction of F_V to V' determines a section $s : V' \to H$ and the universal morphism $H \times \mathbb{P}^1 \to V' \times X$ is smooth over this section by [17, II.3.5.3]. This also imples that $H \to V'$ is smooth along $s(V')$, hence H has a unique irreducible component $W' \subset H$ which contains $s(V')$ and W' is geometrically irreducible. Furthermore, there is an open subset $W \subset W'$, containing $s(V')$ such that the universal morphism $F : W \times \mathbb{P}^1 \to V' \times X \to X$ is smooth. This gives the required morphism. ∎

29 (Proof of (7)). Choose a partial compactification $Y \subset \bar{Y}$ such that h extends to a proper morphism $\bar{h} : \bar{Y} \to X$. By the upper semicontinuity of fiber dimensions, there is a closed subset $T \subset X$ of codimension 2 such that every fiber of h over $X \setminus T$ has the same dimension.

This implies that if B is any irreducible curve and $B \to X \setminus T$ a morphism then every irreducible component of $B \times_X Y$ dominates B. Indeed, the fiber product $B \times_X Y$ is the preimage of the diagonal under the morphism $B \times Y \to X \times X$. Therefore every irreducible component of $B \times_X Y$ has dimension at least $\dim Y - \dim X + 1$. If B maps to $X \setminus T$ then every fiber of $B \times_X Y \to B$ has dimension $\dim Y - \dim X$, hence the claim.

Choose $F : W \times \mathbb{P}^1 \to X$ as in (6). The map F is smooth, so by an easy dimension count, there is an open subset $W^1 \subset W$ such that $F(W^1 \times \mathbb{P}^1) \subset X \setminus T$ (cf. [1, 9]).

Let

$$Y^0 := p^{-1}(X^0) \xrightarrow{q} Z^0 \xrightarrow{r} X^0$$

be as in (9), where Z^0 is normal and irreducible. The map $q_w : Y^0 \times_X \mathbb{P}^1_w \to Z^0 \times_X \mathbb{P}^1_w$ has geometrically irreducible fibers, thus it is enough to prove that $Z^0 \times_X \mathbb{P}^1_w$ is irreducible if $w \in W^0$.

The product $W^1 \times \mathbb{P}^1 \times_X Z^0$ is irreducible by (3) and also smooth, and it has a section over W^1, for instance $w \mapsto (w, 0, F(w, 0))$. So the generic fiber of $W^1 \times \mathbb{P}^1 \times_X Z^0 \to W^1$ is irreducible over $k(W^1)$, smooth and it also has a $k(W^1)$-point. Therefore it is geometrically irreducible. Thus there is an open subset $W^0 \subset W^1$ such that the fibers of $W^1 \times \mathbb{P}^1 \times_X Z^0 \to W^1$ are geometrically irreducible over W^0 (cf. [8, IV.12]). ∎

4. Applications to Non-closed Fields

In this section we derive some consequences of the previous results to fields which are not algebraically closed.

Remark 30 (Fundamental groups over non-closed fields). If k is a field then the connected étale covers of $\operatorname{Spec} k$ are exactly the maps $\operatorname{Spec} K \to \operatorname{Spec} k$ where K/k is a finite separable field extension. This implies that

$$\pi_1(\operatorname{Spec}(k)) = \operatorname{Gal}(k^{sep}/k)$$

where k^{sep} is the separable closure of k. In general, the fundamental group of a k-scheme X_k is made up of the fundamental group of $X_{\bar{k}}$ and the Galois group $\operatorname{Gal}(k^{sep}/k)$. To be precise, if X_k is geometrically connected, then there is an exact sequence

$$1 \to \pi_1(X_{\bar{k}}, \bar{x}) \to \pi_1(X_k, \bar{x}) \to \operatorname{Gal}(k^{sep}/k) \to 1$$

and every point of $X(k)$ defines a splitting of the sequence. See [14, IX.6.1] for details.

This has an important consequence which indicates that surjectivity assertions between fundamental groups are essentially geometric in nature:

Proposition 31. *Let k be a field and $f : Y \to X$ a morphism of geometrically connected k-schemes. Let $\bar{y} \to Y$ be a geometric point. Then the map of algebraic fundamental groups*

$$f_* : \pi_1(Y, \bar{y}) \to \pi_1(X, f(\bar{y}))$$

is surjective (resp. its image has finite index) iff the map of geometric fundamental groups

$$f_* : \pi_1(Y_{\bar{k}}, \bar{y}) \to \pi_1\big(X_{\bar{k}}, f(\bar{y})\big)$$

is surjective (resp. its image has finite index). ∎

For arithmetic applications of these results it is useful to find maps $\mathbb{P}^1 \to X$ defined over the ground field. The current results work very well for certain fields:

32 (Large fields). We are interested in fields K which have the property that on any variety with one smooth K-point the K-points are Zariski dense. Such fields are called *large fields* in [28]. The following are some interesting classes of such fields:

1. Fields complete with respect to a discrete valuation. This in particular includes the finite extensions of the p-adic fields $\mathbb{Q}_p$ and the power series fields $\mathbb{F}_q\big((t)\big)$ over finite fields $\mathbb{F}_q$.
2. More generally, quotient fields of local Henselian domains.
3. $\mathbb{R}$ and all real closed fields.
4. Infinite algebraic extensions of finite fields and, more generally, pseudo algebraically closed fields, cf. [9, Chap. 10]

For large fields, the existence of a smooth k point in $\bar{W}$ in (3) implies that $W(k)$ is dense in W. Using (3) and (5) we obtain the following:

Theorem 33. *Let X be a smooth, projective, SRC variety over a large field k and $x \in X(k)$. Then there is a dominant family of rational curves through x (defined over k)*

$$F : W \times \mathbb{P}^1 \to X, \quad F\big(W \times (0{:}1)\big) = \{x\}$$

with the following properties:

1. *The pullback $F_w^* T_X$ is ample for every $w \in W$.*
2. *Let $X^0 \subset X$ be any open k-subvariety containing x and $\pi_1(X^0, x) \twoheadrightarrow G$ a finite quotient (whose kernel is open). Then there is a dense subset of k-points $W_G^0(k) \subset W$ such that for every $w \in W_G^0(k)$ the composed map of algebraic fundamental groups*

$$\pi_1\big(\mathbb{P}^1_w \cap F^{-1}(X^0), (0{:}1)\big) \to \pi_1(X^0, x) \to G \quad \text{is surjective.}$$ ∎

An arithmetic application of these results is a simple proof of the following theorem, proved in increasing generality in [15, 5, 19]. Moret-Bailly [25, 26] proved analogous results for finite and smooth group schemes in positive characteristic.

Theorem 34. *Let K be a characteristic zero large field, G a linear algebraic group scheme over K and A a principal homogeneous G-space. Then there is an open set $V \subset \mathbb{A}^1_K$ containing 0 and a geometrically irreducible G-torsor $g : V_G \to V$ such that $g^{-1}(0) \cong A$ (as a G-space).*

Proof. Assume that G acts on A from the left and choose an embedding $G \subset GL(n)$ over K. The group $A \times GL(n)$ admits a diagonal left action by G and a right action by $GL(n)$ acting only on $GL(n)$.

The right G-action makes the morphism

$$h : Y^0 := G \backslash (A \times GL(n)) \to G \backslash (A \times GL(n)) / G =: X^0$$

into a G-torsor. Let $G^0 \subset G$ be the connected component of the identity. Then

$$Z^0 := G \backslash (A \times GL(n)) / G^0 \to X^0$$

is a finite étale cover, hence it corresponds to a finite quotient $\pi_1(X^0, x) \to H$ where $x \in X(K)$ is the image of $G \backslash (A \times G)$. Let $X \supset X^0$ be a smooth compactification of X^0. The variety $Z^0_{\bar{K}}$ is isomorphic to $GL(n)$, hence X is unirational and so SRC. The fiber of h over x is isomorphic to A.

Apply (33) to X with the point x to get $f : (0 \in \mathbb{P}^1) \to (x \in X)$ such that $\mathbb{P}^1 \times_X Z^0$ is geometrically irreducible. The fibers of $Y^0 \to Z^0$ are torsors over G^0, hence irreducible. Thus $\mathbb{P}^1 \times_X Y^0$ is also geometrically irreducible. Set $V = f^{-1}(X^0)$ and $V_G = V \times_X Y^0$. ∎

Remark 35. We have used the characteristic zero assumption in two places. The first is the existence of a smooth compactification X. Conjecturally this is always satisfied. A second point is that in general we can conclude that X is SRC only if $h : Y^0 \to X^0$ is separable, that is when G is smooth. The computations of [11, 12] suggest that in general this may be a quite subtle point.

Acknowledgments. I thank E. Szabó for helpful comments and C. Araujo for the nice picture. I am very grateful to J.-L. Colliot-Thélène for a long list of corrections and suggestions which improved the paper enormously. Partial financial support was provided by the NSF under grant number DMS-9970855 and DMS-02-00883.

REFERENCES

[1] C. Araujo and J. Kollár, *Rational Curves on Varieties,* this volume.

[2] A. Borel, *Linear algebraic groups,* 2nd ed. Graduate Texts in Mathematics, vol. 126. Springer (1991).

[3] S. Bosch, W. Lütkebohmert and M. Raynaud, *Néron models,* Ergebnisse der Mathematik und ihrer Grenzgebiete (3) vol. 21, Springer (1990).

[4] F. Campana, On twistor spaces of the class C, *J. Diff. Geom.,* **33** (1991), 541–549.

[5] J.-L. Colliot-Thélène, Rational connectedness and Galois covers of the projective line, *Ann. of Math.,* **151** (2000), 359–373.

[6] O. Debarre, *Variétés rationnellement connexes,* Exposé 905, Séminaire Bourbaki (2001/02).

[7] O. Debarre, *Fano varieties,* this volume.

[8] A. Grothendieck, *Éléments de géométrie algébrique,* Publ. Math. IHES vols. 4, 8, 11, 17, 20, 24, 28, 32 (1960 67).

[9] M. Fried and M. Jarden, *Field arithmetic,* Ergebnisse der Mathematik und ihrer Grenzgebiete (3), vol. 11, Springer (1986).

[10] W. Fulton and R. Pandharipande, Notes on stable maps and quantum cohomology, in: *Algebraic geometry — Santa Cruz 1995* (S. Bloch et al., eds.), Proc. Sympos. Pure Math., vol. 62 (2), Amer. Math. Soc. (Providence, 1997), pp. 45–96.

[11] P. Gille, La R-équivalence sur les groupes algébriques réductifs définis sur un corps global, *Publ. Math. IHES,* **86** (1997), 199–235.

[12] P. Gille, R-équivalence sur les G-revêtements sur les corps locaux non archimédiens, *J. Number Theory,* **91** (2001), 284–292.

[13] M. Goresky and R. MacPherson, *Stratified Morse theory,* Ergebnisse der Mathematik und ihrer Grenzgebiete (3), vol. 14, Springer (1988).

[14] A. Grothendieck, *Revêtements étales et groupe fondamental,* LNM 224, Springer (1971).

[15] D. Harbater, Galois coverings of the arithmetic line, in: *Number Theory Seminar, New York 1984/85,* LNM 1240, Springer (1987), pp. 165–195.

[16] A. J. de Jong and J. Starr, *Every rationally connected variety over a function field has a rational point,* preprint.

[17] J. Kollár, *Rational curves on algebraic varieties,* Ergebnisse der Mathematik und ihrer Grenzgebiete (3), vol. 32, Springer (1996).

[18] J. Kollár, Rationally connected varieties over local fields, *Ann. of Math.,* **150** (1999), 357–367.

[19] J. Kollár, Fundamental groups of rationally connected varieties, *Mich. Math. J.,* **48** (2000), 359–368.

[20] J. Kollár, letter to J. Starr.

[21] J. Kollár, Y. Miyaoka and S. Mori, Rationally Connected Varieties, *J. Alg. Geom.*, **1** (1992), 429–448.

[22] J. Kollár and E. Szabó, Rationally connected varieties over finite fields, *Duke Math. J.*, to appear.

[23] H. Matsumura, *Commutative Algebra*, Benjamin (1980).

[24] A. Mézard, Fundamental group, in: *Courbes semi-stables et groupe fondamental en géométrie algébrique (Luminy, 1998)* (J.-B. Bost et al., eds.), Progress in Math., vol. 187, Birkhäuser, 2000, pp. 141–155.

[25] L. Moret-Bailly, R-équivalence simultanée de torseurs: un complément à l'article de P. Gille, *J. Number Theory*, **91** (2001), 293–296.

[26] L. Moret-Bailly, *Sur la R-équivalence de torseurs sous un groupe fini*, preprint, available at `http://www.maths.univ-rennes1.fr/~moret/`

[27] J. P. Murre, *Lectures on an introduction to Grothendieck's theory of the fundamental group*. Notes by S. Anantharaman. Tata Institute of Fundamental Research Lectures on Mathematics, No. 40. Tata Institute of Fundamental Research (Bombay, 1967).

[28] F. Pop, Embedding problems over large fields, *Ann. of Math*, **144** (1996), 1–34.

János Kollár
Princeton University
Princeton NJ 08544-1000
`kollar@math.princeton.edu`

BOLYAI SOCIETY
MATHEMATICAL STUDIES, 12

Higher Dimensional Varieties and Rational Points, pp. 93–132.

Fano Varieties

OLIVIER DEBARRE

There are few objects canonically attached to a smooth projective algebraic variety X of dimension n. One of them is the invertible sheaf of differential n-forms, or equivalently the corresponding divisor class, for which one chooses a representant K_X, called a *canonical divisor*. One can start a classification according to "how ample" K_X is, and consider the two extreme classes of varieties: those for which K_X is ample, and those for which $-K_X$ is ample. Varieties X for which K_X is ample are called *varieties of general type* and, as their name suggests, there are so many of them that a classification is impossible. In contrast, there are "very few" varieties X for which $-K_X$ is ample; they are called *Fano varieties*. In many cases, such as in Mori's classification program of algebraic varieties known as the Minimal Model Program, it is necessary to allow some kind of singularities, which we describe below.

Our purpose is to survey the state of the present knowledge concerning some of the geometrical properties of Fano varieties.

These are notes from a minicourse given at the Instructional Conference "Higher-Dimensional Varieties and Rational Points" held at the Alfréd Rényi Institute of Mathematics in Budapest, September 2–7, 2001. Szeretném megköszönni a konferencia szervezőinek a meghívást.

1. Definitions and Examples

A (smooth) Fano variety is usually defined as a smooth projective variety defined over an algebraically closed field, whose anticanonical divisor is ample. As explained above, it is necessary to allow some kind of singularities, which we now describe.

Let X be a normal variety defined over an algebraically closed field of characteristic zero such that some multiple of the Weil canonical divisor[1] K_X is a Cartier divisor (we say that K_X is a $\mathbf{Q}$-Cartier divisor). For any smooth variety Y and proper morphism $f : Y \to X$, write

$$K_Y \sim f^*K_X + \sum_i a_i E_i \tag{1}$$

where the E_i are f-exceptional divisors, the a_i are rational numbers, and $\sim$ denotes numerical equivalence.[2] The *discrepancy* of X is the minimum of the a_i for all possible f.

The discrepancy can be computed from any resolution f whose exceptional locus has normal crossings (these resolutions exist by Hironaka's theorem) by the formula

$$\operatorname{discrep}(X) = \begin{cases} \min(1, a_i) & \text{if the } a_i \text{ are all } \geq -1 \\ -\infty & \text{otherwise} \end{cases}$$

([36], Corollary 2.32(2)). It is therefore either $-\infty$ or a rational number in $[-1, 1]$, and is 1 if X is smooth. We say that X has *log terminal singularities* if its discrepancy is > -1.

Definition 1. A Fano variety X is a normal projective variety with log terminal singularities such that the anticanonical divisor $-K_X$ is an ample $\mathbf{Q}$-Cartier divisor.

We will always make the implicit assumption that the characteristic is zero when we consider singular Fano varieties (this is necessary for the definition of log terminal to make sense).

As mentioned in the introduction, Fano varieties, although they are important for many reasons, are quite rare in nature. For example, as we will see later, there are only finitely many deformation types of smooth Fano varieties in any given dimension. Let us give a few examples.

Examples 2. (1) A projective curve is a Fano variety if and only if it is smooth and rational.

[1] For any projective variety X that is nonsingular in codimension 1, we define a canonical (Weil) divisor class K_X as the class of the closure of a canonical divisor of the smooth locus of X (see [27], II, Proposition 6.5(b)).

[2] This means that the degree of both sides on any curve is the same.

(2) A smooth Fano surface is usually called a del Pezzo surface. They have been classified and fall into 10 deformations types, to wit $\mathbf{P}^1 \times \mathbf{P}^1$ and the blow-up of $\mathbf{P}^2$ in at most 8 points. These points need to be in general position (e.g., no three are colinear). Singular del Pezzo surfaces are studied in [21]. A complex normal surface germ has log terminal singularities if and only if it is (analytically) the quotient of a $\mathbf{C}^2$ by a finite subgroup of $\mathrm{GL}(2, \mathbf{C})$.

(3) Smooth Fano threefolds have been classified in characteristic zero: there are 17 families with Picard number 1 ([29], [30],[3] [48][4]), and 88 other families.[5] In positive characteristic, see [47], where Fano threefolds with Picard number 1 are classified, and [41] for additional results.

(4) A finite product of Fano varieties is a Fano variety.

(5) Let Y be a smooth subvariety of $\mathbf{P}^n$, with hyperplane divisor H, such that $K_Y \sim qH$ for some rational number q (this is the case for example when Y is a smooth curve). Assume further that Y is projectively normal, so that the cone X in $\mathbf{P}^{n+1}$ over Y is normal. Let $\tilde{X}$ be the $\mathbf{P}^1$-bundle[6] $\pi : \mathbf{P}(\mathcal{O}_Y \oplus \mathcal{O}_Y(H)) \to Y$ and let Y_0 be the section corresponding the quotient $\mathcal{O}_Y(H)$ of $\mathcal{O}_Y \oplus \mathcal{O}_Y(H)$, so that $Y_0|_{Y_0} \equiv -H$. The cone X is obtained as the contraction $f : \tilde{X} \to X$ of Y_0, and ([27], V, Corollary 2.11)

$$K_{\tilde{X}} \sim -2Y_0 + \pi^*(K_Y - H) = -2Y_0 + \pi^*\big((q-1)H\big)$$

hence

$$K_X \sim (q-1)H$$

(these two Weil divisors coincide outside of the vertex of X). The anticanonical bundle of X is therefore ample if and only if $q < 1$. Writing $K_{\tilde{X}} \sim f^*K_X + aY_0$ as in (1), we get by restricting to Y_0 the relation

$$\big(-2Y_0 + \pi^*\big((q-1)H\big)\big)\big|_{Y_0} = aY_0|_{Y_0}$$

hence $a = -1 - q$. The singularities of X are therefore log terminal if and only if $q < 0$. It follows that X *is a Fano variety if and only if* Y *is a Fano variety.*

[3] Iskovskikh's incomplete classification was completed by Mukai and Umemura in [44] and by Cutkosky in [17].

[4] Iskovskikh's and Shokurov's arguments were greatly simplified by Takeuchi in [50].

[5] Mori and Mukai recently noticed that the family of blow-ups of $\mathbf{P}^1 \times \mathbf{P}^1 \times \mathbf{P}^1$ along a curve of tridegree $(1,1,3)$ is missing from the list in [43].

[6] We follow Grothendieck's notation: for a vector bundle $\mathcal{E}$, the projectivization $\mathbf{P}(\mathcal{E})$ is the space of *hyperplanes* in the fibers of $\mathcal{E}$.

(6) Let X be a normal complete intersection in $\mathbf{P}^{n+s}$ defined by equations of degrees $d_1 \geq \cdots \geq d_s \geq 2$. A canonical divisor is $(d_1 + \cdots + d_s - n - s - 1)H$, where H is a hyperplane section; X is therefore a Fano variety if and only if it has log terminal singularities and $d_1 + \cdots + d_s \leq n + s$. Note that in any given dimension n, there are only finitely many choices for the degrees $d_1, \ldots, d_s$ (because $s \leq \sum(d_i - 1) \leq n$).

(7) A canonical divisor of a normal cyclic covering $X \to \mathbf{P}^n$ of degree $d > 1$ branched along a hypersurface of degree de is $(-n - 1 + (d-1)e)H$, where H is the inverse image of a hyperplane; X is therefore a Fano variety if and only if it has log terminal singularities and $(d-1)e \leq n$ (again, there are only finitely many choices for d and e when n is fixed).

(8) In characteristic zero, a projective variety acted on transitively by a connected linear algebraic group (such as a flag variety) is a smooth Fano variety (see [33], Theorem V.1.4).

(9) If Y is a Fano variety and $D_1, \ldots, D_r$ are nef Cartier divisors on Y such that $-K_Y - D_1 - \cdots - D_r$ is ample, $X = \mathbf{P}\big(\bigoplus_{i=1}^r \mathcal{O}_Y(D_i)\big)$ is a Fano variety. Indeed, if D is a divisor associated with the line bundle $\mathcal{O}_Y(1)$ and π is the canonical map $X \to Y$, one gets as in [27], V, Lemma 2.10,

$$-K_X = rD + \pi^*(-K_Y - D_1 - \cdots - D_r)$$

Since each D_i is nef, the divisor D is nef on X. Since each $-K_Y - D_1 - \cdots - D_r + D_i$ is ample, the divisor $D + \pi^*(-K_Y - D_1 - \cdots - D_r)$ is ample (because a direct sum of nef (resp. ample) line bundles is a vector bundle that has the same property). It follows that $-K_X$ is ample, and the discrepancy of X is equal to the discrepancy of Y. Hence X is a Fano variety.

2. Toric Fano Varieties

Although toric varieties (whose definition is given below in 2.1) are very special among all varieties, they are nevertheless well worth studying for several reasons. First, they provide very intricate examples against which one can test conjectures. For example, they have proved an efficient testing ground for Mori's Minimal Model Program ([45], [52], [14]). The survey [16] (with its very impressive bibliography) shows how active an area of research toric varieties are.

In another direction, the link that the toric construction establishes between the geometry of projective toric varieties and the combinatorial properties of polytopes has been the source of many fruitful interactions between these two fields (see the survey [22] for a sample of results).

Smooth toric Fano varieties are also very special among general smooth Fano varieties, but they again provide many interesting examples. For instance, their anticanonical hyperplane sections are Calabi–Yau varieties and this construction alone provides in the region of 400 million Calabi–Yau threefolds accessible through popular websites. Their birational geometry is also very rich ([13], [15]). Finally, there is hope that certain bounds (rank of the Picard group, degree,...) for smooth toric Fano varieties fall in the right ballpark for smooth Fano varieties in general.

This section is organized as follows. After quickly going through the basics of the toric construction in 2.1 and 2.2, we prove in 2.3 a couple of (nearly optimal) bounds on the Picard number and degree of smooth toric Fano varieties. The first bound implies that in each dimension, there are only finitely many isomorphism types of smooth toric Fano varieties (Corollary 7), a property that is still valid for singular toric Fano varieties whose discrepancy is bounded away from -1 (Corollary 13). This is a result that is conjectured to hold for general Fano varieties.

2.1. Toric varieties

The algebraic torus of dimension n over the algebraically closed field $\mathbf{k}$ is the affine variety

$$T = \operatorname{Spec}\left(\mathbf{k}[U_1, U_1^{-1}, \dots, U_n, U_n^{-1}]\right)$$

A *toric variety* over $\mathbf{k}$ is a normal algebraic variety X defined over $\mathbf{k}$ with an action of T and a dense open orbit isomorphic to T. Toric varieties can all be constructed from very explicit combinatorial data, as we explain below. This makes their study very accessible, and the whole theory is at the same time elementary and very rich.

Affine toric varieties. Let M be a free abelian group of rank n, with dual $N = \operatorname{Hom}_{\mathbf{Z}}(M, \mathbf{Z})$. Set $M_{\mathbf{R}} = M \otimes_{\mathbf{Z}} \mathbf{R}$ and $N_{\mathbf{R}} = N \otimes_{\mathbf{Z}} \mathbf{R}$.

For each cone σ in $N_{\mathbf{R}}$ generated by a finite number of vectors in N and containing no lines, the dual cone

$$\sigma^\vee = \left\{ u \in M_{\mathbf{R}} \mid \langle u, v \rangle \geq 0 \text{ for all } v \in \sigma \right\}$$

is also generated by a finite number of elements of M. Moreover, the semigroup $\sigma^\vee \cap M$ is finitely generated (Gordon's lemma) and we may define an affine variety by

$$X_\sigma = \operatorname{Spec}\left(\mathbf{k}[\sigma^\vee \cap M]\right)$$

If τ is a face of σ, the semigroup $\sigma^\vee \cap M$ is contained in $\tau^\vee \cap M$, so there is a map $X_\tau \to X_\sigma$. In fact, X_τ is a principal open subset of X_σ: if we choose $u \in \sigma \cap M^\vee$ such that $\tau = \sigma \cap u^\perp$, we have $X_\tau = (X_\sigma)_u$.

For example,

$$X_{\{0\}} = \operatorname{Spec}\left(\mathbf{k}[M]\right) = T.$$

The torus T acts on X_σ: the action corresponds to the map

$$\mathbf{k}[\sigma^\vee \cap M] \to \mathbf{k}[\sigma^\vee \cap M] \otimes \mathbf{k}[M]$$

$$u \mapsto u \otimes u$$

on algebras and is compatible with the inclusion $X_\tau \to X_\sigma$ above.

General toric varieties. Now if we consider a *finite* set Δ of cones in N as above, with the property that every face of a cone in Δ is in Δ and the intersection of any two cones in Δ is a face of each (such a set is called a *fan*), we construct an irreducible $\mathbf{k}$-scheme X_Δ with an action of T by gluing the affine varieties X_σ and $X_{\sigma'}$ along the open set $X_{\sigma\cap\sigma'}$, for all σ and σ' in Δ. It turns out that X_Δ is *irreducible, separated and normal* ([22], pp. 21 and 29) and has dimension n. Each X_σ, hence also X_Δ, contain the open set $X_{\{0\}} = T$.

All toric varieties are obtained via this construction.

Examples 3. Let $(e_1, \ldots, e_n)$ be a basis for N and let $U_1, \ldots, U_n$ be the corresponding elements of $\mathbf{k}[M]$.

(1) We have

$$X_{\langle e_1,\ldots,e_n\rangle} = \operatorname{Spec}\left(\mathbf{k}[U_1, \ldots, U_n]\right) = \mathbf{A}^n_{\mathbf{k}}$$

and similarly,

$$X_{\langle e_1,\ldots,e_m\rangle} = \operatorname{Spec}\left(\mathbf{k}[U_1, \ldots, U_m, U_{m+1}, U_{m+1}^{-1}, \ldots, U_n, U_n^{-1}]\right)$$

is isomorphic to the product of $\mathbf{A}^m_{\mathbf{k}}$ with the $(n-m)$-dimensional torus.

(2) The toric variety associated with the set of cones generated by any proper subset of $\{e_0, e_1, \ldots, e_n\}$, where $e_0 = -e_1 - \cdots - e_n$, is isomorphic to the projective space $\mathbf{P}^n_{\mathbf{k}}$. The action of the torus T is given by

$$(t_1, \ldots, t_n) \cdot (x_0, \ldots, x_n) = (x_0, t_1 x_1, \ldots, t_n x_n).$$

Many geometric properties of the variety X_Δ can be read off the fan Δ.

- By Example (1) above, if a cone σ is generated by part of a basis of N, the corresponding variety X_σ is smooth. It turns out that the converse holds: the toric variety X_Δ is smooth if and only if each cone in Δ can be generated by part of a basis of N ([22], p. 29).
- The toric variety X_Δ is projective if and only if Δ is the set of cones spanned by the faces of a polytope[7] Q with vertices in N and the origin in its interior ([22], p. 72). We denote it also by X_Q.
- The toric variety X_Q is $\mathbf{Q}$-factorial (i.e., every Weil divisor is a $\mathbf{Q}$-Cartier divisor) if and only if Q is simplicial (i.e., each facet has n vertices). The Picard group of X_Q is then free abelian of rank ([22], p. 65)

$$\rho_{X_Q} = \operatorname{Card}\big(\{\text{vertices of } Q\}\big) - n \tag{2}$$

- The toric variety X_Q is a Fano variety if the vertices of the polytope Q are primitive vectors in N;[8] in this case, we say that Q is a *Fano polytope.* All toric Fano varieties are obtained in this way. It is a smooth Fano variety if each facet of Q is the convex hull of a basis of N; in this case, we say that Q is a *smooth Fano polytope.* All smooth toric Fano varieties are obtained in this way.

2.2. Toric Fano varieties

Let Q be a Fano polytope in the vector space $N_{\mathbf{R}}$, with vertices primitive vectors in N and associated toric Fano variety X. Its dual

$$P = Q^\vee = \big\{u \in M_{\mathbf{R}} \mid \langle u, v\rangle \geq -1 \text{ for all } v \in Q\big\}$$

[7]A *polytope* is the convex hull Conv (S) of a finite set S. Note that different polytopes may define the same fan.

[8]The condition for the canonical divisor of a toric variety to be $\mathbf{Q}$-Cartier will be worked out in Proposition 12, where it is also proved that the singularities are then automatically log terminal; the rest follows from [22], p. 66, p. 72, and the proposition p. 85.

(this is the opposite of the traditional definition) is a polytope with vertices in $M_{\mathbf{Q}}$ (in M when Q is smooth).

We have ([22], p. 70, Corollary p. 74, p. 75)

$$(-K_X)^n = n!\,\mathrm{vol}\,(P) \tag{3}$$

where the volume is computed with respect to M (i.e., the volume of the cube determined by a basis of M is 1).

The divisor $-mK_X$ is a Cartier divisor for all integers m such that the vertices of mP are in M and for all such positive integers m, we have ([22], p. 111)

$$h^0(X, -mK_X) = \chi(X, -mK_X) = \mathrm{Card}\,(M \cap mP) \tag{4}$$

Examples 4. (1) Up to the action of GL (N), there are only 5 smooth Fano polytopes indimension 2. They are

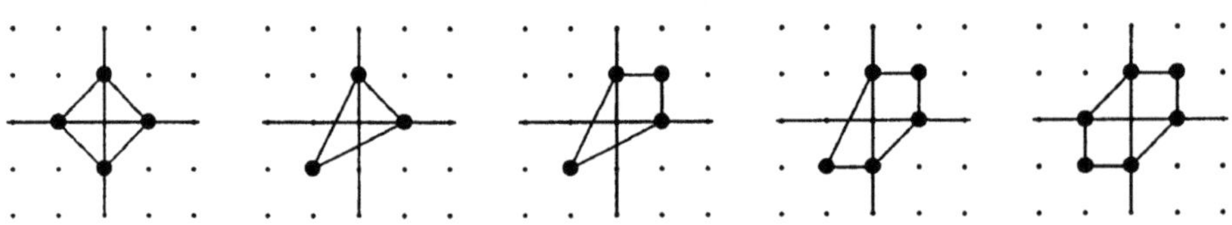

$\mathbf{P}^1 \times \mathbf{P}^1$ $\quad$ $\mathbf{P}^2$ $\quad$ $\mathbf{P}^2$ blown up at a fixed point $\quad$ $\mathbf{P}^2$ blown up at 2 fixed points $\quad$ $\mathbf{P}^2$ blown up at 3 fixed points

(the three fixed points of the action of T on $\mathbf{P}^2$ are the coordinate points $(1,0,0)$, $(0,1,0)$, and $(0,0,1)$). The dual polytopes are

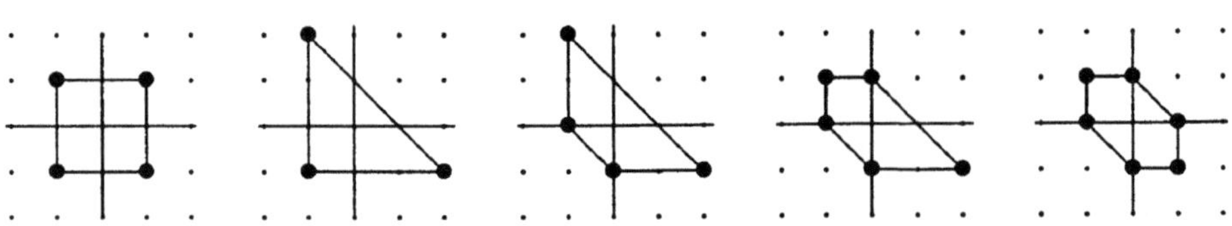

and formulas (3) and (4) are easily checked in these cases.

(2) For any positive integer a, the polytope

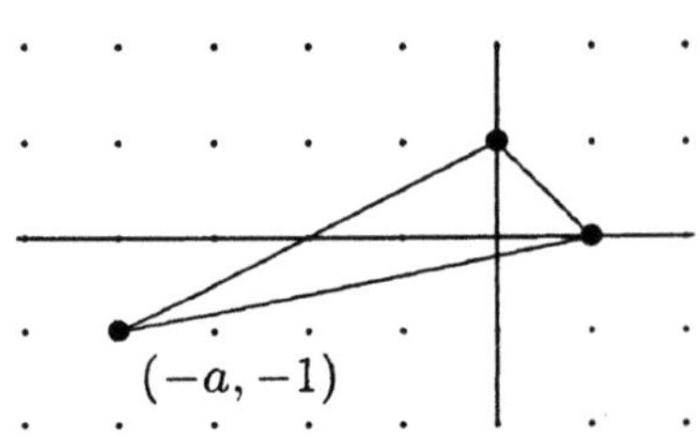

is a Fano polytope which is singular for $a \geq 2$ and whose dual is

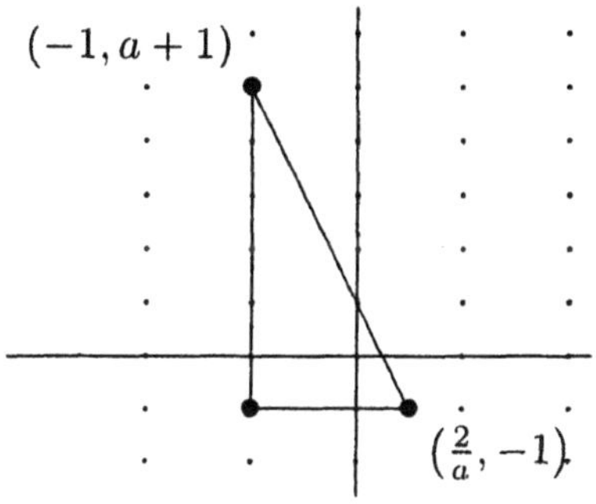

The corresponding Fano variety X_a has a single singular point. Its blow up is the toric variety associated with the fan

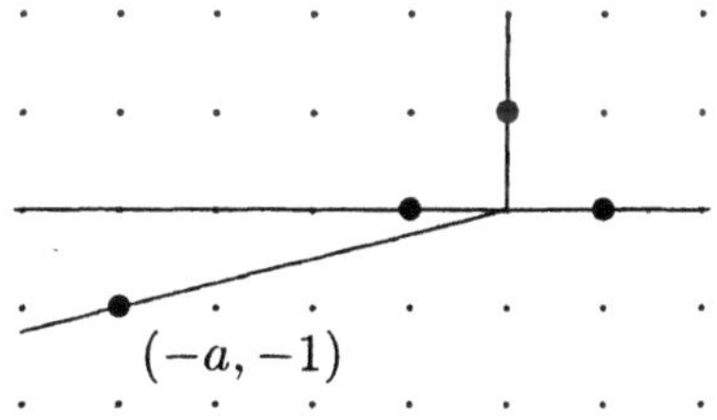

By [22], p. 8, this is the Hirzebruch surface $\mathbf{P}(\mathcal{O}_{\mathbf{P}^1} \oplus \mathcal{O}_{\mathbf{P}^1}(a))$ and X_a is obtained from it by contracting the unique curve with self-intersection $-a$. This is a particular case of Example 2(5), where we computed that the discrepancy of X_a is $-1 + \frac{2}{a}$. The smallest positive integer m such that mK_X is a Cartier divisor is a if a is odd, $a/2$ otherwise.

This shows that there are, in each dimension at least 2, infinitely many isomorphism classes of toric Fano varieties.

2.3. Smooth toric Fano varieties

Let Q be a smooth Fano polytope in $N_{\mathbf{R}}$ with $\rho + n$ vertices, associated smooth Fano variety X, and dual polytope P. From (2), we know that the Picard number of X is ρ. The vertices of Q are in N and the vertices of P are in M.

Remarks 5. (1) There is a one-to-one correspondence between the vertices of Q (resp. of P) and the facets of P (resp. of Q). For any vertex u of P and any vertex v of Q, we have $\langle u, v \rangle \geq -1$ and there is equality if and only if v

is on the facet F_u of Q that corresponds to the vertex u of P (or viceversa). Otherwise, we have $\langle u, v\rangle \geq 0$, because u is in M and v in N.

(2) Let $e_1, \ldots, e_n$ be the vertices of F_u, so that $\langle u, e_i\rangle = -1$ for each i. They form a basis for N (see 2.1) and coordinates will be taken in this basis. Let v^i be the vertex of Q such that the face

$$F_i = \operatorname{Conv}\left(\{e_1, \ldots, e_{i-1}, v^i, e_{i+1}, \ldots, e_n\}\right)$$

is adjacent to F_u. We have $v_i^i < 0$ (because the origin and e_i (resp. and v^i) are on the same side of the affine hyperplane generated by F_i (resp. by F_u)) hence $v_i^i = -1$ since the vertices of F_i form a basis for N. It follows that the affine hyperplane spanned by F_i has equation

$$\langle u, x\rangle + \left(1 + \langle u, v^i\rangle\right) x_i = -1$$

where $\langle u, v^i\rangle \geq 0$ because v^i is not on F_u. Let v be a vertex of Q such that $\langle u, v\rangle = 0$. If $v \neq v^i$, the vertex v is not on F_i hence

$$0 \leq \langle u, v\rangle + \left(1 + \langle u, v^i\rangle\right) v_i = \left(1 + \langle u, v^i\rangle\right) v_i$$

and $v_i \geq 0$. This cannot happen for all i, hence v must be among $v^1, \ldots, v^n$. In particular, there are at most n vertices v of Q such that $\langle u, v\rangle = 0$.

Theorem 6 ([51]). *The Picard number of a smooth toric Fano variety of dimension n is at most $2n^2 - n$.*

Before we begin the proof, let us mention two elementary theorems about the convex hull Conv (S) of a subset S of $\mathbf{R}^n$ (see [18]). The first is Helly's theorem: *any point of* Conv (S) *is in the convex hull of at most $n+1$ points of S.* The second is Steinitz's theorem: *any point of the interior of* Conv (S) *is in the interior of the convex hull of at most $2n$ points of S.*

Proof of the Theorem. Since the origin is in the interior of P, it is in the interior of the convex hull of at most $2n$ vertices by Steinitz's theorem. This means that there is a relation

$$0 = \lambda_1 u_1 + \cdots + \lambda_d u_d$$

with $\lambda_1, \ldots, \lambda_d$ positive, where the vertices $u_1, \ldots, u_d$ of P generate the vector space $M_{\mathbf{R}}$ and $n < d \leq 2n$. Let v be a vertex of Q. Since $u_1, \ldots, u_d$ generate $M_{\mathbf{R}}$ and v is nonzero, we must have $\langle u_i, v\rangle \neq 0$ for at least one i, hence $\langle u_i, v\rangle < 0$ for at least one i. This means that v is on the facet F_{u_i} of Q corresponding to this u_i (Remark 5(1)). Since each facet of Q has n vertices, it follows that $\rho + n \leq nd \leq 2n^2$, hence the theorem. ∎

Corollary 7 ([51]). *In each dimension, there are only finitely many isomorphism types of smooth toric Fano varieties defined over* $\mathbf{k}$.

Proof. It is enough to prove that up to the action of GL (N), there are only finitely many smooth Fano polytopes in $N_{\mathbf{R}}$. We know that the number of vertices of such a polytope Q is at most $2n^2$. Since each facet of Q has n vertices, their number is at most $\binom{2n^2}{n}$, hence the volume of Q is at most $\frac{1}{n!}\binom{2n^2}{n}$. Fix the vertices $e_1, \ldots, e_n$ of a facet of Q and denote by S the simplex Conv $(\{0, e_1, \ldots, e_n\})$. For any other vertex v of Q, the volume of Conv $(S \cup \{v\})$ is at most $\frac{1}{n!}\binom{2n^2}{n}$, hence the absolute value of each coordinate of v in the basis $(e_1, \ldots, e_n)$ is bounded by $\binom{2n^2}{n}$. The integral vector v belongs to some bounded set hence may take only finitely many values (at most $\left(2\binom{2n^2}{n}\right)^n$). ∎

Going back to the Picard number ρ of a smooth toric Fano variety of dimension n, Batyrev conjectured the inequalities[9]

$$\rho \leq \begin{cases} 2n & \text{if } n \text{ is even} \\ 2n-1 & \text{if } n \text{ is odd} \end{cases} \tag{5}$$

for all smooth toric Fano varieties. The inequality $\rho \leq n^2 - n + 1$ for $n \geq 3$ is proved in [51] and using the same ideas, it is possible to (asymptotically) improve their bound.

Theorem 8. *The Picard number of a smooth toric Fano variety of dimension n is at most*

$$2 + 2\sqrt{(n^2-1)(2n-1)}$$

Proof. Let u be a vertex of P. The half line $\mathbf{R}^- u$ meets the boundary of P at a point that belongs to some face of dimension $k \leq n-1$ (choose k minimal) hence is in the relative interior of the convex hull of some vertices $u_1, \ldots, u_d$, with $d \leq k+1$ by Helly's theorem. One can therefore write

$$0 = \lambda u + \lambda_1 u_1 + \cdots + \lambda_d u_d$$

[9] This holds for $n \leq 4$ thanks to Batyrev's classification ([4] and [46]) and for $n = 5$ (where there is no classification) by [15], Theorem 3.2. Equality is attained for n even by $S^{n/2}$ and for n odd by $S^{(n-1)/2} \times \mathbf{P}^1$, where S is $\mathbf{P}^2$ blown up at the 3 fixed points of the torus action (see Example 4(1)).

with $\lambda_1 \geq \cdots \geq \lambda_d > 0$ and $\lambda > 0$. Let $a \in \{0, \ldots, d\}$, let $\mathcal{V}_a(Q)$ be the set of vertices of Q that are on the facets $F_u, F_{u_1}, \ldots, F_{u_a}$ or on a facet adjacent to them, and let u be a vertex of P in the complement $\mathcal{V}_a(Q)^c$ of $\mathcal{V}_a(Q)$ in the set $\mathcal{V}(Q)$ of all vertices of Q. By Remark 5(2), we have

$$\langle u, v\rangle \geq 1 \qquad\qquad \langle u_i, v\rangle \geq 1$$

for $1 \leq i \leq a$, hence

$$-\lambda_{a+1}\langle u_{a+1}, v\rangle - \cdots - \lambda_d\langle u_d, v\rangle = \lambda\langle u, v\rangle + \sum_{i=1}^{a} \lambda_i\langle u, v_i\rangle$$

$$\geq \lambda + \sum_{i=1}^{a} \lambda_i > \sum_{i=1}^{a} \lambda_i$$

This implies that the integer $\langle u_i, v\rangle$ must be equal to -1 for at least $a+1$ indices i in $\{a+1, \ldots, d\}$, i.e., that v must be on at least $a+1$ faces among $F_{u_{a+1}}, \ldots, F_{u_d}$. Consider the set

$$I = \{\, (v, i) \in \mathcal{V}_a(Q)^c \times \{a+1, \ldots, d\} \mid v \in F_{u_i} \}$$

The fiber of i for the second projection $I \to \{a+1, \ldots, d\}$ consists of vertices that are on F_{v_i} but not on the intersection $\bigcap_{i=1}^{d} F_{u_i}$, which is an $(n-k-1)$-dimensional face of Q, hence has $n-k$ vertices. It follows that

$$\operatorname{Card}(I) \leq k(d-a)$$

Since we proved that each fiber of the first projection has at least $a+1$ elements, we get

$$\operatorname{Card}\big(\mathcal{V}_a(Q)^c\big) \leq \frac{\operatorname{Card}(I)}{a+1} \leq \frac{k(d-a)}{a+1}$$

For $i \in \{2, \ldots, a\}$, the intersection $F_{u_1} \cap F_{u_i}$ is a face of Q of dimension at least $n-k-1$, hence contains at least $n-k$ vertices. It follows that

$$\begin{aligned}
\operatorname{Card}\big(\mathcal{V}_a(Q)\big) \leq {} & 2n && \text{vertices on } F_u \text{ and } F_{u_1}\\
& + (a-1)k && \text{vertices on } F_{u_2} \cup \cdots \cup F_{u_a} - F_{u_1}\\
& + (a+1)n && \text{vertices adjacent to } F_u, F_{u_1}, \ldots, F_{u_a}
\end{aligned}$$

hence

$$\begin{aligned}\operatorname{Card}\left(\mathcal{V}(Q)\right) &\leq \max_{1\leq d\leq k+1\leq n}\left(\min_{0\leq a\leq d}\left(2n+(a-1)k+(a+1)n+\frac{k(d-a)}{a+1}\right)\right)\\ &\leq \max_{1\leq d\leq k+1\leq n}\left(\min_{0\leq a\leq d}\left(3n-2k+a(k+n)+\frac{k(d+1)}{a+1}\right)\right)\end{aligned}$$

Taking

$$a=\left[\sqrt{\frac{k(d+1)}{k+n}}\right]$$

we get

$$\begin{aligned}\operatorname{Card}\left(\mathcal{V}(Q)\right) &\leq \max_{1\leq d\leq k+1\leq n}\left(3n-2k+2\sqrt{k(d+1)(k+n)}\right)\\ &= n+2+2\sqrt{(n^2-1)(2n-1)} \qquad \blacksquare\end{aligned}$$

We now turn our attention towards the integer

$$(-K_X)^n = n!\operatorname{vol}(P)$$

often called the *degree* of the smooth Fano variety X. Batyrev gave in [3] an explicit (but very large) bound for $\operatorname{vol}(P)$.

Note that[10]

$$\operatorname{Int}(P)\cap M=\{0\}$$

There is a bound in [37] for the volume of an integral polytope with only one integral interior point (see 2.4); it is doubly exponential in n and is almost optimal in general. For duals of smooth Fano polytopes, we prove that there is a much better bound.

[10]Indeed, any point u of P is in a cone spanned by a facet F of P with vertices $u_1,\dots,u_s$. It can therefore be written as $u=\sum_{i=1}^{s} r_i u_i$ with $r_i\in\mathbf{R}^+$ and $\sum r_i\leq 1$. This facet corresponds to a vertex v of Q that satisfies $\langle u_i,v\rangle=-1$ for all i, hence

$$\langle u,v\rangle=\sum_{i=1}^{s} r_i\langle u_i,v\rangle=-\sum_{i=1}^{s} r_i$$

If u is in M, this is an integer ≥ -1, hence 0 or -1. If it is 0, all the r_i are 0 and $u=0$. If it is -1, the point u is on F.

Theorem 9. *Assume $\rho \geq 2$ and let Q be a smooth Fano polytope in $N_{\mathbf{R}}$ with $\rho + n$ vertices and dual P. We have*

$$\mathrm{vol}\,(P) \leq \left(\frac{(n-1)^{\rho+1}-1}{n-2}\right)^n \leq n^{\rho n}$$

Equivalently, a smooth toric Fano variety X of dimension n and Picard number $\rho \geq 2$ satisfies

$$(-K_X)^n \leq n! n^{\rho n} \tag{6}$$

A complete toric variety X of dimension n with $\rho = 1$ is isomorphic to $\mathbf{P}^n$, hence

$$(-K_X)^n = (n+1)^n$$

Smooth complete toric varieties with $\rho = 2$ have been classified in [32]: they are all projectivizations of a decomposable bundle over a projective space of smaller dimension. A tedious check shows that a smooth toric Fano variety X of dimension n with $\rho = 2$ satisfies

$$(-K_X)^n \leq n^{2n}$$

For any integers $\rho \geq 2$ and $n \geq 4$ such that $\frac{n}{\log n} \geq 2^{\rho-2}$, examples of smooth toric Fano varieties X of dimension n and Picard number ρ with

$$(-K_X)^n \geq \left(\frac{n^\rho}{2^{\rho^2-1}(\log n)^{\rho-1}}\right)^n$$

are constructed in [19], Proposition 5.22, so the bound (6) is not too far off.

The theorem is a consequence of the following two lemmas.

Lemma 10. *Let P be a polytope in $M_{\mathbf{R}}$ whose only integral interior point is the origin. Let b be a positive number such that for each vertex u of P and each vertex v of $P^\vee$, we have*

$$-1 \leq \langle u, v \rangle \leq b$$

Then

$$\mathrm{vol}\,(P) \leq (b+1)^n$$

Proof. This is Theorem 2.5 from [37], but we reproduce the proof for the convenience of the reader. The hypothesis implies $-\frac{1}{b}P \subset P$. Assume to the contrary $\operatorname{vol}(P) > (b+1)^n$ and set

$$P' = \frac{1-\varepsilon}{b+1}P.$$

We have $\operatorname{vol}(P') > 1$ for ε positive small enough hence, by van der Corput's theorem, P' contains distinct points p and p' such that $p - p' \in M$. Since P' is convex and $-\frac{1}{b}P' \subset P'$, the point

$$\frac{p-p'}{b+1} = \frac{p + b\left(-\frac{1}{b}p'\right)}{b+1}$$

is in P', hence $p - p'$ is in $(b+1)P'$, which is contained in the interior of P. This contradicts the fact that the origin is the only integral interior point of P. ∎

Lemma 11. *Let Q be a smooth Fano polytope in $N_{\mathbf{R}}$ with $\rho + n$ vertices. For each vertex u of P and each vertex v of Q, we have*

$$-1 \leq \langle u, v \rangle \leq \frac{n-1}{n-2}\big((n-1)^{\rho} - 1\big)$$

when $\rho > 1$.

Proof. Consider a facet of Q, with vertices a basis $(e_1, \ldots, e_n)$ of N, and the corresponding vertex u of P. It takes the value -1 on each e_i (and nonnegative values on other vertices of Q). Write the distinct values of u on the vertices of Q as an increasing sequence

$$-1 < b_1 < \cdots < b_k$$

with $k \leq \rho$. Let j be an integer in $\{1, \ldots, k\}$ and let v be a vertex of Q such that $\langle u, v \rangle = b_j$.

Consider a minimal subset I of $\{1, \ldots, n\}$ such that v and e_i $(i \in I)$ are *not* all on the same facet of Q. We may and will assume $I = \{1, \ldots, r\}$ for some positive integer r. The vector $v + e_1 + \cdots + e_r$ is in the cone spanned by some facet F of Q. The vertices of F form a basis of N, so we may write[11]

$$v + e_1 + \cdots + e_r = v_1 + \cdots + v_s + e'_1 + \cdots + e'_t$$

where

[11] This is what Batyrev calls a *primitive relation* in [4].

- $v_1, \dots, v_s, e'_1, \dots, e'_t$ are vertices of F (possibly nondistinct);
- $v_1, \dots, v_s \notin \{e_1, \dots, e_n\}$ (so that $\langle u, v_i \rangle \geq 0$);
- $e'_1, \dots, e'_t \in \{e_{r+1}, \dots, e_n\}$, (by minimality of I).

The vector $\frac{1}{r+1}(v + e_1 + \cdots + e_r)$ is in the interior of Q, hence $s + t \leq r$.

If $r = n$, the polytope Q is the standard polytope of $\mathbf{P}^n$ (see Example 3(2)) and $\rho = 1$, so we have $r < n$.

If $s = 0$, we have $\langle u, v \rangle = r - t \leq n - 1$. If $s > 0$, we have for each i in $\{1, \dots, s\}$

$$\langle u, v \rangle - r = \langle u, v_1 \rangle + \cdots + \langle u, v_s \rangle - t \geq \langle u, v_i \rangle - t$$

hence $\langle u, v_i \rangle \leq \langle u, v \rangle - r + t \leq \langle u, v \rangle - s < \langle u, v \rangle$ (in particular, this case cannot happen when $j = 1$, hence $b_1 \leq n - 1$). It follows that $\langle u, v_i \rangle \leq b_{j-1}$ for each i in $\{1, \dots, s\}$, hence

$$\begin{aligned} b_j = \langle u, v \rangle &= \langle u, v_1 \rangle + \cdots + \langle u, v_s \rangle + r - t \\ &\leq s b_{j-1} + n - 1 \\ &\leq (n-1)(b_{j-1} + 1) \end{aligned}$$

From this relation, we get

$$b_j \leq b_1 (n-1)^{j-1} + \frac{n-1}{n-2}\left((n-1)^{j-1} - 1\right) \leq \frac{n-1}{n-2}\left((n-1)^j - 1\right)$$

by induction on j. ■

2.4. Singular toric Fano varieties

As we saw in Example 4(2), there are infinitely many isomorphism types of toric Fano varieties in any given dimension ≥ 2. We show here (Corollary 13) that by bounding the discrepancy away from -1, we get finitely many isomorphism types, as in the smooth case (Corollary 7).

We first show how to compute the discrepancy of a toric variety in terms of the geometry of its fan Δ and the lattice N.

Proposition 12. *Let X be the toric variety associated with a fan Δ in $N_{\mathbf{R}}$. The divisor K_X is a $\mathbf{Q}$-Cartier divisor if and only if, for each maximal* *cone σ in Δ, spanned by primitive vectors $v_1, \ldots, v_s$ in N, there exists an element u_σ of $M_{\mathbf{Q}}$ such that $\langle u_\sigma, v_i \rangle = 1$ for all i. In this case,*

$$\operatorname{discrep}(X) = -1 + \min_{\substack{\sigma \text{ maximal cone in } \Delta \\ v \in \sigma \cap N - \{0, v_1, \ldots, v_s\}}} \langle u_\sigma, v \rangle$$

In particular, X has log terminal singularities.

Proof. We may, by subdividing the cones in Δ, refine Δ into a fan $\tilde{\Delta}$ to obtain a T-equivariant resolution of singularities $f : \tilde{X} \to X$ whose exceptional locus has normal crossings ([22], Proposition p. 48).

Let σ be a maximal cone in Δ, spanned by primitive vectors $v_1, \ldots, v_s$ of N. The cones of $\tilde{\Delta}$ contained in σ are spanned by vectors among $v_1, \ldots, v_s$ and other primitive elements $w_1, \ldots, w_t$ of $\sigma \cap N$. The latter correspond bijectively to the exceptional divisors $E_1, \ldots, E_t$ of f that meet $f^{-1}(X_\sigma)$.

The complement of the dense orbit T in X_σ is the union of T-invariant irreducible hypersurfaces $D_1, \ldots, D_s$ (one for each v_i).

Any element u of M defines a regular function on T, hence a rational function h_u on X_σ whose divisor is $\sum_i \langle u, v_i \rangle D_i$ ([22], p. 61). Since $K_X = -\sum_i D_i$ ([22], Proposition p. 85), mK_X is a Cartier divisor if and only if there exists, for each maximal cone σ in Δ, an element u of M such that $\langle u, v_i \rangle = -m$ for all i ([22], Exercise p. 62).

The coefficient of the divisor $f^*(mK_X)$ on E_j is that of $f^*(\operatorname{div}(h_u))$, to wit $\langle u, w_j \rangle$, whereas the coefficient of $mK_{\tilde{X}}$ is $-m$. With the notation of (1), this means

$$a_j = -1 + \left\langle -\frac{1}{m} u, w_j \right\rangle$$

We have $u_\sigma = -\frac{1}{m} u$. Now each v in $\sigma \cap N$ is in a cone of $\tilde{\Delta}$ hence, since $\tilde{X}$ is smooth, is a linear combination with positive integral coefficients of vectors among $v_1, \ldots, v_s, w_1, \ldots, w_t$. If $v \notin \{0, v_1, \ldots, v_s\}$, the number $\langle u_\sigma, v \rangle$ is at least

$$\min \{ 2, \langle u_\sigma, w_1 \rangle, \ldots, \langle u_\sigma, w_t \rangle \} = 1 + \min \{1, a_1, \ldots, a_t\}$$

This proves the lemma. ∎

Let X be the projective toric variety associated with a polytope Q with vertices in N. By Proposition 12, the divisor K_X is a $\mathbf{Q}$-Cartier divisor if and only if, for each facet of Q, the primitive vectors of N in the direction of the vertices are on an affine hyperplane. Let σ be the cone spanned by this facet and let $\langle u_\sigma, \cdot\rangle = 1$ be the equation of this hyperplane. For any positive integer q, we have

$$\operatorname{discrep}(X) \geq -1 + \frac{1}{q} \Rightarrow \langle u_\sigma, v\rangle \geq \frac{1}{q} \quad \text{for any } v \text{ in } \sigma \cap N - \{0\}$$

For a *Fano* toric variety, the vertices of Q are primitive (see 2.1), so this last relation implies

$$\operatorname{Int}(Q) \cap qN = \{0\}$$

This remark was used in [8] to prove the following.

Corollary 13. *Given positive integers n and q, there are only finitely many isomorphism types of toric Fano varieties of dimension n and discrepancy $\geq -1 + \frac{1}{q}$ defined over* $\mathbf{k}$.

The authors appear to have been unaware of the earlier articles [28] and [37], from which the result follows immediately, with explicit bounds: Theorems 1 and 2 give the bound

$$\operatorname{vol}(Q) \leq q^n \big(7(q+1)\big)^{n2^{n+1}} = c(n, q)$$

for any polytope Q in $\mathbf{R}^n$ such that $\operatorname{Int}(Q) \cap q\mathbf{Z}^n = \{0\}$, and the finiteness of the number of isomorphism classes of such polytopes under the action of $\operatorname{GL}(\mathbf{Z}^n)$. One can also get all kinds of explicit bounds: consider a face of Q with vertices $v_1, \ldots, v_s$ primitive in N and the corresponding vertex u of P. Assume that $v_1, \ldots, v_n$ are linearly independent. We have

$$\delta_u \det_N(v_1, \ldots, v_n) = n! \operatorname{vol}\big(\langle v_1, \ldots, v_n\rangle\big) \leq n! \operatorname{vol}(Q) \leq n! c(n, q)$$

Since $\delta_u u$ is in M, we see that

$$\big(n! c(n, q)\big)! K_X \quad \text{is a Cartier divisor}$$

Let r be the number of facets of Q. Since Q has integral vertices, we have

$$\operatorname{vol}(Q) \geq \frac{r}{n!}$$

If X is $\mathbf{Q}$-factorial, each facet of Q has n vertices (see 2.1), hence the total number $\rho_X + n$ of vertices of Q is at most nr, and

$$\rho_X \leq n n! c(n, q)$$

If mK_X is a Cartier divisor, i.e., if mP is integral, the origin is again the only interior point of mP in mM and Theorem 1 of [37] yields explicit bounds on $(-K_X)^n$ and $h^0(X, -mK_X)$.

3. Rational Curves on Smooth Fano Varieties

We now forget about toric varieties and study smooth Fano varieties in general. Our aim is to show that a result analogous to Corollary 7 holds. The method of course is of course very different, since there is no description of smooth Fano varieties in general (except in dimension at most 3). The main ingredient of this proof is the study of rational curves on smooth Fano varieties (which is of interest by itself) via Mori's bend-and-break lemmas.

3.1. The bend-and-break lemmas

Let $\mathbf{k}$ be an algebraically closed field.

Given varieties X and Y defined over $\mathbf{k}$, with X quasi-projective and Y projective, Grothendieck showed ([24], 4(c)) that morphisms from Y to X are parametrized by a locally Noetherian scheme $\mathrm{Mor}\,(Y, X)$.

Let C be a curve (a projective connected reduced scheme of dimension 1 over $\mathbf{k}$). The scheme $\mathrm{Mor}\,(C, X)$ has in general countably many components, but if we fix an ample divisor H on X and an integer d, the subscheme $\mathrm{Mor}_d(C, X)$ of $\mathrm{Mor}\,(C, X)$ that parametrizes "curves" $f : C \to X$ with H-degree $f_*C \cdot H$ equal to d is quasi-projective over $\mathbf{k}$, and $\mathrm{Mor}\,(C, X)$ is the disjoint union of the $\mathrm{Mor}_d(C, X)$, for all integers d.

The local structure of the scheme $\mathrm{Mor}\,(C, X)$ around the point $[f]$ can be worked out: the Zariski tangent space is isomorphic to

$$H^0\big(C, \mathcal{H}om(f^*\Omega^1_X, \mathcal{O}_C)\big)$$

and, *if X is smooth along* $f(C)$, locally around $[f]$, the scheme $\mathrm{Mor}\,(C, X)$ can be defined by $h^1(C, f^*T_X)$ equations in a nonsingular variety of dimension $h^0(C, f^*T_X)$. In particular, any irreducible component of $\mathrm{Mor}\,(C, X)$ through $[f]$ has dimension at least

$$h^0(C, f^*T_X) - h^1(C, f^*T_X)$$

Since C is a curve, this is equal by the Riemman–Roch theorem to

$$\chi(C, f^*T_X) = -K_X \cdot f_*C + \big(1 - g(C)\big)\dim(X)$$

Fixing a finite subset B of C, we can consider the subscheme

$$\mathrm{Mor}\,\big(C, X; f|_B\big) \quad \text{of} \quad \mathrm{Mor}\,(C, X)$$

that consists of morphisms that coincide with f on B. Fixing the value of a morphism at one point imposes $\dim(X)$ conditions, hence

$$
\begin{aligned}
(7) \qquad & \dim_{[f]} \operatorname{Mor}\big(C, X; f|_B\big) \\
& \geq \dim_{[f]} \operatorname{Mor}(C, X) - \operatorname{Card}(B)\dim(X) \\
& \geq -K_X \cdot f_*C + \big(1 - g(C) - \operatorname{Card}(B)\big)\dim(X)
\end{aligned}
$$

We now prove the first bend-and-break lemma, which can be found in [42] (Theorems 5 and 6). It says that a curve deforming nontrivially, while keeping a point fixed, must break into an effective 1-cycle with a rational component passing through the fixed point.

Proposition 14 (First bend-and-break lemma). *Let X be a projective variety, let $f : C \to X$ be a smooth curve, and let c be a point on C. If $\dim_{[f]} \operatorname{Mor}\big(C, X; f|_{\{c\}}\big) \geq 1$, there exists a rational curve on X through $f(c)$.*

According to (7), *when X is smooth along $f(C)$*, the hypothesis is fulfilled whenever

$$-K_X \cdot f_*C - g(C)\dim(X) \geq 1$$

Proof of the Proposition. We may assume that $f(C)$ is irrational. Let T be the normalization of a 1-dimensional subvariety of $\operatorname{Mor}\big(C, X; f|_{\{c\}}\big)$ passing through $[f]$ and let $\overline{T}$ be a smooth compactification of T. The indeterminacies of the rational evaluation map

$$
\begin{aligned}
\operatorname{ev} : C \times \overline{T} & \dashrightarrow X \\
(x, t) & \longmapsto f_t(x)
\end{aligned}
$$

defined on $C \times T$ can be resolved by blowing up points to get a morphism

$$e : S \xrightarrow{\varepsilon} C \times \overline{T} \overset{\operatorname{ev}}{\dashrightarrow} X$$

If ev is defined at every point of $\{c\} \times \overline{T}$, take an affine open neighborhood U of $f(c)$ in X and an open neighborhood V of c in C such that $\operatorname{ev}(V \times \overline{T})$ is contained in U. For each v in V, the image of $\{v\} \times \overline{T}$ is a complete subvariety of the affine variety U, hence is a point. It follows that ev has infinitely many 1-dimensional fibers, so that its image must be the curve $f(C)$. In particular, for each t in T, the image of f_t must be $f(C)$. But

$f(C)$ being irrational, there are only finitely many morphisms from C to $f(C)$ mapping c to $f(c)$, so this implies that the f_t remain the same, which is absurd.

Hence there exists a point t_0 in $\overline{T}$ such that ev is not defined at (c, t_0). The fiber of t_0 under the projection $S \to \overline{T}$ is the union of the strict transform of $C \times \{t_0\}$ and a (connected) exceptional rational 1-cycle E which is not entirely contracted by e and meets the strict transform of $\{c\} \times \overline{T}$. Since the latter is contracted by e to the point $f(c)$, the rational 1-cycle e_*E passes through $f(c)$. ■

The same proof shows that the proposition still holds when X is a complex compact Kähler variety (because, by a result of Campana, the variety $\mathrm{Mor}\left(C, X; f|_{\{c\}}\right)$ is *Moishezon*, hence contains algebraic curves as soon as its dimension is positive). However, it fails in general for curves on compact complex manifolds.

Once we know there is a rational curve, it may under certain conditions be broken up into several components. More precisely, if it deforms nontrivially while keeping two points fixed, it must break up (into an effective 1-cycle with rational components).

Proposition 15 (Second bend-and-break lemma). *Let X be a projective variety and let $f : \mathbf{P}^1 \to X$ be a rational curve. If*

$$\dim_{[f]}\left(\mathrm{Mor}\left(\mathbf{P}^1, X; f|_{\{0,\infty\}}\right)\right) \geq 2,$$

the 1-cycle $f_*\mathbf{P}^1$ is numerically equivalent to the sum of a least two rational curves whose union is connected and passes through $f(0)$ and $f(\infty)$.

According to (7), *when X is smooth along $f(\mathbf{P}^1)$*, the hypothesis is fulfilled whenever

$$-K_X \cdot f_*\mathbf{P}^1 - \dim(X) \geq 2$$

Proof of the Proposition. The group of automorphisms of $\mathbf{P}^1$ fixing two points is the multiplicative group $\mathbf{G}_m$. Let T be the normalization of a 1-dimensional subvariety of $\mathrm{Mor}\left(\mathbf{P}^1, X; f|_{\{0,\infty\}}\right)$ passing through $[f]$ but not contained in its $\mathbf{G}_m$-orbit. The corresponding map

$$F : \mathbf{P}^1 \times T \to X \times T$$

is finite. Let $\overline{T}$ be a smooth compactification of T, let S be the normalization in the function field of $\mathbf{P}^1 \times T$ of the closure in $X \times \overline{T}$ of the image of F and let

$\overline{F} : S \to X \times \overline{T}$ be the canonical finite morphism.[12] Since $\mathbf{P}^1 \times T$ is already normal, we have $\overline{F}^{-1}(X \times T) = \mathbf{P}^1 \times T$ by uniqueness of normalization, so that there is a commutative diagram

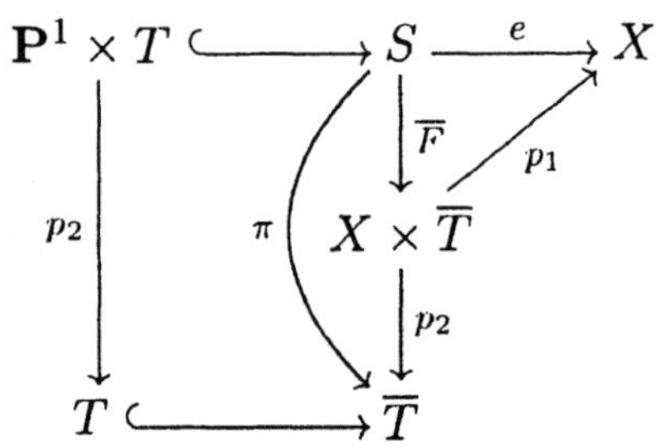

The surface S might not be smooth. On the other hand, we know that no component of a fiber of π is contracted by e (because it would then be contracted by $\overline{F}$).

Since $\overline{T}$ is a smooth curve and S is integral, π is flat ([27], III, Proposition 9.7); hence each fiber C is a 1-dimensional projective scheme without embedded component, whose genus is constant hence equal to 0 ([27], III, Corollary 9.10). In particular, any component C_1 of C_{red} is smooth rational, because $\mathcal{O}_{C_1}$ is a quotient of $\mathcal{O}_C$; hence $H^1(C_1, \mathcal{O}_{C_1})$ is a quotient of $H^1(C, \mathcal{O}_C)$, and therefore vanishes. In particular, if C is integral, it is a smooth rational curve.

Assume all fibers of π are integral. Then S is a (minimal) ruled surface in the sense of [27], V, §2 (Hartshorne assumes that S is smooth, but this hypothesis is not used in the proofs, hence follows from the others). Let T_0 be the closure of $\{0\} \times T$ in S and let T_∞ be the closure of $\{\infty\} \times T$. These sections of π are contracted by e (to $f(0)$ and $f(\infty)$, respectively).

If H is an ample divisor on $e(S)$, which is a surface by construction, we have $(e^*H)^2 > 0$ and $e^*H \cdot T_0 = e^*H \cdot T_\infty = 0$; hence T_0^2 and T_∞^2 are negative by the Hodge index theorem.

However, since T_0 and T_∞ are both sections of π, their difference is linearly equivalent to the pull-back by π of a divisor on $\overline{T}$ ([27], V, Proposition 2.3). In particular,

$$0 = (T_0 - T_\infty)^2 = T_0^2 + T_\infty^2 - 2T_0 \cdot T_\infty < 0$$

[12] This is constructed exactly as the standard normalization (see [27], II, Example 3.8) by patching up the spectra of the integral closures *in* $K(\mathbf{P}^1 \times T)$ of the coordinate rings of affine open subsets of $\overline{F(\mathbf{P}^1 \times T)}$. The fact that $\overline{F}$ is finite follows from the finiteness of integral closure ([27], I, Theorem 3.9A).

which is absurd.

It follows that at least one fiber of π is not integral. Since none of its components is contracted by e, its direct image on X is the required 1-cycle. ■

3.2. Rational curves on smooth Fano varieties

We will apply the bend-and-break lemmas to show that any smooth Fano variety X is covered by rational curves. We start from any curve $f : C \to X$ and want to show, using the estimate (7), that it deforms nontrivially while keeping a point x fixed. In positive characteristic, we use the Frobenius morphism to increase the degree of f without changing the genus of C. The first bend-and-break lemma gives in that case the required rational curve through x. Using the second bend-and-break lemma, we can bound the degree of this curve by a constant depending only on the dimension of X, and this will be essential for the remaining step: reduction of the characteristic zero case to positive characteristic.

Assume for a moment that X and x are defined over $\mathbf{Z}$. For almost all prime numbers p, the reduction of X modulo p is a Fano variety of the same dimension; hence there is a rational curve (defined over the algebraic closure of $\mathbf{F}_p$) through x. This means that the scheme $\bigcup_{d>0} \mathrm{Mor}_d(\mathbf{P}^1, X; 0 \mapsto x)$, which is defined over $\mathbf{Z}$, has a geometric point modulo almost all primes p. Since we can moreover bound the degree d of the curve by a constant independent of p, we are in fact dealing with a quasi-projective scheme, and this implies[13] that it has a point over $\bar{\mathbf{Q}}$, hence over $\mathbf{k}$.

In general, X and x are defined over some finitely generated ring and a similar reasoning yields the existence of a $\mathbf{k}$-point of $\bigcup_{d>0} \mathrm{Mor}_d(\mathbf{P}^1, X; 0 \mapsto x)$, that is, of a rational curve on X through x.

Theorem 16. *Let X be a smooth Fano variety of positive dimension n. Through any point of X there is a rational curve of $(-K_X)$-degree at most $n+1$.*

[13]For a projective scheme, this is because a system of homogeneous polynomial equations with integral coefficients that has, for almost all primes p, a solution modulo p in $\bar{\mathbf{F}}_p$, has a solution in $\bar{\mathbf{Q}}$. One can modify this argument to work for quasi-projective schemes. The proof of this point presented below uses heavier machinery (Chevalley's theorem).

Proof. Let x be a point of X. To construct a rational curve through x, it is enough by the first bend-and-break lemma (Proposition 14) to produce a curve $f : C \to X$ and a point c on C with $\dim_{[f]} \operatorname{Mor}\left(C, X; f|_{\{c\}}\right) \geq 1$ and $f(c) = x$. By the dimension estimate (7), it is enough to have

$$-K_X \cdot f_*C - ng(C) \geq 1$$

Unfortunately, there is no known way to achieve that, except in positive characteristic. Here is how it works.

Assume that the field $\mathbf{k}$ has characteristic $p > 0$. Choose a smooth curve $f : C \to X$ through x and a point c of C such that $f(c) = x$. Consider the ($\mathbf{k}$-linear) Frobenius morphism $C_1 \to C$ ([27], pp. 301–302). It has degree p, but C_1 and C being isomorphic as abstract schemes have the same genus. Iterating the construction, we get a morphism $F_m : C_m \to C$ of degree p^m between curves of the same genus. But

$$-K_X \cdot (f \circ F_m)_* C_m - ng(C_m) = -p^m K_X \cdot f_*C - ng(C)$$

is positive for m large enough. By the first bend-and-break lemma (Proposition 14), there exists a rational curve $f' : \mathbf{P}^1 \to X$, with say $f'(0) = x$. If

$$-K_X \cdot f'_*\mathbf{P}^1 - n \geq 2$$

the scheme $\operatorname{Mor}\left(\mathbf{P}^1, X; f'|_{\{0,1\}}\right)$ has dimension at least 2 at $[f']$. By the second bend-and-break lemma (Proposition 15), one can break up the rational curve $f'(\mathbf{P}^1)$ into at least two rational pieces. The component passing through x has smaller $(-K_X)$-degree, and we can repeat the process as long as $-K_X \cdot \mathbf{P}^1 - n \geq 2$, until we get a rational curve of degree no more than $n + 1$.

This proves the theorem in positive characteristic. Assume now that $\mathbf{k}$ has characteristic 0. Embed X in some projective space, where it is defined by a finite set of equations, and let R be the (finitely generated) subring of $\mathbf{k}$ generated by the coefficients of these equations and coordinates of x. There is a projective scheme $\mathcal{X} \to \operatorname{Spec}(R)$ with an R-point x_R, such that X is obtained from its generic fiber by base change from the quotient field of R to $\mathbf{k}$.

The geometric generic fiber is a Fano variety of dimension n. There is a dense open subset U of $\operatorname{Spec}(R)$ over which $\mathcal{X}$ is flat ([25], Théorème 6.9.1), and even smooth of dimension n ([26], Théorème 12.2.4(iii)). Since ampleness is an open property ([26], Corollaire 9.6.4), we may even, upon shrinking U, assume that the relative dualizing sheaf $\omega_{\mathcal{X}_U/U}$ is ample on each fiber.

It follows that for each maximal ideal $\mathfrak{m}$ of R in U, the (geometric) fiber $X_{\mathfrak{m}}$ is a Fano variety of dimension n. In the future, we will skip these steps when we use this process of reduction to positive characteristic.

Let us take a short break and use a little commutative algebra to show that the finitely generated ring R has the following properties:

- for each maximal ideal $\mathfrak{m}$ of R, the field $R/\mathfrak{m}$ is finite;
- maximal ideals are dense in Spec (R).

The first item is proved as follows. The field $R/\mathfrak{m}$ is a finitely generated $(\mathbf{Z}/\mathbf{Z}\cap\mathfrak{m})$-algebra, hence is finite over the quotient field of $\mathbf{Z}/\mathbf{Z}\cap\mathfrak{m}$ by a theorem of Zariski.[14] If $\mathbf{Z}\cap\mathfrak{m}=0$, the field $R/\mathfrak{m}$ is a finite-dimensional $\mathbf{Q}$-vector space with basis $(e_1,\dots,e_m)$. If $x_1,\dots,x_r$ generate the $\mathbf{Z}$-algebra $R/\mathfrak{m}$, there exists an integer q such that qx_j belongs to $\mathbf{Z}e_1\oplus\cdots\oplus\mathbf{Z}e_m$ for each j. This implies

$$\mathbf{Q}e_1\oplus\cdots\oplus\mathbf{Q}e_m=R/\mathfrak{m}\subset\mathbf{Z}[1/q]e_1\oplus\cdots\oplus\mathbf{Z}[1/q]e_m$$

which is absurd. Therefore, $\mathbf{Z}/\mathbf{Z}\cap\mathfrak{m}$ is finite and so is $R/\mathfrak{m}$.

For the second item, we need to show that the intersection of all maximal ideals of R is $\{0\}$. Let a be a nonzero element of R and let $\mathfrak{n}$ be a maximal ideal of the localization R_a. The field $R_a/\mathfrak{n}$ is finite by the first item; hence its subring $R/R\cap\mathfrak{n}$ is a finite domain, hence a field. Therefore, $R\cap\mathfrak{n}$ is a maximal ideal of R, which is in the open subset Spec (R_a) of Spec (R).

Now back to the proof of the theorem. One can show that there is a quasi-projective scheme

$$\rho:\ \mathrm{Mor}_{\le n+1}(\mathbf{P}^1_R,\mathcal{X};0\mapsto x_R)\to\mathrm{Spec}\,(R)$$

that parametrizes nonconstant morphisms of degree at most $n+1$.

Let $\mathfrak{m}$ be a maximal ideal of R. Since the field $R/\mathfrak{m}$ is finite, hence of positive characteristic, what we just saw implies that the (geometric) fiber over a closed point of the dense open subset U of Spec (R) is nonempty. It follows that the image of ρ, which is a constructible[15] subset of Spec (R) by Chevalley's theorem ([27], II, Example 3.19), contains all closed points of U. It is therefore dense by the second item, hence contains the generic point ([27], II, Example 3.18(b)). This implies that the generic fiber is nonempty. It has therefore a geometric point, which corresponds to a rational curve on

[14]This theorem says that if k is a field and K a finitely generated k-algebra which is a field, K is an algebraic hence finite extension of k; see [38], Theorem 5.2.

[15]Recall that a constructible subset is a finite union of locally closed subsets.

X through x, of degree at most $n+1$, defined over an algebraic closure of the quotient field of R, hence over $\mathbf{k}$.[16] ■

Using a stronger version of the first bend-and-break lemma, one can actually show that *through any point of a Fano*[17] *variety X of positive dimension n, there is a rational curve of $(-K_X)$-degree at most $2n$* (see e.g., [19], Theorem 3.6). This means that X is *uniruled*: there exist a variety Y of dimension $n-1$ and a dominant morphism $\mathbf{P}^1 \times Y \to X$. A variety is *separably uniruled* if there is such a morphism which is in addition separable (i.e., generically smooth). The distinction is only important in positive characteristic. Unfortunately, some Fano varieties are *not* separably uniruled: in characteristic $p > 0$, this is the case for the (smooth Fano) cyclic cover of degree p of $\mathbf{P}^n$, with $n \geq p$, branched along a general hypersurface of degree p ([33], Theorem V.5.11).

There is not much one can do with nonseparably uniruled varieties, hence we will assume in what follows that the characteristic of $\mathbf{k}$ is 0. In this case, on any normal uniruled variety X, there is a *free* rational curve, i.e., a morphism $f : \mathbf{P}^1 \to X$ whose image is contained in the smooth locus of X and such that f^*T_X is generated by global sections.

Free rational curves are essential in many constructions and theorems, such as smoothing results.

3.3. A bound on the degree of smooth Fano varieties in characteristic zero

We want to show that any two points of a smooth Fano variety can be joined by a chain of rational curves (I do not know whether this remains true for a singular Fano variety). The idea for the proof is the following.

First, one constructs, for any normal proper variety X, a kind of quotient $\rho : X \dashrightarrow R(X)$ for the equivalence relation "to be joined by a chain of rational curves;" this we will call the *rational quotient* of X. Then, when X is a smooth Fano variety, a souped-up version of Theorem 16 shows that if

[16] It is important to remark that the "universal" bound on the degree of the rational curve is essential for the proof.

[17] No conditions on the singularities of X are actually necessary beyond its normality. It is interesting to look at Example 2(5): if the base Y of the cone is an abelian variety, the anticanonical divisor of X is $\mathbf{Q}$-Cartier and ample, but the only rational curves that X contains are the lines through its vertex.

$R(X)$ is not a point, there must exist a rational curve on X not contracted by ρ, which is absurd.

As for the construction of the rational quotient (which goes back to [11]), it is no more complicated to consider a family of subvarieties of X given by a diagram

$$\begin{array}{ccc} \mathcal{C} & \xrightarrow{F} & X \\ \big\downarrow{\scriptstyle \pi} & & \\ T & & \end{array} \tag{8}$$

and the associated equivalence relation: two points of X are $\mathcal{C}$-equivalent if they can be connected by a chain of subvarieties $F(\mathcal{C}_{t_1}) \cup \cdots \cup F(\mathcal{C}_{t_m})$. For a given point x of X, the sets $V_m(x)$ of points that can be connected to x by such a chain of fixed length m form an ascending chain of constructible subsets of X whose closures stabilize after at most n steps, where n is the dimension of X. For x general in X, the corresponding sets $\overline{V_n(x)}$ are either disjoint or equal, hence form the fibers of a rational map $\tau : X \dashrightarrow Y$ which is the required quotient for $\mathcal{C}$-equivalence. This process actually works only when F and π are *flat*. For more details, see for example [19], Section 5.2. The precise result is the following.

Proposition 17. *In the diagram (8), assume that π is flat with irreducible fibers and that F is flat. There exists a dense open subset X^0 of X, a variety Y^0 and a morphism $\tau : X^0 \to Y^0$ such that*

(a) *two points of X^0 have the same image by τ if and only if the closures in X of their $\mathcal{C}$-equivalence classes are the same;*

(b) *if x is in X^0, a general point of its fiber $\tau^{-1}(\tau(x))$ can be connected to x by a $\mathcal{C}$-chain of length* $\dim(X) - \dim(Y^0)$.

Smooth uniruled projective varieties with Picard number 1. An infinitesimal calculation shows that free deformations of a free rational curve $\mathbf{P}^1 \to X$ of degree d are parametrized by a smooth quasi-projective variety M and that the evaluation map $\mathrm{ev} : \mathbf{P}^1 \times M \to X$ is smooth (hence flat). There is therefore a quotient $\tau : X^0 \to Y^0$. Assume that X is smooth and has Picard number 1. Standard arguments show that any such morphism must be constant (essentially because any nonzero effective divisor, such as the inverse image of an effective ample divisor on Y^0, is ample). The proposition implies that a general pair of points of X can be connected by a chain of free rational curves of total degree at most $d \dim(X)$. Because we are dealing with free curves, this chain can be smoothed and we have "shown" the following.

Corollary 18. *If a smooth projective variety X with Picard number 1 has a free rational curve of degree d with respect to some ample divisor, there is a free rational curve of degree at most $d \dim(X)$ through a general pair of points of X.*

If X is smooth and separably uniruled with Picard number 1, it has a free rational curve. The corollary applies and produces a rational curve through a general pair of points of X. Varieties with the property that there is a rational curve through a general pair of points are called *rationally connected.*

Smooth Fano varieties with Picard number 1. In characteristic 0, the corollary applies to any smooth Fano variety X with Picard number 1, with $d = \dim(X) + 1$, where the degree is taken with respect to the anticanonical divisor. These varieties are therefore rationally connected. The corollary will also give us a bound on their degree, via the following elementary result.

Proposition 19. *Let X be a projective variety of dimension n, let H be an ample divisor on X, and let d be a nonnegative real number. Let x be a smooth point of X such that a general point of X can be connected to x by an irreducible curve of H-degree at most d. We have $H^n \leq d^n$.*

Proof. By definition of the intersection number, $h^0(X, mH)$ is equivalent to

$$\frac{H^n}{n!} m^n = \frac{\left(m \sqrt[n]{H^n}\right)^n}{n!}$$

for $m \gg 0$. Let $\mathfrak{m}_{X,x}$ be the maximal ideal of $\mathcal{O}_{X,x}$ and let r be a positive integer. Because of the exact sequence

$$0 \to H^0(X, \mathfrak{m}^r_{X,x}(mH)) \to H^0(X, mH) \to H^0(X, \mathcal{O}_{X,x}/\mathfrak{m}^r_{X,x}) \simeq \mathbf{k}^{\binom{n+r-1}{n}}$$

having a point of multiplicity at least r imposes at most $\binom{n+r-1}{n} \sim \frac{r^n}{n!}$ conditions on the elements of the linear system $|mH|$. It follows that given $\varepsilon > 0$, there exist a large positive integer m and a divisor D in $|mH|$ such that

$$\operatorname{mult}_x D \geq m \sqrt[n]{H^n} - m\varepsilon$$

Let C be an irreducible curve of H-degree at most d connecting x to a point outside of D. Since C is not contained in the support of D, we have

$$md \geq C \cdot (mH) = C \cdot D \geq \operatorname{mult}_x D \geq m\sqrt[n]{H^n} - m\varepsilon$$

and the proposition follows by letting ε go to 0. ∎

Corollary 20. *Assume the characteristic is zero. Any smooth Fano variety X of dimension n and Picard number 1 satisfies*

$$(-K_X)^n \leq \big(n(n+1)\big)^n$$

This bound is not too far off (recall $(-K_{\mathbf{P}^n})^n = (n+1)^n$).

The rational quotient and general smooth Fano varieties. To deal with smooth Fano varieties with arbitrary Picard number, we need the construction of the full rational quotient, as explained at the beginning of the section. This requires a lot of technical work and I will only explain the main steps. Say that two points on a variety are $\mathcal{R}$-equivalent if they can be connected by a chain of rational curves. A variety is rationally chain-connected if two general points are $\mathcal{R}$-equivalent.

(1) Extend the construction of Proposition 17 to show that given a normal projective variety X, there exist a normal variety $R(X)$, called *the rational quotient of* X and a rational map $\rho : X \dashrightarrow R(X)$, defined and proper over a dense open subset of X, whose very general fibers are $\mathcal{R}$-equivalence classes.

(2) Extend Theorem 16 to show that for any smooth Fano variety X and nonconstant dominant rational map $\pi : X \dashrightarrow Y$ defined and proper over a dense open subset of X, there exists a rational curve on X that meets but is not contained in a general fiber of π. This implies that a smooth Fano variety is rationally chain-connected. More precisely, any two points of a Fano variety X of dimension n can be connected by a chain of rational curves of total $(-K_X)$-degree at most $(2^n-1)(n+1)$.

(3) In characteristic zero, use smoothing results to show that a rationally chain-connected smooth projective variety is rationally connected. In case of a smooth Fano variety X, any two points can be connected by a single rational curve of $(-K_X)$-degree at most

$$3(2^n-1)(n+1)^{(n+1)(2^n-1)}.$$

Using Proposition 19, we get that in characteristic zero, *any smooth Fano variety X of dimension n is rationally connected and satisfies*

$$(9) \qquad (-K_X)^n \le \left(3(2^n-1)(n+1)^{(n+1)(2^n-1)}\right)^n$$

This bound is probably very far off. In view of (6), it is tempting to conjecture the inequality

$$(-K_X)^n \le \left(n^\rho(n+1)\right)^n$$

for smooth Fano varieties with Picard number ρ (this holds if two general points of X can be joined by a chain of n^ρ free curves of $(-K_X)$-degree at most $n+1$). Getting a bound on ρ in terms of n is another story, although one might maybe expect a linear bound (if X is a del Pezzo surface with Picard number 9, the Fano variety X^m has dimension $n = 2m$ and Picard number $9n/2$).

3.4. Smooth Fano varieties form a bounded family

It is out of the question to prove anything here. We will just give a few references.

There are only finitely many deformation types of Fano varieties X with fixed Hilbert polynomial $P(m) = \chi(X, \mathcal{O}_X(mK_X))$ (see [39]). Using the Hodge index theorem and the Kodaira vanishing theorem, one proves that in characteristic zero, for any ample divisor H on X, the coefficients of the Hilbert polynomial $\chi(X, \mathcal{O}_X(mH))$ are bounded by a function which depends only on the dimension n of X, on H^n and on $K_X \cdot H^{n-1}$ (see [33], Exercise VI.2.15.8). When $H = -K_X$, these numbers are bounded by a function that depends only on n by (9), hence there are finitely many deformation types of Fano varieties in any given dimension. One can even get an effective bound.

Theorem 21. *Let* $\mathbf{k}$ *be an algebraically closed field of characteristic zero. There are at most* $(n+2)^{(n+2)^{n2^{3n}}}$ *deformation types of Fano varieties of dimension* n *defined over* $\mathbf{k}$.

As for singular Fano varieties, it is conjectured that if the dimension n is fixed and the discrepancy bounded away from -1, there should still be only finitely many deformation types.

This was proved in Corollary 13 in the toric case. The point is again to bound $(-K_X)^n$ *and* an integer j such that jK_X is a Cartier divisor. To my knowledge, the best general result in this direction appears in the recent preprint [40] where it is shown that given integers n and j, there are only finitely many deformation types of complex Fano varieties X (in the sense of Definition 1) of dimension n such that jK_X is a Cartier divisor. This is done by bounding $(-K_X)^n$ using techniques that are far beyond the scope of this report.

4. The Differential-Geometric Point of View

If (X, ω) is a compact Kähler manifold, its Ricci form is closed by the Bianchi identity and its cohomology class is the first Chern class of X. Conversely, by Yau's proof of Calabi's conjecture ([53], [54], [9]), given a closed $(1,1)$-form ρ representing $c_1(X)$, there exists a Kähler metric on X whose Ricci form is ρ.

It follows that the smooth complex Fano varieties are exactly the compact Kähler manifolds with positive Ricci curvature ([6], 11.16(ii)). This implies in particular ([6], Theorem 11.26) that they are simply connected (see also 5.1 and 5.2 for more general results).

Compact complex manifolds with a positive Kähler–Einstein metric (i.e., whose Ricci form is a positive (constant) multiple of the Kähler form ω) are trivially Fano varieties, but it is a difficult problem to find conditions on a smooth complex Fano variety so that a Kähler–Einstein metric exists. There are restrictions: for example, the existence of such a metric forces the automorphism group to be reductive[18] ([6], Corollary 11.54) (this excludes for example the projective plane blown up at one or two points). There are also existence results for which the reader is refered to the excellent survey [10] and, in the toric case, to [5].

Classical methods of differential geometry yield a good bound on the degree of compact complex manifolds with a positive Kähler–Einstein metric.

Let X be a compact complex manifold of dimension n with a Kähler–Einstein form ω normalized so that its Ricci form is ω. By [6], 11.5, we

[18] An algebraic group is reductive if it has no nontrivial connected unipotent abelian normal subgroups.

have

$$(-K_X)^n = \operatorname{vol}(X)\,\frac{n!}{(2\pi)^n}$$

From [7], Corollary 4, p. 257, it follows that the volume of X is less than the volume of a sphere of the same dimension and radius $\sqrt{2n-1}$, so that

$$\begin{aligned}(-K_X)^n &\leq \frac{n!}{(2\pi)^n}\big(\sqrt{2n-1}\,\big)^{2n}\,\frac{2^{n+1}\pi^n}{1\cdot 3\cdot \ldots \cdot (2n-1)}\\ &= (2n-1)^n\,\frac{2^{n+1}(n!)^2}{(2n)!}\end{aligned}$$

This bound is equivalent to $2^{n+1}\sqrt{\frac{\pi n}{e}}\,n^n$.

5. Simple Connectedness of Complex Fano Varieties

5.1. Simple connectedness of smooth complex Fano varieties

Any complex complete toric variety is simply connected ([22], p. 56). More generally, any smooth complex projective rationally connected variety (such as a smooth complex Fano variety) is simply connected, as we now explain.

- **A finite connected étale cover of a smooth complex rationally connected variety is trivial.** We will only give the argument for a smooth complex Fano variety X, since this is the only case we will use. By Kodaira's vanishing theorem, $H^m(X, \mathcal{O}_X)$ vanishes for $m > 0$, hence $\chi(X, \mathcal{O}_X) = 1$. If $\pi : \tilde{X} \to X$ is a connected finite étale cover, $\tilde{X}$ is a smooth Fano variety, hence $\chi(\tilde{X}, \mathcal{O}_{\tilde{X}}) = 1$. But $\chi(\tilde{X}, \mathcal{O}_{\tilde{X}}) = \deg(\pi)\,\chi(X, \mathcal{O}_X)$ hence π is an isomorphism.
- **The fundamental group of a rationally connected variety is finite.** Let X be a rationally connected variety. Since a general point x of X can be connected to another general point of X by a rational curve, this means that there is a quasi-projective open subscheme M of $\operatorname{Mor}(\mathbf{P}^1, X; 0 \mapsto x)$ over which the evaluation map $\operatorname{ev} : \mathbf{P}^1 \times M \to X$ is dominant. We use the following fact: *let X and Y be complex algebraic varieties, with Y normal, and let $f : Y \to X$ be a dominant morphism. The image of the induced morphism $\pi_1(f) : \pi_1(Y) \to \pi_1(X)$ has finite*

index. The composition of ev with the injection $\iota : \{0\} \times M \hookrightarrow \mathbf{P}^1 \times M$ is constant, hence

$$\pi_1(\iota) \circ \pi_1(\mathrm{ev}) = 0$$

Since $\mathbf{P}^1$ is simply connected, $\pi_1(\iota)$ is bijective, hence $\pi_1(\mathrm{ev}) = 0$. Since ev is dominant, the lemma implies that the image of $\pi_1(\mathrm{ev})$ has finite index. Therefore, the group $\pi_1(X)$ is finite.

Campana gives in [12] a completely different proof of the simple connectedness of smooth complex rationally connected varieties. He shows more generally that the fundamental group of a compact Kähler manifold is finite if any two points can be connected by a chain of subvarieties whose fundamental group of the normalization has the same property.

In [34], Theorem 4.13, it is proved that the fundamental group of a complex rationally chain-connected normal proper variety is finite (possibly nontrivial if the variety is singular).

5.2. Simple connectedness of complex Fano varieties

This is a particular case of a result of Takayama ([49]) whose proof proceeds along the same steps as in 5.1.

Let X be a complex Fano variety.

- **Any finite connected étale cover of X is trivial.** The argument is the same as in the smooth case, upon replacing Kodaira's vanishing theorem with the Kawamata–Viehweg vanishing theorem (see e.g., [19], Theorem 7.26), which works for varieties with log terminal singularities and yields $H^m(X, \mathcal{O}_X) = 0$ for $m > 0$.
- **The fundamental group of X is finite.** This is the difficult part. The "obvious" idea would be to prove that X is rationally connected (or even only rationally chain-connected), but this is still known only in dimension ≤ 3 by [35].

The argument for the second step is based on completely different analytic techniques. We first explain how the argument works for smooth Fano varieties.

Von Neumann dimensions and Atiyah's L^2 index theorem. First of all, recall that for a vector bundle E with a Hermitian metric on a compact

Kähler manifold X of dimension n, Hodge theory tells us that the vector space $H^q(X, E)$ is isomorphic to the space $\mathcal{H}^{0,q}(X, E)$ of E-valued, $\mathcal{C}^\infty$, *harmonic* forms of type $(0, q)$. In particular, we have

$$\chi(X, \mathcal{E}) = \sum_{q=0}^{n} (-1)^q \dim \mathcal{H}^{0,q}(X, E) \tag{10}$$

where $\mathcal{E}$ is the sheaf of holomorphic sections of E.

If X is not necessarily compact (but still Kähler), we let

$$\mathcal{H}^{0,q}_{(2)}(X, E)$$

be the space of harmonic L^2 forms of type $(0, q)$ with values in E. Although the notation does not take this into account, these spaces depend very much on the choice of the metrics when X is not compact. Also, these spaces are usually infinite-dimensional, so an analog of (10) is not obvious, even to formulate.

Let us go back to the case where X is compact and let $\pi : \tilde{X} \to X$ be its universal cover. We can pull back the Kähler form, the vector bundle E, and its metric to $\tilde{X}$ and consider the vector spaces

$$\mathcal{H}^{p,q}_{(2)}(\tilde{X}, \pi^* E)$$

As it turns out, a lot of Hodge theory can be done on a *complete* Kähler manifold such as $\tilde{X}$. For example, on such a variety, *any harmonic L^2 function is d-closed, hence constant* ([23], § 1). Similarly, any harmonic L^2 form with values in $\pi^* E$ is $\bar{\partial}$- and $\bar{\partial}^*$-closed ([23], § 3) hence elements of $\mathcal{H}^{0,0}_{(2)}(\tilde{X}, \pi^* E)$ are *holomorphic* L^2 sections of $\mathcal{O}_{\tilde{X}}(\pi^* E)$.

There is a way to define, for the spaces $\mathcal{H}^{p,q}_{(2)}(\tilde{X}, \pi^* E)$, or for any closed space H of L^2 sections invariant under the action of $\pi_1(X)$, the von Neumann dimension $\dim_{\pi_1(X)} H$, which is a nonnegative real number that vanishes if and only if $H = 0$ (see, e.g., [34], Chap. 6). The Atiyah index theorem ([2]) states

$$\chi(X, \mathcal{E}) = \sum_{q=0}^{n} (-1)^q \dim_{\pi_1(X)} \mathcal{H}^{0,q}_{(2)}(\tilde{X}, \pi^* E)$$

There is a Kodaira-type vanishing theorem due to Andreotti and Vesentini ([1]) that says

$$\mathcal{H}^{0,q}_{(2)}(\tilde{X}, K_{\tilde{X}} + \pi^* D) = 0 \quad \text{for} \quad q > 0$$

if D is an ample divisor on X. Combining this with the usual Kodaira theorem on X, we get

$$h^0(X, K_X + D) = \dim_{\pi_1(X)} \mathcal{H}^{0,0}_{(2)}(\tilde{X}, K_{\tilde{X}} + \pi^* D)$$

Finiteness of the fundamental group of smooth Fano varieties. Let X be a smooth Fano variety. We can apply the equality above with $D = -K_X$ to get

$$\dim_{\pi_1(X)} \mathcal{H}^{0,0}_{(2)}(\tilde{X}, \mathbf{C}) = 1$$

Therefore, there exists a nonzero harmonic L^2 function on $\tilde{X}$ which is, as we saw above, constant. This implies that $\tilde{X}$ has finite volume. Since this volume is computed with respect to the pull-back of the metric on X, it is equal to the volume of X times the degree of π. It follows that the latter, which is the cardinality of $\pi_1(X)$, is finite.

Finiteness of the fundamental group of Fano varieties. To adapt the proof to the case of an arbitrary Fano variety X, we consider a desingularization $f : Y \to X$ whose exceptional locus has normal crossings and the pull back $\pi : \tilde{Y} = Y \times_X \tilde{X} \to Y$ of the universal cover of X (the map $\pi_1(f)$ is surjective, hence $\tilde{Y}$ is connected). Write

$$K_Y \sim f^* K_X + \sum_i a_i E_i$$

as in (1), with $a_i > -1$ and set

$$\Delta = \sum_{a_i<0} (-a_i) E_i + \sum_{a_i \geq 0} \left(\lceil a_i \rceil - a_i \right) E_i$$

so that

$$K_Y \sim f^* K_X - \Delta + E$$

where $E = \sum_{a_i \geq 0} \lceil a_i \rceil E_i$ is an effective (integral) f-exceptional divisor. Demailly proved in [20] (see also [34], Theorem 11.5) a generalization of the Andreotti–Vesentini vanishing theorem that says in this situation

$$\mathcal{H}^{0,q}_{(2)}(\tilde{Y}, K_{\tilde{Y}} + \pi^*(D + \Delta)) = 0 \quad \text{for} \quad q > 0$$

for any nef and big $\mathbf{Q}$-divisor D on Y such that $D + \Delta$ is integral. Taking $D = -f^* K_X$, we get

$$\mathcal{H}^{0,q}_{(2)}(\tilde{Y}, \pi^* E) = 0 \quad \text{for} \quad q > 0$$

On Y, the Kodaira vanishing theorem is replaced with the Kawamata–Viehweg vanishing theorem. We get as above

$$1 = h^0(Y, E) = \dim_{\pi_1(X)} \mathcal{H}^{0,0}_{(2)}(\tilde{Y}, \pi^* E)$$

hence there is a nonzero *holomorphic* L^2 section s of $\mathcal{O}_{\tilde{Y}}(\pi^* E)$. Since the induced morphism $\tilde{f} : \tilde{Y} \to \tilde{X}$ is birational, this section induces a holomorphic function on the complement of $\tilde{f}(\pi^{-1}(E))$ in $\tilde{X}$, and since this set has codimension at least 2, this function extends to the whole of $\tilde{X}$ because the latter is normal. This means that s vanishes on the divisor $\pi^* E$.

Let t be a section of $\mathcal{O}_Y(E)$ with divisor E. The quotient $s/\pi^* t$ is a holomorphic *function* on $\tilde{Y}$ and a local calculation (see, e.g., [34], Lemma 11.2.2) shows that it is still L^2, hence constant. It follows that $\tilde{Y}$ has finite volume, hence the fundamental group of X is finite.

Conjectures. It is conjectured that

- the smooth locus of a complex Fano variety is simply connected;[19]
- a Fano variety is rationally chain-connected, or even rationally connected.[20]

Example 2(5) shows that normal varieties with discrepancy -1 and ample anticanonical bundle may not be rationally connected (take for Y an abelian variety as in footnote 17). Note however that these examples, as any projective cone in the complex projective space, are simply connected.

[19]This holds in dimension 2 by [55]. The algebraic fundamental group is shown to be finite in [40].

[20]The latter holds in characteristic 0 and dimension ≤ 3 by [35].

REFERENCES

[1] A. Andreotti and E. Vesentini, Carleman estimates for the Laplace–Beltrami equation in complex manifolds, *Publ. Math. IHES,* **25** (1965), 81–130.

[2] M. F. Atiyah, Elliptic operators, discrete groups and von Neumann algebras, *Astérisque,* **32–33** (1976), 43–72.

[3] V. Batyrev, Boundedness of the degree of multidimensional toric Fano varieties (Russian), *Vestnik Moskov. Univ. Ser. I Mat. Mekh.,* **37** (1982), 22–27. English translation: *Moscow Univ. Math. Bull.,* **37** (1982), 28–33.

[4] V. Batyrev, On the classification of toric Fano 4-folds, *J. Math. Sci. (New York),* **94** (1999). 1021–1050.

[5] V. Batyrev, and E. Selivanova, Einstein–Kähler metrics on symmetric toric Fano manifolds, *J. reine angew. Math.,* **512** (1999), 225–236.

[6] A. Besse, *Einstein Manifolds,* Ergebnisse der Mathematik und ihrer Grenzgebiete (3), vol. 10, Springer-Verlag (New York, 1987).

[7] R. Bishop and R. Crittenden, *Geometry of Manifolds,* Pure and Applied Mathematics, vol. 15, Academic Press (New York, 1964).

[8] A. Borisov and L. Borisov, Singular toric Fano varieties, (in Russian) *Mat. Sb.,* **183** (1992), 134–141. English transl.: *Russian Acad. Sci. Sb. Math.,* **75** (1993), 277–283.

[9] J.-P. Bourguignon, Premières formes de Chern des variétés kählériennes compactes, d'après E. Calabi, T. Aubin et S. T. Yau, *Séminaire Bourbaki, 1977/78,* Exp. 507, LNM 710, Springer-Verlag (New York, 1979).

[10] J.-P. Bourguignon, Métriques d'Einstein–Kähler sur les variétés de Fano: obstructions et existence (d'après Y. Matsushima, A. Futaki, S. T. Yau, A. Nadel et G. Tian), *Séminaire Bourbaki, 1996/97,* Exp. 830, Astérisque **245** (1997), 277–305.

[11] F. Campana, Coréduction algébrique d'un espace analytique compact faiblement kählérien, *Invent. Math.,* **63** (1981), 187–223.

[12] F. Campana, Remarques sur le revêtement universel des variétés kählériennes compactes, *Bull. Soc. Math. France,* **122** (1994), 255–284.

[13] C. Casagrande, On the birational geometry of toric Fano 4-folds, *C. R. Acad. Sci. Paris,* **332** (2001), 1093–1098.

[14] C. Casagrande, Contractible classes in toric varieties, e-print `math.AG/0111332`, to appear in *Math. Zeit.,* **243** (2003), 99–126.

[15] C. Casagrande, Toric Fano varieties and birational morphisms, e-print `math.AG/0112007`, to appear in *Int. Math. Res. Not.*

[16] D. Cox, Update on toric geometry, in: *Geometry of Toric Varieties* (L. Bonavero and M. Brion, eds.), Séminaires et Congrès, vol. 6., Soc. Math. France (2002), pp. 1–41.

[17] S. D. Cutkosky, On Fano 3-folds, *Manuscripta Math.,* **64** (1989), 189–204.

[18] L. Danzer, B. Grünbaum and V. Klee, Helly's theorem and its relatives, *Proc. Sympos. Pure Math.*, vol. 7, Amer. Math. Soc., Providence, RI (1963), pp. 101–180.

[19] O. Debarre, *Higher-Dimensional Algebraic Geometry,* Universitext, Springer (2001).

[20] J.-P. Demailly, Singular Hermitian metrics on positive line bundles, *Complex Algebraic Varieties,* LNM 1507, Springer (1992), pp. 87–104.

[21] M. Demazure et al. (eds.) *Séminaire sur les Singularités des Surfaces,* Centre de Mathématiques de l'École Polytechnique, Palaiseau, 1976–1977, LNM 777 (Berlin, 1980).

[22] W. Fulton, *Introduction to toric varieties,* Annals of Mathematics Studies, vol. 131, Princeton University Press (1993).

[23] M. Gromov, Kähler hyperbolicity and L_2-Hodge theory, *J. Diff. Geom.,* **33** (1991), 263–292.

[24] A. Grothendieck, Techniques de construction et théorèmes d'existence en géométrie algébrique IV: les schémas de Hilbert, *Séminaire Bourbaki, 1960/61,* Exp. 221, reprinted in Astérisque hors série **6,** Soc. Math. France (1997).

[25] A. Grothendieck, Éléments de Géométrie Algébrique IV, 2, *Publ. Math. IHES,* vol. 24 (1965).

[26] A. Grothendieck, Éléments de Géométrie Algébrique IV, 3, *Publ. Math. IHES,* vol. 28 (1966).

[27] R. Hartshorne, *Algebraic Geometry,* Graduate Texts in Mathematics, vol. 52, Springer (1977).

[28] D. Hensley, Lattice vertex polytopes with interior lattice points, *Pacific J. Math.,* **105** (1983), 183–191.

[29] V. Iskovskikh, Fano 3-folds I (Russian), *Izv. Akad. Nauk SSSR Ser. Mat.,* **41** (1977), 516–562. English translation: *Math. USSR Izvestiya,* **11** (1977), 485–527.

[30] V. Iskovskikh, Fano 3-folds II (Russian), *Izv. Akad. Nauk SSSR Ser. Mat.,* **42** (1978), 506–549. English translation: *Math. USSR Izvestiya,* **12** (1978), 469–506.

[31] V. Iskovskikh and Yu. Prokhorov, Fano Varieties (Russian), in: *Algebraic geometry V* (A. Parshin and I. Shafarevich, eds.), Ross. Akad. Nauk, Vseross. Inst. Nauchn. i Tekhn. Inform., Moscow. English translation: Encyclopaedia Math. Sci., vol. 47, Springer-Verlag (Berlin, 1999), pp. 1–245.

[32] P. Kleinschmidt and B. Sturmfels, Smooth toric varieties with small Picard number are projective, *Topology,* **30** (1991), 289–299.

[33] J. Kollár, *Rational Curves on Algebraic Varieties,* Ergebnisse der Mathematik und ihrer Grenzgebiete (3), vol. 32, Springer (1996).

[34] J. Kollár, *Shafarevich maps and automorphic forms,* Princeton University Press (1995).

[35] J. Kollár, Y. Miyaoka and S. Mori, Rationally connected varieties, *J. Algebraic Geom.,* **1** (1992), 429–448.

[36] J. Kollár and S. Mori, *Birational Geometry of Algebraic Varieties,* with collaboration of C. H. Clemens and A. Corti, Cambridge Tracts in Mathematics, vol. 134, Cambridge University Press (1998).

[37] J. Lagarias and G. Ziegler, Bounds for lattice polytopes containing a fixed number of interior points in a sublattice, *Canad. J. Math.,* **43** (1991), 1022–1035.

[38] H. Matsumura, *Commutative ring theory,* Second edition, Cambridge Studies in Advanced Mathematics, vol. 8, Cambridge University Press (1989).

[39] T. Matsusaka, On canonically polarized varieties (II), *Amer. J. Math.,* **92** (1970), 283–292.

[40] J. McKernan, Boundedness of log terminal Fano pairs of bounded index, e-print `math.AG/0205214`.

[41] G. Megyesi, Fano threefolds in positive characteristic, *J. Algebraic Geom.,* **7** (1998), 207–218.

[42] S. Mori, Projective manifolds with ample tangent bundles, *Ann. of Math.,* **110** (1979), 593–606.

[43] S. Mori and S. Mukai, Classification of Fano threefolds with $b_2 \geq 2$, *Manuscripta Math.,* **36** (1981), 147–162.

[44] S. Mukai and H. Umemura, Minimal rational threefolds, *Algebraic geometry (Tokyo/Kyoto, 1982)* LNM 1016, Springer (1983), pp. 490–518.

[45] M. Reid, Decomposition of toric morphisms, *Arithmetic and geometry* (M. Artin, J. Tate, eds.) Progr. Math., vol. 36, Birkhäuser, Boston (1983), Vol. II, 395–418.

[46] H. Sato, Toward the classification of higher-dimensional toric Fano varieties, *Tôhoku Math. J.,* **52** (2000), 383–413.

[47] N. Shepherd-Barron, Fano threefolds in positive characteristic, *Compositio Math.,* **105** (1997), 237–265.

[48] V. Shokurov, The existence of a line on Fano varieties (Russian), *Izv. Akad. Nauk SSSR Ser. Mat.,* **43** (1979), 922–964. English translation: *Math. USSR Izvestiya,* **14** (1980), 395–405.

[49] S. Takayama, Simple Connectedness of weak Fano varieties, *J. Algebraic Geom.,* **9** (2000), 403–407.

[50] K. Takeuchi, Some birational maps of Fano 3-folds, *Compositio Math.,* **71** (1989), 265–283.

[51] V. Voskresenskiĭ and A. Klyachko, Toric Fano varieties and systems of roots (Russian) *Izv. Akad. Nauk SSSR Ser. Mat.,* **48** (1984), 237–263. English translation: *Math. USSR Izv.,* **24** (1985), 221–244.

[52] J. Wiśniewski, Toric Mori Theory and Fano Manifolds, in: *Geometry of Toric Varieties* (L. Bonavero and M. Brion, eds.), Séminaires et Congrès, vol. 6, Soc. Math. France (2002), pp. 249–272.

[53] S. T. Yau, On Calabi's conjecture and some new results in algebraic geometry, *Proc. Nat. Acad. Sci. USA,* **74** (1977), 1798–1799.

[54] S. T. Yau, On the Ricci curvature of a compact Kähler manifold and the complex Monge–Ampère equation, I, *Comm. Pure and Appl. Math.*, **31** (1978), 339–411.

[55] D. Q. Zhang, Algebraic surfaces with nef and big anti-canonical divisor, *Math. Proc. Cambridge Philos. Soc.*, **117** (1995), 161–163.

Olivier Debarre
Université Louis Pasteur
7, rue René Descartes
67084 Strasbourg
France

debarre@math.u-strasbg.fr

BOLYAI SOCIETY
MATHEMATICAL STUDIES, 12

Higher Dimensional Varieties and Rational Points, pp. 133–167.

Families of Varieties of General Type: the Shafarevich Conjecture and Related Problems

SÁNDOR J. KOVÁCS*

Consider the curves on the (x, y)-plane parametrized by λ and given by the equation,

$$y^2 = x^5 - 5\lambda x + 4\lambda.$$

An easy computation shows that these curves are non-singular for $\lambda \in \mathbb{C} \setminus \{0, 1\}$. On the other hand the projective closure of each of these curves is singular at the point $[0 : 1 : 0]$. Blowing up this point and another one infinitely near this one yields a resolution, C_λ. The collection of C_λ forms a family of smooth projective curves over the base $\mathbb{C} \setminus \{0, 1\}$.

Exercise 0.1. Prove that the projection of C_λ onto the $y = 0$ axis is the only degree 2 morphism onto $\mathbb{P}^1$ that C_λ admits. (Hint: See [28], IV.5.3). Conclude that for $\lambda_1 \neq \lambda_2$, $C_{\lambda_1} \not\simeq C_{\lambda_2}$.

A family of curves is called isotrivial if any two general members are isomorphic. For example the blow up of the projective plane at a single point, considered as a $\mathbb{P}^1$-bundle over $\mathbb{P}^1$, is an isotrivial family since all of its members are isomorphic to the projective line. Since the genus of a curve in a family is constant, and there is only one curve of genus zero, one cannot expect much more in this case. However, for higher genus curves one can have non-isotrivial families as the above example shows.

Exercise 0.2. Compute the genus of C_λ. Find another non-isotrivial family of smooth curves of the same genus over $\mathbb{C} \setminus \{0, 1\}$. Make a guess: Is there an infinite sequence of pairwise different families of that kind?

At the 1962 ICM in Stockholm, Shafarevich conjectured that there exist only finitely many isomorphism classes of non-isotrivial families of smooth

*Supported in part by NSF Grants DMS-0196072, DMS-0092165, and a Sloan Research Fellowship.

projective curves of a given genus over a given base curve. Furthermore, if there is such a family, then the base curve satisfies a certain hyperbolic condition. (For definitions and a more precise formulation, see Section §1). The conjecture was confirmed by Parshin [57] for the case of a compact base and by Arakelov [3] in general. It was recently generalized to families of higher dimensional varieties. This generalization is our main topic.

It will be advantageous to work with a compactification of the family. Considering families over a compact base curve B naturally leads to a slightly different view on the problem. Instead of smooth families over a non-compact base, we are looking at arbitrary families over a compact base, and consider the locus over which the family is smooth. It is fairly obvious that these are equivalent situations. An arbitrary family over a compact base gives a smooth family over some open subset and a smooth family over an open curve can be extended to a (not necessarily smooth) family over the projective closure of the curve cf. [28], III.9.8.

One could ask what can be said about the singular members of the family. On the simplest level, how many are there? In fact, Szpiro has asked this: What is the lower bound on the number of singular members if $B \simeq \mathbb{P}^1$? This is the same as asking how close the base of the smooth family can be to being compact. Beauville [6] gave the following answer to Szpiro's question: there are always at least three singular members and there are families with exactly three. In fact, Beauville's proof also shows that there is at least one singular member if the base curve is elliptic. In short, $2g(B) - 2 + \delta > 0$, where $g(B)$ is the genus of the (compact) base curve and δ is the number of singular members of the family. In other words, the base of a smooth family must be hyperbolic.

Notice that the existence of Kodaira surfaces shows that there are families over high genus curves without any singular fibers. For the construction of Kodaira surfaces see for example [5], V.14. An interesting point to mention is that even though we do not expect general obstacles for higher genus bases, the known examples are limited in terms of the possible base curves. Taking repeated general hyperplane sections in a moduli space of curves, one can produce "examples" over compact curves, however, this still does not produce a method to find a family over a given curve. It would be interesting to see explicit examples that are significantly different from the Kodaira construction. Having said that, we return to the opposite end of the problem, the focus of this article.

Recently, Catanese and Schneider [8] asked whether something similar to the Szpiro–Beauville bound holds for families with higher dimensional fibers. The conjecture of Shokurov [59] translates to the same question: Is it true that for a family of varieties of general type, $\delta \geq 3$ if $g(B) = 0$ and $\delta \geq 1$ if $g(B) = 1$? Equivalently, is $2g(B) - 2 + \delta > 0$? In effect we are asking whether "hyperbolicity of the base" holds in higher dimensions.

The wide range of applications this question relates to is interesting to note: Catanese and Schneider wanted to use this statement to obtain good estimates for the size of the automorphism group of a variety of general type, while Shokurov needed it for proving quasi-projectivity of certain moduli spaces. Hyperbolicity for families of curves was used by Moret-Bailly [53] and de Jong and Oort [12] to extend those families over codimension 2 subsets.

The structure of the article is as follows. In the first two sections, the original Shafarevich conjecture is discussed in the function field and in the number field case. Section §3 is a brief account of hyperbolicity in general. Section §4 is a short look at a few vanishing theorems. Section §5 contains a sketch of how to prove hyperbolicity of the base for families of curves. Section §6 is a short introduction to the theory of moduli of curves. Beginning with Section §7, we turn to a more detailed discussion of the tools and ideas involved. Sections §7 and §8 contain preliminary steps to Section §9. Weak boundedness, introduced in Section §9, turns out to be one of the most important notions we encounter. In Section §10 we turn to higher dimensional generalizations of the Shafarevich problem. Section §11 contains a brief discussion of positivity of push-forwards and Section §12 is a tour of more vanishing theorems. Section §13 deals with a further generalization, which ultimately leads to studying singular families. Section §14 touches on the subject of families over higher dimensional bases.

Acknowledgements. My work on the subject and consequently this article owes a great deal to János Kollár and Eckart Viehweg. Their ideas and philosophy are present throughout.

I would like to thank Ron Irving, Grisha Stewart, Zack Treisman, and Michael Van Opstall for numerous suggestions that helped improve the presentation of the article.

Finally, I would like to dedicate this article to József Pelikán, whose abstract algebra and topics courses were very influential during my undergraduate years at Eötvös University.

§1. The Function Field Case

The following notation will be preserved throughout the entire article:

Notation 1.1. Let B be a smooth projective curve of genus g over an algebraically closed field k of characteristic 0 and $\Delta \subset B$ a finite subset. In the sequel B and Δ will be fixed.

We will consider *families* over B, i.e., flat projective morphisms $f : X \to B$ with connected fibers, where X is a smooth projective variety over k.

For a morphism $Y \to S$ and a base change morphism $T \to S$, the symbol Y_T will denote $Y \times_S T$. In particular, for $Y = X$, $S = B$ and $b \in B$ we write $X_b = f^{-1}(b)$. In addition, if $T = \operatorname{Spec} F$, then Y_T will also be denoted by Y_F.

Definition 1.2. A family $f : X \to B$ is *isotrivial* if $X_a \simeq X_b$ for general points $a, b \in B$. The family $f : X \to B$ will be called *admissible* (with respect to (B, Δ)) if it is not isotrivial and Δ contains the discriminant locus of f, i.e., the map $f : X \setminus f^{-1}(\Delta) \to B \setminus \Delta$ is smooth.

Our starting point is the aforementioned conjecture of Shafarevich:

1.3. Shafarevich's Conjecture. Let (B, Δ) be fixed and $q \geq 2$ an integer.

(1.3.1) There exist only finitely many isomorphism classes of admissible families of curves of genus q.

(1.3.2) If $2g - 2 + \#\Delta \leq 0$, then there exist no such families.

Exercise 1.4. Prove that the inequality in (1.3.2) can be satisfied only if B is either a rational or an elliptic curve:

$$2g - 2 + \#\Delta \leq 0 \quad \Leftrightarrow \quad \begin{cases} g = 0 \quad \text{and} \quad \#\Delta \leq 2, \\ g = 1 \quad \text{and} \quad \Delta = \emptyset. \end{cases}$$

Shafarevich showed a special case of (1.3.2): There exist no smooth families of curves of genus q over $\mathbb{P}^1$. (1.3.1) was confirmed by Parshin [57] for $\Delta = \emptyset$ and by Arakelov [3] in general.

Our main goal is to generalize this statement to higher dimensional families. In order to do that we will have to reformulate the statement as Parshin and Arakelov did, but before doing so, we need a little bit of background on deformations and parameter spaces.

1.5. Deformations. In general, a deformation of an object means to include that object in a family. There is a potentially confusing point here. Our main objects of study are families, that is, deformations of their members. However, we do not want to consider our families as deformations. We want to look at deformations *of* these families. This may be a bit harder to imagine first, but it works just the same way as deformations of other objects. In addition, we want to fix the base of these families, so we are interested in deformations leaving the base fixed, which makes both the notation and the theory easier.

A *deformation* of a family $\phi : Y \to S$ with the base fixed is a family $\mathfrak{Y} \to S \times T$, where T is connected and such that for some $t_0 \in T$ we have $(\mathfrak{Y}_{t_0} \to S \times \{t_0\}) \simeq (Y \to S)$:

$$\begin{array}{ccc} Y \simeq \mathfrak{Y}_{t_0} & \longrightarrow & \mathfrak{Y} \\ \phi \downarrow & & \downarrow \\ S \simeq S \times \{t_0\} & \longrightarrow & S \times T. \end{array}$$

We say that two families $Y_1 \to S$ and $Y_2 \to S$ have the same *deformation type* if they can be deformed into each other, i.e., if there exists a connected T and a family $\mathfrak{Y} \to S \times T$ such that for some $t_1, t_2 \in T$,

$$(\mathfrak{Y}_{t_i} \to S \times \{t_i\}) \simeq (Y_i \to S) \quad \text{for} \quad i = 1, 2.$$

We will consider deformations of admissible families. It will be advantageous to restrict to deformations of the family over $B \setminus \Delta$. Doing so potentially allows more deformations than over the original base B: it can easily happen that a deformation over $B \setminus \Delta$, that is, a family $\mathfrak{X} \to (B \setminus \Delta) \times T$, cannot be compactified to a (flat) family over $B \times T$, because the compactification may contain fibers of higher than expected dimension. This however, will not cause any problems because of the nature of our inquiry cf. (1.7).

Parameter Spaces. Given a family $\phi : Y \to S$, we say that S *parametrizes* the members of the family. If the members of the family ϕ form a class, $\mathfrak{C}$, of varieties, then we say that S is a *parameter space* for the class $\mathfrak{C}$. Note that we do not require that the members of $\mathfrak{C}$ appear only once in the family.

Exercise 1.6. Show that $\mathbb{P}^5$ is a parameter space for plane conics.

A very useful parameter space is the *Hilbert scheme*, a parameter space for subschemes of $\mathbb{P}^n$. The Hilbert scheme of $\mathbb{P}^n$ decomposes as the disjoint union of Hilbert schemes of subschemes with a given Hilbert polynomial. The components of this union are projective schemes (in particular of finite type). When one is hoping to parametrize the members of a class of varieties, then the most likely way to succeed is to try to find the parameter space as a subscheme of an appropriate Hilbert scheme. For more on Hilbert schemes see [37] and [66]. We will return to their importance and some applications in Section §6.

For more on parameter spaces see [24], Lectures 4, 21.

1.7. With regard to the Shafarevich conjecture, Parshin made the following observation. In order to prove that there are only finitely many admissible families, one can try to proceed the following way. Instead of aiming for the general statement, first try to prove that there are only finitely many deformation types. This is expected to be somewhat easier, because there are ways to parametrize deformations, and the fact that there are only finitely many types translates to the parameter space being of finite type. The next step then is to prove that admissible families are rigid, that is, they do not admit non-trivial deformations. Notice that if we prove these statements for families over $B \setminus \Delta$, then they also follow for families over B. Now since every deformation type contains only one family, and since there are only finitely many deformation types, the original statement follows.

The following is the reformulation of the Shafarevich conjecture that was used by Parshin and Arakelov.

1.8. Reformulation. Let $q \geq 2$ be an integer. Recall that (B, Δ) is fixed.

(1.8.1) Boundedness (**B**): There exist only finitely many deformation types of admissible families, i.e., admissible families of curves of genus q are parametrized by $\mathbb{T}$, a scheme of finite type.

(1.8.2) Rigidity (**R**): There exist no non-trivial deformations of admissible families of curves of genus q, i.e., $\dim \mathbb{T} = 0$.

(1.8.3) Hyperbolicity (**H**): No admissible families of curves of genus q exist if $2g - 2 + \#\Delta \leq 0$, i.e., $\mathbb{T} \neq \emptyset \ \Rightarrow \ 2g - 2 + \#\Delta > 0$.

Remark 1.8.4. As we discussed above, (**B**) and (**R**) together imply (1.3.1) and (**H**) is clearly equivalent to (1.3.2).

§2. The Number Field Case

The number field version of Shafarevich's conjecture played a very important role in Faltings' proof of the Mordell conjecture. This section is a brief detour to this very exciting area, but it is disconnected from the rest of the article. The reader should feel free to skip this section and continue with the next one.

Definition 2.1. Let $(R, \mathfrak{m})$ be a DVR, $F = \mathrm{Frac}(R)$, and C a smooth projective curve over F. C is said to have *good reduction over* R if there exists a scheme Z, smooth and projective over $\operatorname{Spec} R$, such that $C \simeq Z_F$,

$$\begin{array}{ccccc} C & \xrightarrow{\simeq} & Z_F & \longrightarrow & Z \\ & & \downarrow & & \downarrow \\ & & \operatorname{Spec} F & \longrightarrow & \operatorname{Spec} R \end{array}$$

Definition 2.2. Let R be a Dedekind ring, $F = \mathrm{Frac}(R)$, and C a smooth projective curve over F. C has *good reduction at the closed point* $\mathfrak{m} \in \operatorname{Spec} R$ if it has good reduction over $R_{\mathfrak{m}}$.

2.3. Shafarevich Conjecture. Let $q \geq 2$ be an integer.

(2.3.1) Let F be a number field, $R \subset F$ the ring of integers of F, and $\Delta \subset \operatorname{Spec} R$ a finite set. Then there exists only finitely many smooth projective curves over F of genus q that have good reduction outside Δ.

(2.3.2) There are no smooth projective curves of genus q over $\operatorname{Spec} \mathbb{Z}$.

Remark 2.3.3. The Shafarevich Conjecture in the number field case has been confirmed: (2.3.1) by Faltings [20] and (2.3.2) by Fontaine [21].

One can reformulate (1.3.1) to resemble the above statement:

2.4. Shafarevich Conjecture (function field case, version 2). Let $q \geq 2$ be an integer, $F = K(B)$ the function field of B. We may assume that Δ is not empty, i.e., that $B \setminus \Delta$ is affine. Let R be the subring of F such that $B \setminus \Delta = \operatorname{Spec} R$. Then there exist only finitely many smooth projective non-isotrivial curves over F of genus q having good reduction over all closed points of $\operatorname{Spec} R$.

Definition 2.5. If C is a smooth projective curve over F (an arbitrary field), then there exists a morphism $C \to \operatorname{Spec} F$. Sections, $\operatorname{Spec} F \to C$, of this morphism correspond in a one-to-one manner to *F-rational points* of C, points that are defined over the field F. F-rational points of C will be denoted by $C(F)$.

Example 2.5.1.

(1) The $\mathbb{R}$-rational points of the curve $x^2 + y^2 - z^2 = 0$ form a circle, its $\mathbb{C}$-rational points form a sphere.

(2) The curve $x^2 + y^2 + z^2 = 0$ has no $\mathbb{R}$-rational points.

(3) Let C_n be the curve defined by the equation $x^n + y^n - z^n = 0$. If $n \geq 3$, then according to Wiles' Theorem, which is also known as Fermat's Last Theorem,

$$C_n(\mathbb{Q}) = \begin{cases} \{[1:0:1],[0:1:1],[1:-1:0]\}, & \text{if } n \text{ is odd,} \\ \{[1:0:1],[0:1:1],[1:0:-1],[0:1:-1]\} & \text{if } n \text{ is even.} \end{cases}$$

As mentioned earlier, Faltings used (2.3) to prove:

2.6. Mordell Conjecture. Let F be a number field and C a smooth projective curve of genus $q \geq 2$ defined over F. Then $C(F)$ is finite.

The function field version of this conjecture was proved earlier by Manin [51]:

2.7. Mordell Conjecture for function fields. Let F be a function field (i.e., the function field of a variety over k, where k is an algebraically closed field of characteristic 0) and let C be a smooth projective non-isotrivial curve over F of genus $q \geq 2$. Then $C(F)$ is finite.

Remark 2.7.1. The essential case to settle is when $\operatorname{tr.deg}_k F = 1$, i.e., $F = K(B)$, where B is a smooth projective curve over k.

Exercise 2.8. Let B be a smooth projective curve over k and $F = K(B)$ its function field. Let C be a smooth projective curve over F, and let $f : X \to B$ be the "closure" of $f_0 : C \to \operatorname{Spec} F$. Prove that the following statements are equivalent.

(1) There exists an F-rational point of C;

(2) There exists a $\operatorname{Spec} F$-section of f_0;

(3) There exists a B-section of f.

Shafarevich's conjecture implies Mordell's in both the function field and the number field case by an argument due to Parshin:

2.9. Parshin's Covering Trick. For every $P \in C(F)$ or equivalently, for every section $X \overset{\sigma_P}{\leftrightarrows} B$, there exists a finite cover of X, $W_P \xrightarrow{\pi_P} X$ such that

- the degree of π_P is bounded in terms of q,
- the projection $W_P \to B$ is smooth over $B \setminus \Delta$,
- the map π_P is ramified exactly over the image of σ_P,
- the genus of the fibers of $W_P \to B$ is bounded in terms of q.

For details on the construction of the covers, $W_P \xrightarrow{\pi_P} X$, see [50], IV.2.1.

The last ingredient of the proof is:

Theorem 2.10. (de Franchis [11]) *Let C and D be smooth projective curves of genus at least two. Then there exist only finitely many dominant rational maps $D \to C$.*

The Shafarevich Conjecture implies that there are only finitely many different W_P's. Viewing W_P and X as curves over F, de Franchis' theorem implies that a fixed W_P can admit only finitely many different maps to X.

Since those maps are ramified exactly over the image of the corresponding σ_P, this means that there are only finitely many σ_P's, i.e., $C(F)$ is finite, and therefore Mordell's conjecture follows from that of Shafarevich.

We end our little excursion to the number field case here. In the rest of the article we work in the function field case and use the notation and assumptions of (1.1).

§3. Hyperbolicity

Definition 3.1. A complex analytic space X is called *Brody hyperbolic* if every holomorphic map $\mathbb{C} \to X$ is constant.

Remark 3.1.1. Another important, related notion is *Kobayashi hyperbolicity.* For its definition and relation to Brody hyperbolicity the reader is referred to [49].

Exercise 3.2. Let T be a complex torus. Prove that if X is Brody hyperbolic, then every holomorphic map $\mathbb{C}^* \to X$ is constant, and every holomorphic map $T \to X$ is constant.

We would like to define the algebraic analogue of hyperbolicity motivated by this observation. Algebraic maps are more restrictive than holomorphic ones. For instance the universal covering map, $\mathbb{C} \to E$, of an elliptic curve, E, is not algebraic. In particular, excluding algebraic maps from $\mathbb{C}$ to X does not exclude maps from E to X.

The following is a working definition. For a notion that resembles Kobayashi hyperbolicity one would need a different definition. This should perhaps be called "algebraic Brody hyperbolicity" to emphasize that fact. Nevertheless, the following is the way it is currently used:

Definition 3.3. An algebraic variety X is called *algebraically hyperbolic* if

- every regular map $\mathbb{A}^1 \setminus \{0\} \to X$ is constant, and
- every regular map $A \to X$ is constant, where A is an abelian variety.

Next we recall some basic, but very important notions.

Definition 3.4. Let $\mathcal{L}$ be a line bundle on X. It is said to be *generated by global sections* if for every point $x \in X$ there exists a global section $\sigma_x \in H^0(X, \mathcal{L})$ such that the germ σ_x generates the stalk $\mathcal{L}_x$ as an $\mathcal{O}_X$-module. If $\mathcal{L}$ is generated by global sections, then the global sections define a morphism $\phi_{\mathcal{L}} : X \to \mathbb{P}^N = \mathbb{P}\big(H^0(X, \mathcal{L})\big)$. $\mathcal{L}$ is called *semi-ample* if $\mathcal{L}^m$ is generated by global sections for $m \gg 0$. $\mathcal{L}$ is called *ample* if it is semi-ample and $\phi_{\mathcal{L}^m}$ is an embedding for $m \gg 0$. A line bundle $\mathcal{L}$ on X is called *big* if the global sections of $\mathcal{L}^m$ define a rational map $\phi_{\mathcal{L}^m} : X \dashrightarrow \mathbb{P}^N$ such that X is birational to $\phi_{\mathcal{L}^m}(X)$ for $m \gg 0$. Note that in this case $\mathcal{L}^m$ is not necessarily generated by global sections, so $\phi_{\mathcal{L}^m}$ is not necessarily defined everywhere.

A smooth projective variety, X, is of *general type* if ω_X is big. It is easy to see that this condition is invariant under birational equivalence between smooth projective varieties. An arbitrary projective variety is of *general type* if so is a desingularization of it.

It is expected that varieties of general type are close to being hyperbolic. Lang has made several conjectures describing this phenomenon more concretely [50], Ch. I §3. One of those conjectures is the following:

3.5. Lang's Conjecture. Let X be a projective variety. Then X is algebraically hyperbolic if and only if every subvariety of X is of general type.

§4. Vanishing Theorems

Vanishing theorems have played a central role in algebraic geometry for the last couple of decades, especially in classification theory. Kollár [34] gives an introduction to the basic use of vanishing theorems as well as a survey of results and applications available at the time. For more recent results one should consult [15], [18], [38], [45], [60]. Because of the availability of those surveys, we will only recall statements that are important for the present article.

In nearly any discussion of vanishing theorems, one must start with the fundamental vanishing theorem of Kodaira.

Theorem 4.1. (Kodaira [33]) *Let X be a smooth complex projective variety and $\mathcal{L}$ an ample line bundle on X. Then*

$$H^i(X, \omega_X \otimes \mathcal{L}) = 0 \quad \text{for} \quad i > 0.$$

This has been generalized in several ways, but as noted above we will restrict to a select few. The original statement of Kodaira was generalized to allow semi-ample and big line bundles in place of ample ones by Grauert and Riemenschneider.

Theorem 4.2. (Grauert–Riemenschneider [22]) *Let X be a smooth complex projective variety and $\mathcal{L}$ a semi-ample and big line bundle on X. Then*

$$H^i(X, \omega_X \otimes \mathcal{L}) = 0 \quad \text{for} \quad i > 0.$$

Remark 4.2.1. "Semi-ample" was later replaced by "nef" in the statement by Kawamata and Viehweg ([29], [63]).

Akizuki and Nakano extended Kodaira's vanishing to include other exterior powers of the sheaf of differential forms:

Theorem 4.3. (Akizuki–Nakano [2]) *Let X be a smooth complex projective variety and $\mathcal{L}$ an ample line bundle on X. Then*

$$H^q(X, \Omega^p_X \otimes \mathcal{L}) = 0 \quad \text{for} \quad p + q > \dim X.$$

4.4. Ramanujam [58] gave a simplified proof for (4.3) and showed that it does not hold if one only requires $\mathcal{L}$ to be semi-ample and big.

§5. Families of Curves

In this section we are going to give an indication how to prove hyperbolicity for a smooth family of curves over a compact base.

Lemma 5.1. *Let $f : X \to B$ be a smooth family of curves of genus $q \geq 2$. If $\omega_{X/B}$ is ample, then $\omega_{X/B} \otimes f^*\omega_B^{-1}$ is not ample.*

Proof. Since f is smooth then there exists a short exact sequence of locally free sheaves,

$$0 \to f^*\omega_B \to \Omega_X \to \omega_{X/B} \to 0.$$

Twisting this short exact sequence by $\omega_{X/B}$, one obtains another one:

$$0 \to \omega_X \to \Omega_X \otimes \omega_{X/B} \to \omega_{X/B}^2 \to 0.$$

Since $\omega_{X/B}$ is ample, $H^2(X, \Omega_X \otimes \omega_{X/B}) = 0$ by (4.3). Hence the map $H^1(X, \omega_{X/B}^2) \to H^2(X, \omega_X)$ is surjective. The latter group is non-zero, so the former is non-zero as well. However, by (4.1), this implies that indeed $\omega_{X/B} \otimes f^*\omega_B^{-1}$ cannot be ample. ■

Exercise 5.2. Let $f : X \to B$ be a smooth family of curves of genus $q \geq 2$. Prove that if $\omega_{X/B}$ is ample, then $g(B) > 1$. Notice that this is a step toward (**H**). To prove (**H**), one can proceed in a similar way using more delicate vanishing theorems.

To finish the proof one needs to prove that for a non-isotrivial family $\omega_{X/B}$ is ample. As we will discuss later, this ampleness follows from positivity results on the push-forwards of powers of $\omega_{X/B}$. See §11 for more details.

To argue that it is actually reasonable to expect that $\omega_{X/B}$ is ample, let us restrict to the case of $g(B) = 1$. We are trying to prove that in this case no smooth non-isotrivial f exists. The fibers of f are smooth curves of genus at least two, and the base is of genus one, so X cannot contain any rational curves. Therefore X is a non-ruled minimal surface and thus some power of its canonical bundle is generated by global sections. By adjunction, the restriction of $\omega_{X/B}$ to a fiber is the canonical bundle of the fiber, hence it is positive on it. Therefore the morphism induced by the global sections of $\omega_{X/B}$ cannot contract any fibers.

Now there are two possibilities. If the image of this morphism is a curve, then all fibers map isomorphically onto this curve. This implies that the fibers are pairwise isomorphic, contradicting the non-isotriviality of f. If the image of the morphism is a surface, then X is of general type and the only way ω_X may not be ample is if it contracts some rational curves. However, we have already established that X does not contain any rational curves, so $\omega_X = \omega_{X/B}$ must be ample. This completes the proof of (**H**) in the case of $g(B) = 1$. $\mathbb{P}^1$ can be covered by an elliptic curve, hence a smooth family over $\mathbb{P}^1$ induces one over the elliptic curve. This implies that no smooth non-isotrivial f exists over $\mathbb{P}^1$ either.

§6. Moduli Spaces of Curves

The notion of parameter spaces was invented in and thus to some extent belongs to the 19th century, a time when varieties were meant to be embedded in projective space and families of varieties were families of embedded varieties.

The 20th century saw a dramatic change in the way we think of varieties. They are no longer subobjects, they exist without a necessary reference to an ambient space. Some of them do not even admit an embedding to projective space, but even those that do are preferred to be viewed abstractly. Correspondingly, we would like to parametrize isomorphism classes without respect to any embedding. For instance, while smooth plane conics are parametrized by a Zariski open subset of $\mathbb{P}^5$, they are all isomorphic to $\mathbb{P}^1$, so up to isomorphism the parameter space is a single point. Of course, the way to parametrize the possible degree two embeddings of $\mathbb{P}^1$ into $\mathbb{P}^2$, that is smooth plane conics, should not be forgotten, but in order to follow our principle of regarding varieties in an abstract way we need a corresponding notion of parametrization.

Roughly speaking, a *moduli space* is a parameter space whose points are in one–to–one correspondence with isomorphism classes of the members of a class of varieties (or schemes).

Recall that k is an algebraically closed field of characteristic 0. Let

$$\mathcal{M}_q(k) = \{\text{smooth projective curves of genus } q, \text{ defined over } k\} \big/ \simeq .$$

By definition, this is only a set, but the theory of moduli tells us that we can make it an algebraic scheme in a natural way:

Theorem 6.1. (Mumford [54]) *For $q \geq 2$ there exists a quasi-projective scheme $\mathfrak{M}_q$ and a natural bijection between $\mathcal{M}_q(k)$ and the set of k-points of $\mathfrak{M}_q$.*

"Natural" here means that every flat family $f : X \to B$ of smooth projective curves of genus q induces a morphism from the base of the family to $\mathfrak{M}_q$, $\eta_f : B \to \mathfrak{M}_q$, such that each point is mapped to the isomorphism class of the fiber over the point, that is, for all $b \in B$, $\eta_f(b) = [X_b]$. A scheme with this property is called a *coarse moduli scheme*, or a *moduli space*. The $\mathfrak{M}_q$ in the theorem is the moduli space of smooth projective curves of genus q.

In order to study compact families, we would like to work with a projective compactification: $\mathfrak{M}_q \subseteq \overline{\mathfrak{M}}_q$. However, in order for this to be useful, the compactification should also be natural. In other words, it should be the moduli space for an extended class of varieties. This leads us to consider non-smooth families of curves as well.

Definition 6.2. A curve singularity is called a *normal crossing* if it is locally analytically isomorphic to the singularity of the plane curve $xy = 0$. A reduced projective curve C is *stable* if it has only normal crossing singularities and ω_C is ample.

The extended moduli problem studies

$$\overline{\mathcal{M}}_q(k) = \{\text{stable curves of arithmetic genus } q, \text{ defined over } k\} \big/ \simeq$$

and we have a statement similar to (6.1):

Theorem 6.3. (Deligne–Mumford [13], Knudsen–Mumford [32]) *For $q \geq 2$ there exists a projective coarse moduli scheme $\overline{\mathfrak{M}}_q$ for $\overline{\mathcal{M}}_q$, such that $\mathfrak{M}_q \subseteq \overline{\mathfrak{M}}_q$ is an open subscheme.*

Hence for every family $f : X \to B$ of stable curves of genus q there exists a morphism $\eta_f : B \to \overline{\mathfrak{M}}_q$ such that for all $b \in B$, $\eta_f(b) = [X_b]$.

6.4. Moduli spaces and Hilbert schemes. The Hilbert scheme, $\mathrm{Hilb}_p \mathbb{P}^N$, parametrizes subschemes of $\mathbb{P}^N$ with fixed Hilbert polynomial p. The Hilbert scheme, $\mathrm{Hilb}\,\mathbb{P}^N$, parametrizes all subschemes of $\mathbb{P}^N$. It is the disjoint union of $\mathrm{Hilb}_p \mathbb{P}^N$ as p runs through all Hilbert polynomials.

Hilbert schemes belong to the best kind of parameter spaces. There exists a universal family of the schemes parametrized by $\mathrm{Hilb}\,\mathbb{P}^N$ such that any family of subschemes of $\mathbb{P}^N$ can be obtained from the universal family by pull-back.

Hilbert schemes play a very important role in the construction of moduli spaces. The main idea is the following.

We want to find a parameter space for all smooth (or stable) projective curves of a given genus. The first difficulty is to try to get a hold on all of them at the same time. This problem is solved by a certain boundedness property: For a curve C of genus $q \geq 2$, the sheaf of differential forms, ω_C, is an ample line bundle and one can prove that ω_C^3 is generated by global sections and defines an embedding of C into $\mathbb{P}^N$ for $N = 5q - 7$. From the construction it is clear that all smooth projective curves of genus q are represented as a curve in $\mathbb{P}^N$ with the same Hilbert polynomial. The next step is to verify that the points of the corresponding Hilbert scheme parametrizing our curves form a locally closed subscheme, $\mathcal{H}$, of the ambient Hilbert scheme.

This scheme is still not exactly that we want because the same curve is represented by several points. Even though the sheaf of differentials is unique, giving a universal choice, the embedding defined by the global sections is unique only up to an isomorphism of the target projective space. The morphism depends on the actual generators we choose among the sections. In other words, $\mathrm{PGL}_n(k)$ acts on $\mathcal{H}$ and the orbits of this action correspond to the isomorphism classes of smooth projective curves of genus q.

Therefore the desired moduli space is given as the $\mathrm{PGL}_n(k)$ quotient of $\mathcal{H}$. The fact that this quotient exists as a scheme and has reasonably nice properties is a highly non-trivial fact, but fortunately Mumford's geometric invariant theory tells us that this is the case. Hence the desired moduli space exists.

The naturality of moduli spaces implies that a smooth family of curves of genus q, $f : X \to B$, induces a map $\eta_f : B \to \mathfrak{M}_q$. One might hope that the converse is true, as well, namely that maps into $\mathfrak{M}_q$ induce families. In that case families would be in one-to-one correspondence with maps from the base of the family to the moduli space and $\mathfrak{M}_q$ would be a *fine moduli space*. This, however, turns out to be too much to ask for.

Of course, this is only the tip of an iceberg. The theory of moduli of curves is a highly developed beautiful subject. The reader is encouraged to continue discovering this topic. A very readable and up to date account is [25].

We now return to Shafarevich's Conjecture and examine the conditions listed in (1.8).

§7. Boundedness

The aim of this section is to sketch a proof of the following statement.

Theorem 7.1. *In order to prove* (**B**), *it suffices to show that for some* $m > 1$ *the degree of* $f_*\omega^m_{X/B}$ *is bounded, the bound depending on* $B \setminus \Delta$ *and* q, *but not on* f.

This section may be hard to digest for the beginner. Anyone willing to accept the above statement should feel free to continue with the next chapter.

Proof. (Sketch) First recall that condition (**B**) says that admissible families are parametrized by a scheme of finite type.

The naturality of moduli spaces implies that a family $f : X \to B$ induces a map $\eta_f : B \setminus \Delta \to \mathfrak{M}_q$, and since B is a smooth curve, η_f induces another map $\bar{\eta}_f : B \to \overline{\mathfrak{M}}_q$. Unfortunately, as we discussed in the previous section, maps do not necessarily induce families so we cannot parametrize our families this way.

However, we can proceed a different way. One can prove that there exist $m, N > 1$ such that for any admissible family $f : X \to B$, the global sections of $\omega^m_{X/B}$ define a map to $\mathbb{P}^N$ such that it embeds each smooth fiber of f as a degree d curve in $\mathbb{P}^N$. Here d is the degree of $\omega^m_{X/B}$ restricted to a fiber of f. This induces a map, ζ_f, from the base, $B \setminus \Delta$, to the Hilbert scheme of degree d curves in $\mathbb{P}^N$, $\operatorname{Hilb}_d \mathbb{P}^N$. This Hilbert scheme admits a natural map to $\overline{\mathfrak{M}}_q$ and the composition of this natural map with ζ_f is η_f.

This allows one to parametrize admissible families by the maps $\zeta_f : B \setminus \Delta \to \operatorname{Hilb}_d \mathbb{P}^N$. Maps can be characterized by their graphs. The closure of the graph of ζ_f is a curve contained in $B \times \mathrm{Hilb}_d \mathbb{P}^N$ such that the first projection maps it isomorphically onto B. Hence the parameter space we are looking for is a subscheme of the Hilbert scheme of $B \times \mathrm{Hilb}_d \mathbb{P}^N$ cf. [37].

The Hilbert scheme is an infinite union of schemes of finite type, the components corresponding to the different Hilbert polynominals. The parametrizing scheme being of finite type means that all such graphs should be contained in only finitely many components. That is, there should be only finitely many Hilbert polynomials that can actually occur. All of these curves are isomorphic to B and their image in $\overline{\mathfrak{M}}_q$ is non-trivial, so the only relevant coefficient of the Hilbert polynomial is determined by the degree of a fixed ample line bundle pulled back from $\overline{\mathfrak{M}}_q$.

Therefore, if $\mathcal{L}$ is a fixed ample line bundle on $\overline{\mathfrak{M}}_q$, then to prove (**B**), it suffices to prove that $\deg \bar{\eta}_f^* \mathcal{L}$ is bounded on B, the bound depending on B, Δ and q, but not on f. The construction of $\overline{\mathfrak{M}}_q$ produces natural ample line bundles: for p sufficiently large and divisible there exist line bundles, $\lambda_m^{(p)}$ on $\overline{\mathfrak{M}}_q$, such that for a family of stable curves $f : X \to B$, if $\bar{\eta}_f : B \to \overline{\mathfrak{M}}_q$ is the induced map, then

$$\left(\det\left(f_* \omega_{X/B}^m\right)\right)^p = \bar{\eta}_f^* \lambda_m^{(p)}.$$

This statement is a consequence of the construction of $\lambda_m^{(p)}$ and the fact that one uses the powers of the relative dualizing sheaf to embed the members of the family into projective space. The only missing information we need is that $\lambda_m^{(p)}$ is ample for $m > 1$. This was proved by Arakelov. Later on we will need the analogous statement for a family of higher dimensional varieties. Fortunately that is also available due to work of Kawamata, Kollár, and Viehweg cf. (9.2). Therefore, to prove (**B**), it suffices to show that $\deg f_* \omega_{X/B}^m$ is bounded on B for some $m > 1$. ∎

Remark 7.1.1. The notation $\lambda_m^{(p)}$ might strike the reader as being odd. The point of this notation is the following. When this line bundle is pulled back to B, then it is a pth power, but originally it may not be. We are considering maps defined by sections of $\left(\det(f_* \omega_{X/B}^m)\right)^p$ for m, p large and divisible enough. We have no reason to expect this sheaf in general to have any sections for $p = 1$, so $\det(f_* \omega_{X/B}^m)$ is not expected to be the pull-back of a line bundle from the moduli space.

§8. Hyperbolicity Revisited

The first row in the following diagram is the statement of condition (**H**). The last row shows equivalent conditions for both the assumption and the conclusion.

$$\begin{array}{ccc} 2g-2+\#\Delta \leq 0 & \Longrightarrow & \nexists\, f : X \to B \text{ admissible} \\ \Updownarrow & & \Updownarrow \\ B\setminus\Delta = \begin{cases} \mathbb{P}^1 \\ \mathbb{A}^1 \\ \mathbb{A}^1\setminus\{0\} \\ \text{elliptic curve} \end{cases} & & \begin{array}{c} \nexists\, B\setminus\Delta \to \mathfrak{M}_q \\ \text{non-constant,} \\ \text{induced by} \\ \text{a family.} \end{array} \end{array}$$

Therefore proving (**H**) is equivalent to proving that there does not exist a non-constant morphism, induced by a family, of the form $\mathbb{A}^1\setminus\{0\} \to \mathfrak{M}_q$ or $E \to \mathfrak{M}_q$, where E is an arbitrary elliptic curve. This resembles hyperbolicity notions we encountered earlier, justifying the name. It is convenient at this point to make a new definition.

Definition 8.1. $\mathfrak{M}_q$ is *modular hyperbolic* if every morphism $B \to \mathfrak{M}_q$ induced by a family is constant for $B = \mathbb{A}^1\setminus\{0\}$ or B is abelian variety.

Remark 8.1.1. Essentially, this means that we require the moduli stack to be hyperbolic. For details on stacks see [13], [4], [48].

8.2. The previous observation can be now stated as follows: If $\mathfrak{M}_q$ is modular hyperbolic, then (**H**) holds.

§9. Weak Boundedness

In this section we will discuss a property that implies both (**B**) and (**H**) and add it to the list of (1.8).

(9.1) Weak Boundedness (**WB**) For an admissible family $f : X \to B$, the degree of $f_*\omega^m_{X/B}$ is bounded above in terms of $g(B), \#\Delta$, $g(X_{\text{gen}})$, m. In particular, the bound is independent of f.

Using this new notion we can rephrase (7.1):

Theorem 9.2. **(WB) ⇒ (B)**

The traditional proof of hyperbolicity for curves proceeds via some form of weak boundedness. The key point is that the upper bound obtained on $\deg f_*\omega^m_{X/B}$ has the form of $(2g(B)-2+\#\Delta)\cdot c(g(B),\#\Delta, g(X_{\text{gen}}), m)$ where $c(g(B),\#\Delta, g(X_{\text{gen}}), m) > 0$. This proves hyperbolicity; since $\det f_*\omega^m_{X/B}$ is ample, its degree is positive, so any upper bound of it is positive as well.

In higher dimensions, the bounds obtained are not always in this form. However, perhaps somewhat surprisingly, hyperbolicity follows from the fact of weak boundedness, not from the explicit bound.

Theorem 9.3. **(WB) ⇒ (H)**

A more precise and somewhat more general formulation is the following:

Theorem 9.4. ([46], 0.9) *Let $\mathfrak{F}$ be a collection of smooth varieties of general type, B a smooth projective curve and $\Delta \subset B$ a finite subset of B. Let*

$$\mathcal{F}am(B,\Delta,\mathfrak{F}) = \{\, f : X \to B \mid X \text{ is smooth, } f \text{ is flat and } f^{-1}(t) \in \mathfrak{F} \text{ for all } t \in B \setminus \Delta \}.$$

Assume that there exist $M, m \in \mathbb{N}$ such that for all $(f : X \to B) \in \mathcal{F}am(B,\Delta,\mathfrak{F})$

$$\deg\left(f_*\omega^m_{X/B}\right) \le M,$$

and that $\mathcal{F}am(B,\Delta,\mathfrak{F})$ contains non-isotrivial families. Then $2g(B)-2+\#\Delta > 0$.

Proof. First consider the case $g(B) = 1$ and assume that $\Delta = \emptyset$. Let $\tau : B \to B$ be a finite étale endomorphism of degree > 1. Then $(f_\tau : X_\tau \to B_\tau) \in \mathcal{F}am(B,\Delta,\mathfrak{F})$ and, by assumption,

$$\deg\tau \cdot \deg\left(f_*\omega^m_{X/B}\right) = \deg\left(f_{\tau *}\omega^m_{X_\tau/B_\tau}\right) \le M.$$

Since $\deg\tau$ can be arbitrary large and $\deg\left(f_*\omega^m_{X/B}\right) > 0$ by [35], this leads to contradiction.

The case of $g(B) = 0$, $\#\Delta \le 2$ works similarly. One may assume that $\#\Delta = 2$. Again, there exists a finite endomorphism $\tau : B \to B$ of degree > 1 such that τ is smooth over $B \setminus \Delta$ and completely ramified over Δ: assume that $\Delta = \{0, \infty\}$ and consider the map induced by $z \mapsto z^n$. At this point we would like to apply the same argument as above, but before we can do that we have to deal with the singularities this procedure produces. In fact we have to start by applying stable reduction to the family ([25], 3.47), and then the singularities we end up working with are sufficiently mild to allow us to use the above argument to finish the proof. For more details see [46], 0.9. ■

These statements suggest that **(WB)** is a condition worth studying.

Next we turn to higher dimensional generalizations. As a first step, one can try to keep the base of the family be a curve and allow higher dimensional fibers. Independently or simultaneously, one can study families over higher dimensional bases. Finally, generalizing the conditions on the fibers naturally leads to the study of families with singular fibers. We will discuss the state of affairs in all of these directions.

§10. Higher Dimensional Fibers

The first task is to state what we are after, so we need to generalize both the statement and the conditions. The condition that a curve has genus at least 2, i.e., our assumption that $g(X_{\mathrm{gen}}) \geq 2$, is equivalent to the condition that $\omega_{X_{\mathrm{gen}}}$ is ample. In higher dimensions, the role of the genus is played by the Hilbert polynomial, so fixing $g(X_{\mathrm{gen}})$ will be replaced by fixing $h_{\omega_{X_{\mathrm{gen}}}}$, the Hilbert polynomial of $\omega_{X_{\mathrm{gen}}}$.

10.1. Therefore we have the following starting data:

(10.1.1) a fixed smooth curve B of genus g,

(10.1.2) a fixed finite subset $\Delta \subset B$, and

(10.1.3) a fixed polynomial h.

Definition 10.2. An *admissible family* is a non-isotrivial family $f : X \to B$, such that X is a smooth projective variety and for all $b \in B \setminus \Delta$, the variety X_b is smooth and projective with ω_{X_b} ample and $h_{\omega_{X_b}} = h$. Two such families are *equivalent* if they are isomorphic over $B \setminus \Delta$.

Having made this definition, the various parts of Shafarevich's conjecture make sense in any dimension.

10.3. Higher Dimensional Shafarevich Conjecture.

(B) Admissible families are parametrized by a scheme $\mathbb{T}$ of finite type.

(R) $\dim \mathbb{T} = 0$.

(H) $\mathbb{T} \neq \emptyset \Rightarrow 2g - 2 + \#\Delta > 0.$

(WB) For an admissible family $f : X \to B$, the degree of $f_*\omega^m_{X/B}$ is bounded in terms of $g, \#\Delta, h$ and m.

Next we are going to try to determine which of these statements remain true in higher dimensions.

10.4. (R)

Let $Y \to B$ be an arbitrary non-isotrivial family of curves of genus ≥ 2, and C a smooth projective curve of genus ≥ 2. Then $f : X = Y \times C \to B$ is an admissible family, and a deformation of C gives a deformation of f. Therefore (**R**) fails as stated.

Question 10.5. Under what additional conditions does (**R**) hold?

10.6. (B), (H), and (**WB**).

Several results have been obtained in this regard recently.

Migliorini [52] showed that for families of minimal surfaces a somewhat weakened hyperbolicity statement holds, namely that $\delta \geq 1$ if $g \leq 1$. The author showed the same in [39] for families of minimal varieties of arbitrary dimension, later proved (**H**) for families of minimal surfaces in [40], and then (**H**) for families of canonically polarized varieties in [43]. Most recently, Viehweg and Zuo [70] proved the analytic version of (**H**), i.e., Brody hyperbolicity.

Bedulev and Viehweg [7] proved that (**B**) holds for families of surfaces of general type and that (**WB**) (and in some cases (**B**)) holds for families of canonically polarized varieties. As a byproduct of their proof they also obtained that (**H**) holds in these cases.

Viehweg and Zuo [68] extended (**H**) and (**WB**) to families of varieties of general type and of varieties admitting a good minimal model. In [46] the author obtained similar results with different methods allowing the fibers to have rational Gorenstein singularities, but restricting to the case of families of minimal varietes of general type.

(**WB**) implies (**B**) as in the curve case cf. [69], 6.2. The proof of (9.4) works in any dimension, so (**WB**) implies (**H**) as well. These suggest that (**WB**) is again the "right" condition to study.

Currently the best known results are the following:

Theorem 10.7. (Bedulev–Viehweg [7]) *Let* $f : X \to B$ *be an admissible family with* B, Δ, h *fixed. Let* $\delta = \#\Delta$, $g = g(B)$, *and* $n = \dim X_{\rm gen} = \dim X - 1$. *If* $f_*\omega^m_{X/B} \neq 0$, *then there exists a positive integer,* $e = e(m, h)$ *such that*

$$\deg f_*\omega^m_{X/B} \leq m \cdot e \cdot \operatorname{rk} f_*\omega^m_{X/B} \cdot \big(n(2g - 2 + \delta) + \delta\big).$$

Theorem 10.8. (Viehweg–Zuo [68], Kovács [46]) *A similar statement holds if instead of assuming* $X_{\rm gen}$ *smooth and* $\omega_{X_{\rm gen}}$ *ample, we assume that* $f : X \to B$ *is not birationally isotrivial, and*

- [68], [46] $X_{\rm gen}$ *is smooth of general type, or*
- [68] $X_{\rm gen}$ *smooth, has a good minimal model and* $\kappa(X_{\rm gen}) \geq 0$, *or*
- [46] $X_{\rm gen}$ *has rational Gorenstein singularities and* $\omega_{X_{\rm gen}}$ *is ample.*

Remark 10.9. One can ask whether the Shafarevich problem holds for families of other type of varieties. There are some known results in this setting as well.

Faltings [19] studied the Shafarevich problem for families of abelian varieties and proved that (**B**) holds, while (**R**) fails in general. He also gave an equivalent condition for (**R**) to hold in this case.

Oguiso and Viehweg [56] considered (**H**) for families of non-general type surfaces. Their work combined with the previous results show that (**H**) holds for families of minimal surfaces of non-negative Kodaira dimension.

§11. Positivity of Push-forwards

One of the most important ingredients in the proofs of the known results is an appropriate variant of a fundamental positivity result.

Definition 11.1. A locally free sheaf, $\mathcal{E}$, is *ample* if $\mathcal{O}_{\mathbb{P}(\mathcal{E})}(1)$ on $\mathbb{P}(\mathcal{E})$ is ample.

Theorem 11.2. (Kawamata [30], Kollár [35], [36], Viehweg [64], [65]) *Let* $f : X \to B$ *be an admissible family and* $m > 1$. *If* $f_*\omega^m_{X/B} \neq 0$, *then* $f_*\omega^m_{X/B}$ *is ample on* B.

Corollary 11.3. *Let* $f : X \to B$ *be an admissible family and* $m > 1$. *If* $f_*\omega^m_{X/B} \neq 0$, *then* $\deg f_*\omega^m_{X/B} > 0$.

The methods used to prove (11.2) give a more precise estimate of the positivity of these push-forwards:

Theorem 11.4. (Esnault–Viehweg [17], 2.4) *Let $f : X \to B$ be an admissible family, and $\mathcal{M}$ a line bundle on B. Assume that there is an $m > 1$ such that $\deg \mathcal{M} < \deg f_*\omega^m_{X/B}$. Set $r = \operatorname{rk} f_*\omega^m_{X/B}$. Then there exists a positive integer $e = e(m,h)$, such that $(f_*\omega^m_{X/B})^{\otimes e\cdot r} \otimes \mathcal{M}^{-1}$ is ample on B.*

Corollary 11.5. (for $\Delta = \emptyset$) *Let $\mathcal{N}$ be a line bundle on B such that $\deg \mathcal{N}^{m\cdot e\cdot r} < \deg f_*\omega^m_{X/B}$. Then $\omega_{X/B} \otimes f^*\mathcal{N}^{-1}$ is ample on X.*

Proof. (Sketch) Since $\left(f_*(\omega^m_{X/B} \otimes f^*\mathcal{N}^{-m})\right)^{\otimes e\cdot r} \simeq (f_*\omega^m_{X/B})^{\otimes e\cdot r} \otimes \mathcal{N}^{-m\cdot e\cdot r}$, (11.4) implies that $f_*(\omega^m_{X/B} \otimes f^*\mathcal{N}^{-m})$ is ample on B. Furthermore, $\omega^m_{X/B} \otimes f^*\mathcal{N}^{-m}|_{X_{\mathbf{gen}}} \simeq \omega^m_{X_{\mathbf{gen}}}$ is ample on X_{gen}. Hence $\omega_{X/B} \otimes f^*\mathcal{N}^{-1}$ is ample both "horizontally" and "vertically", so it is ample. For details about the last step see [46], 7.6. ■

This allows us to reduce the proof of (**WB**) to finding an appropriate line bundle:

11.6. Plan for proving (**WB**):

- Find an $\mathcal{N}$ depending only on B and Δ such that $\omega_{X/B} \otimes f^*\mathcal{N}^{-1}$ is not ample on X.
- Then by (11.5) we have that $\deg \mathcal{N}^{m\cdot e\cdot r} \not< \deg f_*\omega^m_{X/B}$. In other words $\deg f_*\omega^m_{X/B} \leq m \cdot e \cdot r \cdot \deg \mathcal{N}$.

11.7. We will find such an $\mathcal{N}$ using vanishing theorems. The main idea is the following: we want to find a line bundle such that twisting with the relative dualizing sheaf does not yield an ample line bundle. Ample line bundles appear in many vanishing theorems, so one way to prove that a given line bundle is not ample is to prove that a cohomology group does not vanish that would if the line bundle were ample. Next we are going to look at the needed vanishing theorems.

§12. Logarithmic Vanishing Theorems

In order to proceed we will need more delicate vanishing theorems than before. Our starting point is the theorem of Esnault and Viehweg that extends (4.3) to sheaves of logarithmic differential forms:

Theorem 12.1. (Esnault–Viehweg [17], 6.4) *Let X be a smooth complex projective variety and $\mathcal{L}$ an ample line bundle on X. Further let D be a normal crossing divisor on X. Then*

$$H^q\big(X, \Omega^p_X(\log D) \otimes \mathcal{L}\big) = 0 \quad \textit{for} \quad p+q > \dim X.$$

12.2. Extending the known vanishing theorems in a different direction, Navarro-Aznar *et al.* proved a version of the Kodaira–Akizuki–Nakano vanishing theorem for singular varieties that implies our previous statements: (4.1), (4.2), and (4.3) cf. [55] in [22]. We will return to this statement later.

As mentioned earlier, in order to prove **(WB)**, we need a suitable vanishing theorem. The following is a somewhat weaker statement than that is really needed, but shows the main idea of the proof and how to apply it.

Theorem 12.3. [40], [43] *Let $f : X \to B$ be a family such that B is a smooth projective curve. Assume that $D = f^*\Delta$ is a normal crossing divisor. Let $n = \dim X_{\text{gen}}$ and $\mathcal{L}$ an ample line bundle on X such that $\mathcal{L} \otimes f^*\omega_B(\Delta)^{-n}$ is also ample. Then*

$$H^{n+1}\big(X, \mathcal{L} \otimes f^*\omega_B(\Delta)\big) = 0.$$

Proof. After taking exterior powers of the sheaves of logarithmic differential forms, one has the following short exact sequence for each $p = 1, \dots, n+1$:

$$0 \longrightarrow \Omega^{p-1}_{X/B}(\log D) \otimes f^*\omega_B(\Delta) \longrightarrow \Omega^p_X(\log D) \longrightarrow \Omega^p_{X/B}(\log D) \longrightarrow 0.$$

Define $\mathcal{L}_p = \mathcal{L} \otimes f^*\omega_B(\Delta)^{1-p}$ for $p = 0, \dots, n+1$. Then the above short exact sequence yields:

$$0 \longrightarrow \Omega^{p-1}_{X/B}(\log D) \otimes \mathcal{L}_{p-1} \longrightarrow \Omega^p_X(\log D) \otimes \mathcal{L}_p \longrightarrow \Omega^p_{X/B}(\log D) \otimes \mathcal{L}_p \longrightarrow 0.$$

$\mathcal{L}_p$ is ample for $p = 1, \dots, n+1$, since either $\omega_B(\Delta)$ or $\omega_B(\Delta)^{-1}$ is nef, so by (12.1) $H^{n+1-(p-1)}(X, \Omega^p_X(\log D) \otimes \mathcal{L}_p) = 0$ (recall that $\dim X = n+1$). Hence the map

$$H^{n+1-p}\big(X, \Omega^p_{X/B}(\log D) \otimes \mathcal{L}_p\big) \longrightarrow H^{n+1-(p-1)}\big(X. \Omega^{p-1}_{X/B}(\log D) \otimes \mathcal{L}_{p-1}\big)$$

is surjective for $p = 1, \ldots, n+1$. Observe that these maps form a chain as p runs through $p = n+1, n, \ldots, 1$. Hence the composite map

$$H^0\big(X, \Omega^{n+1}_{X/B}(\log D) \otimes \mathcal{L}_{n+1}\big) \longrightarrow H^{n+1}(X, \mathcal{L}_0)$$

is also surjective. However, $\Omega_{X/B}(\log D)$ is of rank n, so $\Omega^{n+1}_{X/B}(\log D) = 0$, and therefore $H^{n+1}(X, \mathcal{L}_0) = H^{n+1}(X, \mathcal{L} \otimes f^*\omega_B(\Delta)) = 0$ as well. ■

12.4. We are finally able to prove (**WB**), at least for $\Delta = \emptyset$, by combining positivity and vanishing: (11.3) and (11.5) with $\mathcal{N} = \mathcal{O}_B$ imply that $\omega_{X/B}$ is ample. Since

$$H^{n+1}(X, \underbrace{\omega_{X/B} \otimes f^*\omega_B}_{\omega_X}) \neq 0,$$

this and (12.3) imply that $\omega_{X/B} \otimes f^*\omega_B^{-n}$ cannot be ample. Then (11.5) with $\mathcal{N} = f^*\omega_B^n$ implies that

$$\deg f_*\omega^m_{X/B} \leq \deg f^*\omega_B^{n \cdot m \cdot e \cdot r} = m \cdot e \cdot r \cdot \dim X_{\text{gen}} \cdot (2g-2).$$

Remark 12.4.1. For a complete proof of (**WB**) without the assumption $\Delta = \emptyset$, see [7], [46], or [69].

Next we will look at other generalizations.

§13. Smooth Fibers that are Minimal of General Type and Singular Fibers

Viehweg and Zuo [69], respectively Kovács [46] present different approaches to this case. Here we discuss the latter. For a survey on the former, the reader is referred to Viehweg [67].

The following principle has been applied with great success in birational geometry.

Principle 13.1. Studying an ample line bundle on a singular variety is similar to studying a semi-ample and big line bundle on a smooth variety.

13.2. The usual situation is the following. Our goal is to prove a statement for a pair, $(X, \mathcal{L})$, where X is possibly singular, and $\mathcal{L}$ is ample on X. Instead of working on X we work on a desingularization $f : Y \to X$, and consider the semi-ample and big line bundle $\mathcal{K} = f^*\mathcal{L}$. A prominent example of this trick is the use of the Kawamata–Viehweg vanishing theorem in the Minimal Model Program.

13.3. Here we will turn the situation upside-down. Our goal is a statement for $(Y, \mathcal{K})$, where Y is smooth and $\mathcal{K}$ is semi-ample and big on Y. Instead of working on Y we construct a pair $(X, \mathcal{L})$ and a map $f : Y \to X$, where X is possibly singular, $\mathcal{L}$ is ample on X, f is birational, and $\mathcal{K} = f^*\mathcal{L}$.

The motivation for this approach is that we would like to extend the results to the case when $\omega_{X_{\mathrm{gen}}}$ is only semi-ample and big. One crucial ingredient of the proof is an appropriate version of the Kodaira–Akizuki–Nakano vanishing theorem (4.3). However, as Ramanujam (4.4) pointed out, (4.3) fails if the line bundle in question is only assumed to be semi-ample and big. On the other hand, Navarro-Aznar *et al.* proved a singular version (12.2), so one hopes that this way the proof can be made to work.

13.4. Derived categories. In order to state the singular version of the Kodaira–Akizuki–Nakano vanishing theorem, we need to use derived categories. The reader unfamiliar with the basics may wish to consult [26] and [9] for definitions and details.

A derived category is essentially the unified approach to deal with resolutions and derived functors simultaneously. The following dictionary shows a correspondence between derived categorical notions and what they generalize.

THIS	$\rightsquigarrow$	GENERALIZES THIS
"complex"	$\rightsquigarrow$	"sheaf"
"hypercohomology"	$\rightsquigarrow$	"cohomology"
$\mathbb{H}^i(X, A)$	$\rightsquigarrow$	$H^i(X, A)$
"distinguished triangle"	$\rightsquigarrow$	"short exact sequence"
$A \longrightarrow B \longrightarrow C \xrightarrow{+1}$	$\rightsquigarrow$	$0 \to A \to B \to C \to 0$

13.5. Du Bois's complex. We also need Du Bois's generalized De Rham complex. The original construction of Du Bois's complex, $\underline{\Omega}^{\cdot}_X(\log D)$, is based on simplicial resolutions. The reader interested in the details is referred to the original article [14]. Note also that a simplified construction was later obtained in [23] via the general theory of cubic resolutions. An easily accessible introduction can be found in [61].

The word "hyperresolution" will refer to either simplicial or cubic resolution. Formally, the construction of $\underline{\Omega}^{\cdot}_X(\log D)$ is the same regardless which resolution is used and no specific aspects of either resolution will be used.

The following definition is included to make sense of the statements of some of the forthcoming theorems. It can be safely ignored if the reader is not interested in the detailed properties of Du Bois's complex and is willing to accept that it is a very close analogue of the De Rham complex of smooth varieties.

Definition 13.6. Let X be a complex scheme and D a closed subscheme whose complement in X is dense. Then $(X_., D_.) \to (X, D)$ is a *good hyperresolution* if $X_. \to X$ is a hyperresolution, and if $U_. = X_. \times_X (X \setminus D)$ and $D_. = X_. \setminus U_.$, then D_i is a divisor with normal crossings on X_i for all i.

Let X be a complex scheme of dimension n. Let $D_{\text{filt}}(X)$ denote the derived category of filtered complexes of $\mathcal{O}_X$-modules with differentials of order ≤ 1 and $D_{filt,coh}(X)$ the subcategory of $D_{\text{filt}}(X)$ of complexes $K^{\cdot}$, such that for all i, the cohomology sheaves of $Gr^i_{\text{filt}} K^{\cdot}$ are coherent cf. [14], [23]. Let $D(X)$ and $D_{\text{coh}}(X)$ denote the derived categories with the same definition except that the complexes are assumed to have the trivial filtration. The superscripts $+, -, b$ carry the usual meaning (bounded below, bounded above, bounded). Isomorphism in these categories is denoted by $\simeq_{\text{qis}}$. If $K^{\cdot}$ is a complex in any of the above categories, then $h^i(K^{\cdot})$ denotes the i-th cohomology sheaf of $K^{\cdot}$. In particular, every sheaf is naturally a complex with $h^i = 0$ for $i \neq 0$.

The right derived functor of an additive functor F, if it exists, is denoted by RF and R^iF is short for $h^i \circ RF$. $\mathbb{H}^i$, $\mathbb{H}^i_Z$, and $\mathcal{H}^i_Z$ denote $R^i\Gamma$, $R^i\Gamma_Z$, and $R^i\mathcal{H}_Z$ respectively, where Γ is the functor of global sections, Γ_Z is the functor of global sections with support in the closed subset Z, and $\mathcal{H}_Z$ is the functor of local sections with support in the closed subset Z. Note that according to this terminology, if $\phi : Y \to X$ is a morphism and $\mathcal{F}$ is a coherent sheaf on Y, then $R\phi_*\mathcal{F}$ is the complex whose cohomology sheaves give the usual higher direct images of $\mathcal{F}$.

Theorem 13.7. (Du Bois [14], 6.3, 6.5) *Let X be a proper complex scheme of finite type and D a closed subscheme whose complement is dense in X.*

Then there exists a unique $\underline{\Omega}^{\cdot}_X(\log D) \in \mathrm{Ob}\left(D_{\mathrm{filt}}(X)\right)$ *with the following properties, using the notation:*

$$\underline{\Omega}^p_X(\log D) := Gr^p_{\mathrm{filt}}\, \underline{\Omega}^{\cdot}_X(\log D)[p].$$

(13.7.1) *Let* $j : X \setminus D \to X$ *be the inclusion map. Then*

$$\underline{\Omega}^{\cdot}_X(\log D) \simeq_{\mathrm{qis}} Rj_*\mathbb{C}_{X\setminus D}.$$

(13.7.2) $\underline{\Omega}^{\cdot}_{(\_)}(\log(\_))$ *is functorial, i.e., if* $\phi : Y \to X$ *is a morphism of proper complex schemes of finite type, then there exists a natural map* ϕ^* *of filtered complexes*

$$\phi^* : \underline{\Omega}^{\cdot}_X(\log D) \to R\phi_*\underline{\Omega}^{\cdot}_Y(\log \phi^* D)$$

Furthermore, $\underline{\Omega}^{\cdot}_X(\log D) \in \mathrm{Ob}\left(D^b_{filt,coh}(X)\right)$ *and if* ϕ *is proper, then* ϕ^* *is a morphism in* $D^b_{filt,coh}(X)$.

(13.7.3) *Let* $U \subseteq X$ *be an open subscheme of* X. *Then*

$$\underline{\Omega}^{\cdot}_X(\log D)\big|_U \simeq_{\mathrm{qis}} \underline{\Omega}^{\cdot}_U\left(\log D|_U\right).$$

(13.7.4) *There exists a spectral sequence degenerating at* E_1 *and abutting to the singular cohomology of* $X \setminus D$:

$$E_1^{pq} = \mathbb{H}^q\left(X, \underline{\Omega}^p_X(\log D)\right) \Rightarrow H^{p+q}(X \setminus D, \mathbb{C}).$$

(13.7.5) *If* $\varepsilon_\cdot : (X_\cdot, D_\cdot) \to (X, D)$ *is a good hyperresolution, then*

$$\underline{\Omega}^{\cdot}_X(\log D) \simeq_{\mathrm{qis}} R\varepsilon_{\cdot *}\Omega^{\cdot}_{X_\cdot}(\log D_\cdot).$$

In particular, $h^i\left(\underline{\Omega}^p_X(\log D)\right) = 0$ *for* $i < 0$.

(13.7.6) *There exists a natural map,* $\mathcal{O}_X \to \underline{\Omega}^0_X(\log D)$, *compatible with* (13.7.2).

(13.7.7) *If* X *is smooth and* D *is a normal crossing divisor, then*

$$\underline{\Omega}^{\cdot}_X(\log D) \simeq_{\mathrm{qis}} \Omega^{\cdot}_X(\log D).$$

In particular,

$$\underline{\Omega}^p_X(\log D) \simeq_{\mathrm{qis}} \Omega^p_X(\log D).$$

(13.7.8) *If* $\phi : Y \to X$ *is a resolution of singularities, then*

$$\underline{\Omega}^{\dim X}_X(\log D) \simeq_{\mathrm{qis}} R\phi_*\omega_Y(\phi^* D).$$

Naturally, one may choose $D = \emptyset$ and then it is simply omitted from the notation. The same applies to $\underline{\Omega}^p_X := Gr^p_{\mathrm{filt}}\, \underline{\Omega}^{\cdot}_X[p]$. We are now able to state the singular version of the Kodaira–Akizuki–Nakano vanishing theorem.

Theorem 13.8. (Navarro-Aznar [55], [23]) *Let X be a complex projective variety and $\mathcal{L}$ an ample line bundle on X. Then*

$$\mathbb{H}^q(X, \underline{\Omega}_X^p \otimes \mathcal{L}) \quad \textit{for} \quad p + q > \dim X.$$

Since Du Bois's complex agrees with the De Rham complex for smooth varieties, this theorem reduces to the Kodaira–Akizuki–Nakano theorem in the smooth case. However, this theorem is still not strong enough in our original situation if $\Delta \neq \emptyset$. We need a singular version of Esnault–Viehweg's logarithmic vanishing theorem (12.1).

Theorem 13.9. [46] *Let X be a complex projective variety and $\mathcal{L}$ an ample line bundle on X. Further let D be a normal crossing divisor on X. Then*

$$\mathbb{H}^q(X, \underline{\Omega}_X^p(\log D) \otimes \mathcal{L}) = 0 \quad \textit{for} \quad p + q > \dim X.$$

To adapt the proof of (**WB**) to the singular case we need a singular version of (12.3). Besides the above vanishing theorem we also need an analogue of the sheaf of relative logarithmic differentials.

Theorem-Definition 13.10. [46] *Let $f : X \to B$ be a morphism between complex varieties such that $\dim X = n + 1$ and B is a smooth curve. Let $\Delta \subsetneq B$ be a finite set and $D = f^*\Delta$. For every nonnegative integer p there exists a complex $\underline{\Omega}_{X/B}^p(\log D) \in \mathcal{O}bj\left(D(X)\right)$ with the following properties.*

(13.10.1) *The natural map $\wedge_p$ factors through $\underline{\Omega}_{X/B}^p(\log D) \otimes f^*\omega_B(\Delta)$, i.e., there exist maps:*

$$w_p'' : \underline{\Omega}_X^p(\log D) \otimes f^*\omega_B(\Delta) \to \underline{\Omega}_{X/B}^p(\log D) \otimes f^*\omega_B(\Delta) \quad \textit{and}$$
$$w_p' : \underline{\Omega}_{X/B}^p(\log D) \otimes f^*\omega_B(\Delta) \to \underline{\Omega}_X^{p+1}(\log D)$$

such that $\wedge_p = w_p' \circ w_p''$.

(13.10.2) *If $w_p = w_p'' \otimes id_{f^*\omega_B(\Delta)^{-1}} : \underline{\Omega}_X^p(\log D) \to \underline{\Omega}_{X/B}^p(\log D)$, then*

$$\underline{\Omega}_{X/B}^p(\log D) \otimes f^*\omega_B(\Delta) \xrightarrow{w_p'} \underline{\Omega}_X^{p+1}(\log D) \xrightarrow{w_{p+1}} \underline{\Omega}_{X/B}^{p+1}(\log D) \xrightarrow{+1}$$

is a distinguished triangle in $D(X)$.

(13.10.3) *w_p is functorial, i.e., if $\phi : Y \to X$ is a B-morphism, then there are natural maps in $D(X)$ forming a commutative diagram:*

$$\begin{array}{ccc} \underline{\Omega}^p_X(\log D) & \longrightarrow & \underline{\Omega}^p_{X/B}(\log D) \\ \downarrow & & \downarrow \\ R\phi_*\underline{\Omega}^p_Y(\log \phi^* D) & \longrightarrow & R\phi_*\underline{\Omega}^p_{Y/B}(\log \phi^* D). \end{array}$$

(13.10.4) *If f is smooth over $B \setminus \Delta$, then $\underline{\Omega}^p_{X/B}(\log D) \simeq_{\mathrm{qis}} \Omega^p_{X/B}(\log D)$.*

(13.10.5) *$\underline{\Omega}^r_{X/B}(\log D) = 0$ for $r > n$ and if f is proper, then $\underline{\Omega}^p_{X/B}(\log D) \in \mathcal{O}bj\left(D^b_{\mathrm{coh}}(X)\right)$ for every p.*

Using these objects one can make the proof work to obtain the following theorem. It is in a non-explicit form. For more precise statements see [46], (7.8), (7.10), (7.11), (7.13).

Theorem 13.10. *Fix B, $\Delta \subset B$. Then weak boundedness holds for families of canonically polarized varieties with rational Gorenstein singularities over $B \setminus \Delta$ with fixed Hilbert polynomial admitting a simultaneous resolution of singularities. In particular, $2g - 2 + \#\Delta > 0$ for these families by* (9.4).

As a corollary, one obtains weak boundedness for non-birationally-isotrivial families of minimal varieties of general type.

§14. Higher Dimensional Bases

The remaining generalization is to allow B to have arbitrary dimension. Let B be a smooth projective variety, $\Delta \subset B$ a divisor with normal crossings and h a polynomial. The definition of an admissible family is formally the same as in (10.2).

By [66], there exists a coarse moduli scheme, $\mathfrak{M}_h$, parametrizing such X_b's. Hence for an admissible family, $f : X \to B$, there exists an associated map, $\eta_f : B \setminus \Delta \to \mathfrak{M}_h$, defined by $b \mapsto [X_b]$.

Since B is now allowed to be higher dimensional, the notion of isotriviality is no longer the best one to consider. Observe that f is isotrivial if and only if η_f is constant. Saying that f is not isotrivial would allow the family to be isotrivial in certain directions. What we want to assume is that the family "truly" changes in any direction on B. To express this we define the family's variation in moduli.

Definition 14.1. (Viehweg [64], [65], Kollár [35]) $\operatorname{Var} f := \dim(\operatorname{Im} \eta) \leq \dim B$.

We are interested in the case $\operatorname{Var} f = \dim B$. In (8.2), we observed that hyperbolicity follows if we know that $\mathfrak{M}_h$ is modular hyperbolic. In fact, for hyperbolicity over a 1-dimensional base, we only needed the corresponding property of $\mathfrak{M}_h$ for curves. However, we would also like to know that every morphism $A \to \mathfrak{M}_h$ induced by a family is constant, where A is an arbitrary abelian variety. This follows from the next theorem.

Theorem 14.2. [41], [43] *$\mathfrak{M}_h$ is modular hyperbolic.*

Remark 14.2.1. This statement also follows from boundedness by an argument similar to the one used in the proof of (9.4).

As before, this implies that if $f : X \to \mathbb{P}^1$ is an admissible family, then $\#\Delta > 2$. More generally, for an admissible family $f : X \to \mathbb{P}^m$ with $\operatorname{Var} f = m$, this implies that $\deg \Delta > 2$. However, we expect that in this case $\deg \Delta$ should be larger than $m + 1$.

Theorem 14.3. (Viehweg–Zuo [69]) $\deg \Delta > m + 1$, *i.e., $\omega_{\mathbb{P}^m}(\Delta)$ is ample.*

Remark 14.3.1. Viehweg and Zuo actually prove a lot more than this in [69]. Please see the article for details.

It is now natural to suspect that a more general statement should hold. The following statement to this effect is part of a more general conjecture of Viehweg [67].

14.4. Viehweg's Conjecture. If $f : X \to B$ is an admissible family, then $\omega_B(\Delta)$ is big.

This is known to be true for families of curves by [67], 2.6 and for $B = \mathbb{P}^n$ and various other cases by [69]. However, this question is far from being completely settled.

References

[1] U. Angehrn and Y.-T. Siu, Effective freeness and point separation for adjoint bundles, *Invent. Math.*, **122** (1995), 291–308.

[2] Y. Akizuki and S. Nakano, Note on Kodaira-Spencer's proof of Lefschetz theorems, *Proc. Jap. Acad.*, **30** (1954), 266–272.

[3] S. Arakelov, Families of algebraic curves with fixed degeneracies, *Izv. A. N. SSSR*, **35** (1971), 1269–1293.

[4] M. Artin, Versal deformations and algebraic stacks, *Invent. Math.*, **27** (1974), 165–189.

[5] W. Barth, C. Peters and A. Van de Ven, *Compact Complex Surfaces,* Ergebnisse der Mathematik und ihrer Grenzgebiete (3), vol. 4., Springer (1984).

[6] A. Beauville, Le nombre minimal de fibres singulières d'une courbe stable sur $\mathbb{P}^1$, *Astérisque,* **86** (1981), 97–108.

[7] E. Bedulev and E. Viehweg, On the Shafarevich conjecture for surfaces of general type over function fields, *Invent. Math.*, **139** (2000), 603–615.

[8] F. Catanese and M. Schneider, Polynomial bounds for abelian groups of automorphisms, *Compositio Math.*, **97** (1995), 1–15.

[9] B. Conrad, *Grothendieck Duality and Base Change,* LNM 1750, Springer (2000).

[10] M. A. A. de Cataldo, Vanishing via lifting to second Witt vectors and a proof of an isotriviality result, *J. Algebra,* **219** (1999), 255–265.

[11] M. de Franchis, Un teorema sulle involuzioni irrazionali, *Rend. Circ. Mat. Palermo,* **36** (1913), 368. See also in "Collected works of Michele de Franchis", Edited by C. Ciliberto and E. Sernesi., *Rend. Circ. Mat. Palermo (2) Suppl.,* **27** (1991), 523.

[12] A. J. de Jong and F. Oort, On extending families of curves, *J. Algebraic Geom.*, **6** (1997), 545–562.

[13] P. Deligne and D. Mumford, The irreducibility of space of curves of given genus, *Publ. Math. IHES,* **36** (1969), 75–110.

[14] Ph. Du Bois, Complex de De Rham filtré d'une variété singulière, *Bull. Soc. Math. France,* **109** (1981), 41–81.

[15] L. Ein, Multiplier ideals, vanishing theorems and applications, in: *Algebraic Geometry, Santa Cruz 1995* (S. Bloch et al., eds.), Proc. Symp. Pure Math., vol. 62 (1997), pp. 203–219.

[16] H. Esnault and E. Viehweg, Logarithmic de Rham complexes and vanishing theorems, *Invent. Math.,* **86** (1986), 161–194.

[17] H. Esnault and E. Viehweg, Effective bounds for semipositive sheaves and for the height of points on curves over complex function fields, *Compositio Math.*, **76** (1990), 69–85.

[18] H. Esnault and E. Viehweg, *Lectures on vanishing theorems,* DMV Seminar, vol. 20, Birkhäuser (1992).

[19] G. Faltings, Arakelov's Theorem for abelian varieties, *Invent. Math.*, **73** (1983), 337–348.

[20] G. Faltings, Endlichkeitssätze für abelsche Varietäten über Zahlkörpern, *Invent. Math.*, **73** (1983), 349–366.

[21] J.-M. Fontaine, Il n'y a pas de variété abélienne sur $\mathbb{Z}$, *Invent. Math.*, **81** (1985), 515–538.

[22] H. Grauert and O. Riemenschneider, Verschwindungssätze für analytische Kohomologiegruppen auf komplexen Räumen, *Invent. Math.*, **11** (1970), 263–292.

[23] F. Guillén, V. Navarro-Aznar, P. Pascual-Gainza and F. Puerta, *Hyperrésolutions cubiques et descente cohomologique*, LNM 1335, Springer (1988).

[24] J. Harris, *Algebraic Geometry: A First Course*, Graduate Texts in Mathematics, vol. 133, Springer (1992).

[25] J. Harris and I. Morrison, *Moduli of curves*, Graduate Texts in Mathematics, vol. 187, Springer (New York, 1998).

[26] R. Hartshorne, *Residues and Duality*, LNM 20, Springer (1966).

[27] R. Hartshorne, On the De Rham cohomology of algebraic varieties, *Publ. Math. IHES*, **45** (1975), 5–99.

[28] R. Hartshorne, *Algebraic Geometry*, Graduate Texts in Mathematics, vol. 52, Springer (1977).

[29] Y. Kawamata, A generalisation of Kodaira-Ramanujam's vanishing theorem, *Math. Ann.*, **261** (1982), 43–46.

[30] Y. Kawamata, Kodaira dimension of algebraic fiber spaces over curves, *Invent. Math.*, **66** (1982), 57–71.

[31] Y. Kawamata, K. Matsuda and K. Matsuki, Introduction to the Minimal Model Problem, in: *Algebraic Geometry* (Sendai, 1985) (T. Oda, ed.), Adv. Stud. Pure Math., vol. 10, Kinokuniya - North-Holland (1987), pp. 283–360.

[32] F. F. Knudsen and D. Mumford, The projectivity of the moduli space of stable curves. I. Preliminaries on "det" and "Div", *Math. Scand.*, **39** (1976), 19–55.

[33] K. Kodaira, On a differential geometric method in the theory of analytic stacks, *Proc. Nat. Acad. USA*, **39** (1953), 1268–1273.

[34] J. Kollár, Vanishing theorems for cohomology groups, in: *Algebraic Geometry Bowdoin 1985*, Proc. Symp. Pure Math., vol. 46 (1987), pp. 233–243.

[35] J. Kollár, Subadditivity of the Kodaira dimension: Fibers of general type, in: *Algebraic Geometry* (Sendai, 1985), (T. Oda, ed.), Advanced Studies in Pure Math., vol. 10, Kinokuniya-North Holland (1987), pp. 361–398.

[36] J. Kollár, Projectivity of Complete Moduli, *J. Diff. Geom.*, **32** (1990), 235–268.

[37] J. Kollár, *Rational Curves on Algebraic Varieties*, Ergebnisse der Mathematik und ihrer Grenzgebiete (3), vol. 32, Springer (1996).

[38] J. Kollár, Singularities of pairs, in: *Algebraic Geometry Santa Cruz 1995*, Proc. Symp. Pure Math., vol. 62 (1997), pp. 221–287.

[39] S. J. Kovács, Smooth families over rational and elliptic curves, *J. Algebraic Geom.*, **5** (1996), 369–385.

[40] S. J. Kovács, On the minimal number of singular fibers in a family of surfaces of general type, *J. reine angew. Math.*, **487** (1997), 171–177.

[41] S. J. Kovács, Families over a base with a birationally nef tangent bundle, *Math. Ann.*, **308** (1997), 347–359.

[42] S. J. Kovács, Relative De Rham complexes for non-smooth morphisms, in: *Birational Algebraic Geometry* (Y. Kawamata, V. Shokurov eds.), Contemp. Math., vol. 207 (1997), pp. 89–100.

[43] S. J. Kovács, Algebraic hyperbolicity of fine moduli spaces, *J. Algebraic Geom.*, **9** (2000), 165–174.

[44] S. J. Kovács, A characterization of rational singularities, *Duke Math. J.*, **102** (2000), 187–191.

[45] S. J. Kovács, Rational, log canonical, Du Bois singularities II: Kodaira vanishing and small deformations, *Compositio Math.*, **121** (2000), 297–304.

[46] S. J. Kovács, Logarithmic vanishing theorems and Arakelov-Parshin boundedness for singular varieties, *Compositio Math.*, **131** (2002), 291–317.

[47] S. J. Kovács, Vanishing theorems, boundedness and hyperbolicity over higher dimensional bases, Proc. AMS (to appear).

[48] G. Laumon and L. Moret-Bailly, *Champs algébriques,* Ergebnisse der Mathematik und ihrer Grenzgebiete (3), vol. 39, Springer (2000).

[49] S. Lang, *Introduction to complex hyperbolic spaces,* Springer (1987).

[50] S. Lang, *Survey of Diophantine Geometry,* Springer (1997).

[51] Ju. I. Manin, Rational points on algebraic curves over function fields, *Izv. Akad. Nauk SSSR Ser. Mat.*, **27** (1963), 1395–1440.

[52] L. Migliorini, A smooth family of minimal surfaces of general type over a curve of genus at most one is trivial, *J. Algebraic Geom.*, **4** (1995), 353–361.

[53] L. Moret-Bailly, Un théorème de pureté pour les familles de courbes lisses, *C.R. Acad. Sci. Paris Sér. I Math.*, **300** (1985), 489–492.

[54] D. Mumford, *Geometric invariant theory,* Ergebnisse der Mathematik und ihrer Grenzgebiete, vol. 34, Springer (1965).

[55] V. Navarro-Aznar, Théorèmes d'annulation, in: *Hyperrésolutions cubiques et descente cohomologique*, LNM 1335, Springer (1988), pp. 133–160.

[56] K. Oguiso and E. Viehweg, On the isotriviality of families of elliptic surfaces, *J. Algebraic Geom.*, **10** (2001), 569–598.

[57] A. Parshin, Algebraic curves over function fields, *Izv. A.N. SSSR,* **32** (1968), 1145–1170.

[58] C. P. Ramanujam, Remarks on the Kodaira vanishing theorem, *J. Indian Math. Soc.*, **36** (1972), 41–51.

[59] V. V. Shokurov, Letters of a bi-rationalist I. A projectivity criterion, in: *Birational Algebraic Geometry,* (Y. Kawamata, V. Shokurov, eds.), Contemp. Math., vol. 207 (1997), pp. 143–152.

[60] K. E. Smith, Vanishing, singularities and effective bounds via prime characteristic local algebra, in: *Algebraic Geometry Santa Cruz 1995,* Proc. Symp. Pure Math., vol. 62 (1997), pp. 289–325.

[61] J. H. M. Steenbrink, Vanishing theorems on singular spaces, *Astérisque,* **130** (1985), 330–341.

[62] S. L. Tan, The minimal number of singular fibres of a semistable curve over $\mathbb{P}^1$, *J. Algebraic Geom.,* **4** (1995), 591–596.

[63] E. Viehweg, Vanishing theorems, *J. reine angew. Math.,* **335** (1982), 1–8.

[64] E. Viehweg, Weak positivity and the additivity of Kodaira dimension for certain fibre spaces, in: *Algebraic Varieties and Analytic Varieties, Advanced Studies in Pure Math.,* vol. 1, North-Holland (1983), pp. 329–353.

[65] E. Viehweg, Weak positivity and the additivity of the Kodaira dimension II., in: *Classification of algebraic and analytic manifolds,* Progress in Math., vol. 39, Birkhäuser (1983), pp. 567–590.

[66] E. Viehweg, *Quasi-Projective Moduli of Polarized Manifolds,* Ergebnisse der Mathematik und ihrer Grenzgebiete (3), vol. 30, Springer (1995).

[67] E. Viehweg, *Positivity of direct image sheaves and applications to families of higher dimensional manifolds* (2000) Lecture notes, "School on Vanishing Theorems and Effective Results in Algebraic Geometry", ICTP, Trieste.

[68] E. Viehweg and K. Zuo, On the isotriviality of families of projective manifolds over curves, *J. Algebraic Geom.,* **10** (2001), 781–799.

[69] E. Viehweg and K. Zuo, Base spaces of non-isotrivial families of smooth minimal models, *Complex geometry* (Göttingen, 2000), 279–328, Springer (Berlin, 2002).

[70] E. Viehweg and K. Zuo, On the Brody hyperbolicity of moduli spaces for canonically polarized manifolds, preprint.

[71] Q. Zhang, Global holomorphic one-forms on projective manifolds with ample canonical bundles, *J. Alg. Geom.,* **6** (1997), 777–787.

Sándor J. Kovács

Department of Mathematics
University of Washington
Seattle, WA 98195
U.S.A.

`kovacs@math.washington.edu`

Part II: Arithmetic

BOLYAI SOCIETY
MATHEMATICAL STUDIES, 12

Higher Dimensional Varieties and Rational Points, pp. 171–221.

Points rationnels sur les fibrations

JEAN-LOUIS COLLIOT-THÉLÈNE

Introduction

On trouvera ici une description de résultats récents sur les points rationnels des variétés algébriques sur un corps de nombres, obtenus par la méthode de la descente et la méthode des fibrations. Je rappelle quelques notions de base, mais renvoie le lecteur à mon précédent rapport [4] pour une description détaillée de la plupart des résultats obtenus avant 1996.

Le texte comprend trois parties de nature assez différente.

Au chapitre 1, après avoir donné des démonstrations simplifiées d'un certain nombre de résultats de base sur l'obstruction de Brauer–Manin ("lemme formel" de D. Harari), on décrit la descente sur une variété ouverte (utilisée pour la première fois dans [12]) et, simplifiant légèrement l'argument de Heath-Brown et Skorobogatov [24], on l'applique à l'étude des solutions de l'équation

$$t^a(1-t)^b = \varepsilon.N_{K/k}(\mathbf{x})$$

où a et b sont des entiers naturels, $\varepsilon \in k$ est une constante non nulle, t varie dans k et $\mathbf{x}$ dans un corps K extension finie de k, et $N_{K/k}$ est la norme de K à k. Dans ce chapitre on s'est efforcé de donner des démonstrations complètes.

Tel n'est pas le cas au chapitre 2, où l'on évoque les résultats obtenus par une méthode inventée par Swinnerton-Dyer [35] et perfectionnée dans [14], qui vise à assurer l'existence de (beaucoup de) points rationnels sur une variété que l'on peut fibrer sur la droite projective, les fibres étant des espaces homogènes principaux de variétés abéliennes. Le principe de la méthode est décrit, mais le détail de l'exécution finale est très élaboré et ne saurait être complètement détaillé. On énonce ensuite les principaux résultats obtenus par cette méthode. Un résultat particulièrement frappant a été ainsi obtenu par Swinnerton-Dyer ([39]) : supposant les groupes de Tate–Shafarevich de courbes elliptiques finis (une conjecture bien connue), il montre que le principe de Hasse vaut pour les hypersurfaces cubiques diagonales de dimension au moins 3 sur le corps des rationnels.

La finitude des groupes de Tate–Shafarevich des variétés abéliennes sur un corps de nombres a un analogue, dû à Tate, sur les corps de fonctions $\mathbf{F}_q(C)$ d'une variable sur un corps fini $\mathbf{F}_q$. Des cas particuliers de cet analogue sont connus. Au chapitre 3, qui est original, je montre que ces cas particuliers suffisent pour établir de façon *inconditionnelle* l'analogue fonctionnel du résultat de Swinnerton-Dyer : *Sur un corps de fonctions* $\mathbf{F}_q(C)$ *avec* q *impair et* $q \equiv 2 \bmod 3$, *le principe de Hasse vaut pour les hypersurfaces cubiques diagonales de dimension au moins* 3. La lecture de ce chapitre présuppose celle de [39].

Je remercie David Harari, Alexei Skorobogatov et Peter Swinnerton-Dyer pour leurs commentaires sur des versions préliminaires de ce texte.

Commençons par quelques rappels (pour plus de détails, voir [4]).

Accouplement de Brauer–Manin

Le groupe de Brauer (cohomologique) d'un schéma X est le groupe

$$\mathrm{Br}\,(X) = H^2_{\text{ét}}(X, \mathbf{G}_m).$$

On supposera le lecteur familier avec la série de trois textes de Grothendieck [16], ou du moins avec les propriétés de base du groupe de Brauer (voir [15], §1). Etant donnés k un corps, $\overline{k}$ une clôture séparable de k, Γ le groupe de Galois de $\overline{k}$ sur k, puis X une k-variété (c'est-à-dire un k-schéma séparé de type fini), et $\overline{X} = X \times_k \overline{k}$, on note $\mathrm{Br}_1(X) = \mathrm{Ker}\left[\mathrm{Br}\,(X) \to \mathrm{Br}\,(\overline{X})\right]$. On a la suite exacte naturelle

$$H^2(\Gamma, \overline{k}[X]^*) \to \mathrm{Br}_1(X) \to H^1(\Gamma, \mathrm{Pic}\,(\overline{X})),$$

où $\overline{k}[X]^*$ désigne le groupe des unités de $\overline{X}$ et $\mathrm{Pic}\,(\overline{X})$ est le groupe de Picard de $\overline{X}$.

Rappelons la définition de l'obstruction de Brauer–Manin (pour plus de détails, voir par exemple [4], §1). Soit k un corps de nombres. Notons Ω l'ensemble des places de k. Rappelons ce qu'est l'ensemble $X(\mathbf{A}_k)$ des adèles d'une k-variété X. On choisit un modèle $\mathcal{X}/O$ de X au-dessus d'un ouvert Spec (O) du spectre de l'anneau des entiers du corps de nombres k, c'est-à-dire un schéma $\mathcal{X}$ plat et de type fini sur O, de fibre générique X/k. On note S l'ensemble fini des places de k non dans cet ouvert. L'ensemble

$$X(\mathbf{A}_k) \subset \prod_{v \in \Omega} X(k_v)$$

est la réunion, pour T variant parmi les sous-ensembles finis de Ω avec $S \subset T$, des sous-ensembles $\prod_{v \in T} X(k_v) \times \prod_{v \notin T} \mathcal{X}(O_v)$ de $\prod_{v \in \Omega} X(k_v)$. Cette définition ne dépend pas du choix du modèle $\mathcal{X}$. On munit $X(\mathbf{A}_k)$ de la topologie de produit restreint, dont une base d'ouverts est formée des ensembles de la forme $\prod_{v \in T} U_v \times \prod_{v \notin T} \mathcal{X}(O_v)$, avec T comme ci-dessus et U_v ouvert de $X(k_v)$. Si X est propre sur k, l'espace $X(\mathbf{A}_k)$ est simplement le produit topologique des $X(k_v)$ pour v dans Ω.

La théorie du corps de classes permet de définir un accouplement (dit de Brauer–Manin)

$$X(\mathbf{A}_k) \times \mathrm{Br}\,(X) \to \mathbf{Q}/\mathbf{Z}$$

entre l'ensemble des adèles de X, supposé non vide, et le groupe de Brauer $\mathrm{Br}\,(X)$ de X. A une adèle $\{P_v\} \in X(\mathbf{A}_k)$ et à $A \in \mathrm{Br}\,(X)$ on associe la somme (finie) $\sum_{v \in \Omega} A(P_v)$. Ici $A(P_v) \in \mathrm{Br}\,(k_v)$ est identifié à un élément de $\mathbf{Q}/\mathbf{Z}$ par la théorie du corps de classes local. Pour $B \subset \mathrm{Br}\,(X)$, on note $X(\mathbf{A}_k)^B$ le fermé de $X(\mathbf{A}_k)$ sur lequel les éléments de B s'annulent. On note $X(\mathbf{A}_k)^{\mathrm{Br}} = X(\mathbf{A}_k)^{\mathrm{Br}\,(X)}$.

Comme remarqué par Manin, la théorie du corps de classes global (loi de réciprocité) montre que l'image diagonale de $X(k)$ dans $X(\mathbf{A}_k)$ est incluse dans $X(\mathbf{A}_k)^{\mathrm{Br}}$. En particulier, si $X(\mathbf{A}_k)^{\mathrm{Br}} = \emptyset$, alors $X(k) = \emptyset$: c'est l'obstruction de Brauer–Manin au principe de Hasse. L'adhérence de $X(k)$ dans $X(\mathbf{A}_k)$ pour la topologie adélique est aussi contenue dans $X(\mathbf{A}_k)^{\mathrm{Br}}$. Il est intéressant de trouver des classes de variétés pour lesquelles cette dernière inclusion est une bijection : on dit alors que l'obstruction de Brauer–Manin au principe de Hasse et à l'approximation faible est la seule pour ces variétés. Lorsque X est projective et que le quotient $\mathrm{Br}\,(X)/\,\mathrm{Br}\,(k)$ est fini, cette propriété implique l'approximation faible-faible : il existe un ensemble fini de places S_0 de k tel que pour tout ensemble fini T de places disjoint de S_0, l'image diagonale de $X(k)$ dans $\prod_{v \in T} X(k_v)$ est dense.

Il était déraisonnable d'espérer que cette obstruction au principe de Hasse fût la seule pour toutes les variétés (géométriquement connexes) projectives et lisses. C'est cependant récemment que Skorobogatov [33] a exhibé le premier exemple de telle variété X satisfaisant $X(\mathbf{A}_k)^{\mathrm{Br}} \neq \emptyset$ mais néanmoins $X(k) = \emptyset$. La variété considérée est une surface bielliptique. De même, il était déraisonnable d'espérer que cette obstruction fût, pour de telles variétés, la seule obstruction à l'approximation faible. Inspiré par [33], Harari [19] montra comment fabriquer de nombreux exemples de k-variétés projectives et lisses X pour lesquelles le groupe de Brauer $\mathrm{Br}(X)$ ne suffit pas à rendre compte du défaut d'approximation faible. L'obstruction qu'il utilise repose sur l'existence de revêtements finis étales galoisiens $Y \to X$, et plus généralement de G-torseurs sur X, pour G un k-groupe algébrique fini. Lorsque le groupe G est abélien, on peut réinterpréter l'obstruction en termes du groupe de Brauer de X, mais il n'en est pas ainsi en général. D'où de nombreux exemples lorsque le groupe fondamental géométrique de X n'est pas abélien. Puis Harari et Skorobogatov [21] montrèrent que l'exemple de [33] peut s'interpréter de ce point de vue. (Harari [20] montre par contre qu'on ne peut espérer fabriquer de tels exemples au moyen de torseurs sous un k-groupe G linéaire connexe.)

Il reste néanmoins des classes très intéressantes de variétés pour lesquelles on espère que l'obstruction de Brauer–Manin au principe de Hasse et à l'approximation faible est la seule. Je renvoie en particulier le lecteur à la conjecture 1 de [4] (p. 25) (le commentaire suivant cette conjecture n'a plus lieu d'être, le travail cité dans ce commentaire étant erroné). Sansuc et moi-même avons conjecturé que l'obstruction de Brauer–Manin est la seule pour les surfaces (géométriquement) rationnelles ; on peut se demander si plus généralement il en est ainsi pour les variétés (géométriquement) rationnellement connexes (pour la définition de ces dernières, voir le texte de C. Araujo et J. Kollár dans ce volume).

Fibrations

Soient k un corps de caractéristique zéro, X une k-variété projective et lisse, géométriquement intègre, et $p : X \to \mathbf{P}^1_k$ un k-morphisme dominant. Etant donné un point $M \in \mathbf{P}^1_k$, de corps résiduel $k(M)$, on note $X_M/k(M)$ la fibre de p en M. On note η le point générique de $\mathbf{P}^1_k$; son corps résiduel est le corps des fonctions $K = k(\mathbf{P}^1)$ de $\mathbf{P}^1_k$. La fibre générique X_η/K de p est lisse. La fibre générique géométrique est $X_\eta \times_K K_s$, où K_s est une clôture algébrique de K. On supposera cette fibre géométrique intègre. On dira alors que p est une *fibration*.

On va s'intéresser aux fibrations p satisfaisant la propriété : pour tout point fermé $M \in \mathbf{P}^1_k$, la fibre $X_M/k(M)$ possède une composante de multiplicité 1 (en d'autres termes toute fibre géométrique possède une composante de multiplicité 1, en d'autres termes encore la fibration p est localement scindée pour la topologie étale sur $\mathbf{P}^1_k$).

On sera amené à considérer le sous-groupe $\mathrm{Br}_{\mathrm{vert}}(X/\mathbf{P}^1_k)$ du groupe de Brauer de X, groupe qu'on appellera "groupe de Brauer vertical" (par rapport à la fibration p). C'est par définition le groupe des éléments de $\mathrm{Br}\,(X)$ dont la restriction à la fibre générique $X_\eta/k(\mathbf{P}^1)$ provient de $\mathrm{Br}\,\big(k(\mathbf{P}^1)\big)$. Sous l'hypothèse de multiplicité 1 faite sur les fibres, le quotient $\mathrm{Br}_{\mathrm{vert}}(X/\mathbf{P}^1_k)/\,\mathrm{Br}\,(k)$ est fini.

Pour de telles fibrations, lorsque k est un corps de nombres, on essaye de trouver des points rationnels sur la k-variété X en se servant des propriétés arithmétiques des fibres de p au-dessus des points k-rationnels.

L'expérience montre que la difficulté de cette entreprise croît avec l'invariant $\delta = \delta(p)$ d'une telle fibration ([32], [4]). Rappelons-en la définition. C'est la somme des degrés $\big[k(M) : k\big]$ des points fermés $M \in \mathbf{P}^1_k$ tels que la fibre $X_M/k(M)$ ne soit pas "scindée", c'est-à-dire ne possède pas de composante de multiplicité 1 géométriquement intègre sur $k(M)$.

Théorème 0.1. *Soit $p : X \to \mathbf{P}^1_k$ une fibration comme ci-dessus. Dans chacun des cas suivants, l'obstruction de Brauer–Manin au principe de Hasse est la seule pour X :*

(i) [17] [18] *On a $\delta(p) \leq 1$, la fibre générique X_η de p satisfait $H^i(X_\eta, O_{X_\eta}) = 0$ pour $i = 1, 2$, le groupe de Néron–Severi de la fibre générique géométrique est sans torsion, et l'on suppose que l'obstruction de Brauer–Manin est la seule pour les fibres lisses de p au-dessus des points rationnels.*

(ii) [13] [12] *On a $\delta(p) \leq 2$, et l'on suppose la validité du principe de Hasse pour les fibres de p au-dessus des points rationnels.*

(iii) [15] [13] *(énoncé conditionnel) Le principe de Hasse vaut pour les fibres de p au-dessus des points rationnels ; pour tout point fermé $M \in \mathbf{P}^1_k$, il existe une composante intègre Y de multiplicité 1 de $X_M/k(M)$ telle que la clôture intégrale de $k(M)$ dans le corps des fonctions de Y soit une extension abélienne de k ; l'entier $\delta(p)$ est quelconque, mais l'on suppose la validité de l'hypothèse de Schinzel.*

L'hypothèse de Schinzel est une extension audacieuse du théorème de Dirichlet sur les premiers dans une progression arithmétique, dont c'est le seul cas connu. Un cas particulier de cette hypothèse est la conjecture des nombres premiers jumeaux.

L'énoncé (iii) ci-dessus est un cas particulier du résultat suivant.

Théorème 0.2 [13]. *Soit $p : X \to \mathbf{P}^1_k$ une fibration comme ci-dessus. Supposons que pour tout point fermé $M \in \mathbf{P}^1_k$ la fibre $X_M/k(M)$ possède une composante Y de multiplicité 1 telle que la clôture intégrale de $k(M)$ dans le corps des fonctions de Y est une extension abélienne de $k(M)$. Supposons qu'il existe une adèle $\{P_v\} \in X(\mathbf{A}_k)$ orthogonale à $\mathrm{Br}_{\mathrm{vert}}(X/\mathbf{P}^1_k)$ pour l'accouplement de Brauer–Manin. Supposons satisfaite l'une des deux hypothèses :*

(a) *pour tout point fermé $M \in \mathbf{P}^1_k$ sauf au plus 2 points rationnels, la fibre X_M possède une composante de multiplicité 1 géométriquement intègre sur $k(M)$ (en particulier $\delta(p) \leq 2$) ;*

(b) *$\delta(p)$ est quelconque, mais l'on admet l'hypothèse de Schinzel.*

Alors il existe une infinité de points $M \in \mathbf{P}^1(k)$ tels que $X_M(\mathbf{A}_k) \neq \emptyset$. Plus précisément, l'ensemble des points $M \in \mathbf{P}^1(k)$ dont la fibre X_M est lisse et satisfait $X_M(\mathbf{A}_k) \neq \emptyset$ est dense dans $p\big(X(\mathbf{A}_k)^{\mathrm{Br}_{\mathrm{vert}}}\big) \subset \mathbf{P}^1(\mathbf{A}_k)$.

La démonstration de (a) est un cas particulier de celle de (b) : elle utilise le théorème de la progression arithmétique.

Le théorème 0.2 ramène dans une certaine mesure l'étude du principe de Hasse pour X à celui du principe de Hasse pour les fibres de p. Dans ce rapport, on s'intéresse au cas où la fibre générique de p est une compactification lisse d'un espace homogène sous un groupe algébrique connexe.

Un cas est particulièrement agréable : c'est celui où cette fibre générique est une variété projective espace homogène d'un groupe linéaire connexe. Dans ce cas les fibres au-dessus de points rationnels satisfont le principe de Hasse (Harder). Sous les hypothèses du théorème on conclut que l'obstruction de Brauer–Manin verticale est la seule obstruction pour la k-variété X.

Un autre cas intéressant (et difficile) est le suivant. Considérons une extension finie de corps de nombres K/k, une base ω_i, $i = 1, \ldots, n$ de K sur k, et un polynôme non nul $P(t) \in k[t]$. Notons $N_{K/k}$ la norme de K à k. L'hypersurface affine définie par l'équation

$$P(t) = N_{K/k}\Big(\sum_{i=1}^{n} x_i\omega_i\Big)$$

dans l'espace affine $\mathbf{A}_k^{n+1}$ admet un modèle projectif et lisse X équipé d'une fibration $p : X \to \mathbf{P}_k^1$ étendant la projection sur la coordonnée t. Une fibre générale de p est une compactification lisse d'un espace homogène principal sous le k-tore $R^1_{K/k}\mathbf{G}_m$ des éléments de K^* de norme 1. L'invariant $\delta = \delta(p)$ est inférieur ou égal (et souvent égal) au nombre de racines distinctes de P, augmenté de 1. Sauf si $P(t)$ n'a qu'une seule racine, on aura donc en général $\delta(p) \geq 3$, et on ne pourra pas appliquer le théorème 0.1 (ii). On peut néanmoins conjecturer que dans ce cas l'adhérence de $X(k)$ dans $X(\mathbf{A}_k)$ coïncide avec $X(\mathbf{A}_k)^{\mathrm{Br}}$. C'est connu dans les cas suivants :

(i) $[K : k] = 2$ et $P(t)$ est de degré 3 ou 4 ([11]) ($\delta = 4$).

(ii) $[K : k] = 3$ et $P(t)$ est de degré au plus 3 ([9]) ($\delta = 3$).

(iii) L'extension K/k est cyclique et l'on accepte l'hypothèse de Schinzel (application du théorème 0.2). Ici δ est quelconque.

(iv) Le corps k est le corps des rationnels, l'extension K/k est quelconque, $P(t) = \varepsilon t^a(1-t)^b$ avec $\varepsilon \in k^*$, et $(a, b) = 1$ ([24], théorème 1.7 ci-après). Ici $\delta = 3$.

Dans chacun des trois premiers cas, d'une part $\mathrm{Br}_{\mathrm{vert}}(X/\mathbf{P}_k^1) = \mathrm{Br}\,(X)$, d'autre part le principe de Hasse vaut pour les fibres. Il n'en est pas de même dans le cas (iv).

Chapitre 1: La descente ouverte

Accouplement de Brauer–Manin et variétés ouvertes

Proposition 1.1. *Soient k un corps de nombres et X une k-variété algébrique. Soit $\mathrm{Spec}\,(O)$ un ouvert du spectre de l'anneau des entiers de k et $\mathcal{X}/O$ un O-schéma de type fini, de fibre générique $X/\mathrm{Spec}\,(k)$. Etant donné un sous-ensemble fini $B \subset \mathrm{Br}\,(X)$, pour presque toute place $v \in \mathrm{Spec}\,(O)$, on a $\alpha(M_v) = 0$ pour tout $\alpha \in B$ et tout $M_v \in \mathcal{X}(O_v)$.*

Démonstration. Quitte à restreindre $\mathrm{Spec}\,(O)$, on peut supposer que chaque $\alpha \in B$ provient d'un élément $\mathcal{A} \in \mathrm{Br}\,(\mathcal{X})$ (passage à la limite en cohomologie étale). La nullité du groupe de Brauer $\mathrm{Br}\,(O_v)$ de l'anneau des entiers de O_v (anneau des entiers du complété k_v pour v place finie) permet de conclure.

Proposition 1.2. *Soient k un corps de nombres et X une k-variété algébrique propre. Soit $\alpha \in \mathrm{Br}(X)$. Pour presque toute place v de k, pour tout $M_v \in X(k_v)$, on a $\alpha(M_v) = 0$.*

Démonstration. Il existe un ouvert $\mathrm{Spec}(O)$ du spectre de l'anneau des entiers de k et $\mathcal{X}/O$ un O-schéma propre de fibre générique X/k, tel que de plus α provienne d'un élément $\mathcal{A} \in \mathrm{Br}(\mathcal{X})$. Pour toute place $v \in \mathrm{Spec}(O)$, on a $X(k_v) = \mathcal{X}(O_v)$ (propreté). Pour $M_v \in X(k_v)$, notons $\tilde{M}_v$ le point correspondant de $\mathcal{X}(O_v)$. L'élément $\alpha(M_v) \in \mathrm{Br}(k_v)$ est l'image de $\mathcal{A}(\tilde{M}_v) \in \mathrm{Br}(O_v)$, et $\mathrm{Br}(O_v) = 0$.

L'énoncé suivant, connu et utilisé auparavant dans des cas particuliers, est dû à D. Harari [17]. J'en donne ici une démonstration simplifiée.

Théorème 1.3 (Harari, [17], Théorème 2.1.1 p. 226). *Soient k un corps de nombres et X une k-variété lisse connexe. Soit $\mathcal{X}/O$ un modèle entier de X au-dessus d'un ouvert* $\mathrm{Spec}(O)$ *du spectre de l'anneau des entiers de k. Soit $U \subset X$ un ouvert de X. Pour tout élément $\alpha \in \mathrm{Br}(U)$ qui n'est pas dans* $\mathrm{Br}(X) \subset \mathrm{Br}(U)$, *il existe une infinité de places v de k pour lesquelles il existe $M_v \in U(k_v) \cap \mathcal{X}(O_v)$ avec $\alpha(M_v) \neq 0$.*

Démonstration. Il existe une sous-variété fermée intègre $Z \subset X$, de codimension 1, de point générique ζ, telle que le résidu $\partial_\zeta(\alpha) \in H^1(k(Z), \mathbf{Q}/\mathbf{Z})$ au point générique de Z soit non nul. Quitte à remplacer X par un ouvert, on peut supposer que Z est lisse sur k, que α n'a qu'un seul résidu non trivial sur X, à savoir celui en ζ, et que le résidu $\partial_\zeta(\alpha)$ appartient à $H^1(Z, \mathbf{Q}/\mathbf{Z}) \subset H^1(k(Z), \mathbf{Q}/\mathbf{Z})$ (la cohomologie utilisée est la cohomologie étale). Quitte à restreindre Z, on peut supposer que Z est un revêtement fini d'un espace affine $\mathbf{A}^d_k$. Soit Z_1/Z le revêtement fini (cyclique) défini par $\partial_\zeta(\alpha)$. Le théorème d'irréductibilité de Hilbert, appliqué au revêtement composé $Z_1/\mathbf{A}^d_k$, assure l'existence de points k-rationnels de $\mathbf{A}^d_k$ dont la fibre est intègre. Choisissons un tel k-point. Son image réciproque par l'appplication $Z \to \mathbf{A}^d_k$ est un point fermé $P \in Z$ tel que $\partial_\zeta(\alpha)(P) \in H^1(k(P), \mathbf{Q}/\mathbf{Z})$ est non nul.

On peut compléter l'équation locale de $Z \subset X$ en P par un système régulier de paramètres de l'anneau local régulier $O_{X,P}$. On trouve ainsi une courbe fermée intègre $C \subset X$, passant par P, lisse en P, et transverse à Z en P. Quitte à restreindre encore X, on peut supposer C/k lisse et $Z \cap C = P$.

Les réductions opérées assurent que α appartient à $\mathrm{Br}\,(X\setminus Z)$. Ainsi α induit un élément $\alpha_C \in \mathrm{Br}\,(C \setminus P)$. De la transversalité de C et Z on tire l'égalité

$$\partial_P(\alpha_C) = \partial_\zeta(\alpha)(P) \in H^1\big(k(P), \mathbf{Q}/\mathbf{Z}\big).$$

Ainsi $\partial_P(\alpha_C) \neq 0$. Comme l'inclusion $C \subset X$ s'étend en une inclusion de modèles entiers sur un ouvert convenable du spectre de l'anneau des entiers de k, on est réduit à établir l'énoncé pour une k-courbe lisse connexe.

On considère dorénavant une k-courbe lisse connexe (donc intègre) C, un point fermé $P \in C$, on note $U = C \setminus P$, et on se donne $\alpha \in \mathrm{Br}\,(U)$ de résidu $\chi = \partial_P(\alpha) \in H^1\big(k(P), \mathbf{Q}/\mathbf{Z}\big)$ non nul. Soit $r > 1$ l'ordre de $\partial_P(\alpha)$. On a donc $\partial_P(\alpha) \in H^1\big(k(P), \mathbf{Z}/r\big) \subset H^1\big(k(P), \mathbf{Q}/\mathbf{Z}\big)$. Quitte à restreindre C, on peut supposer C affine, soit $C = \mathrm{Spec}\,(A)$, et P donné par l'annulation de $f \in A$. Soit A_{hs} l'hensélisé de A en P. La restriction $H^1(A_{hs}, \mathbf{Z}/r) \to H^1\big(k(P), \mathbf{Z}/r\big)$ est un isomorphisme. Il existe donc un ouvert étale connexe $q : \mathrm{Spec}\,(B) \to \mathrm{Spec}\,(A)$, soit $q : D \to C$, tel que $q : Q = q^{-1}(P) \to P$ soit un isomorphisme, et tel que χ soit la restriction de $\xi \in H^1(D, \mathbf{Z}/r)$. Soit $V = D \setminus Q$. Soit $(f, \xi) \in \mathrm{Br}\,(V)$ la classe obtenue par cup-produit de la classe de $f \in k[V]^*/k[V]^{*r} \subset H^1(V, \mu_r)$ avec la classe $\xi \in H^1(D, \mathbf{Z}/r)$. La différence $\beta = \alpha_D - (f, \xi) \in \mathrm{Br}\,(V)$ a un résidu trivial en Q, elle appartient donc à $\mathrm{Br}\,(D)$.

Il existe un ouvert $\mathrm{Spec}\,(O)$ du spectre de l'anneau des entiers de k, une courbe affine relative $\mathcal{C}/\,\mathrm{Spec}\,(O)$ (plate et de type fini), une courbe affine relative $\mathcal{D}/\,\mathrm{Spec}\,(O)$ (plate et de type fini), un morphisme $\mathcal{D} \to \mathcal{C}$, étendant la situation $D/C/\,\mathrm{Spec}\,(k)$, tels que f provienne d'un élément, encore noté f, de l'anneau de $\mathcal{C}$, que $\xi \in H^1(D, \mathbf{Z}/r)$ soit la restriction d'un élément $\xi \in H^1(\mathcal{D}, \mathbf{Z}/r)$, que $\beta \in \mathrm{Br}\,(D)$ soit la restriction d'un élément $\beta \in \mathrm{Br}\,(\mathcal{D})$. On peut en outre supposer que les fermés de $\mathcal{D}$ et de $\mathcal{C}$ définis par $f = 0$ sont intègres et isomorphes via la projection $\mathcal{D} \to \mathcal{C}$, et que la projection naturelle de ces fermés vers $\mathrm{Spec}\,(O)$ est finie et étale. Soit R l'anneau dont le spectre est le fermé défini par $f = 0$ dans $\mathcal{C}$ (et dans $\mathcal{D}$). Le corps des fractions de R est le corps $k(P)$.

Le théorème de Tchebotarev assure l'existence d'une infinité de places v de k telles qu'il existe une place w de $k(P)$ avec $k_v \simeq k(P)_w$ (i.e. w est de degré 1 sur v) et w inerte dans l'extension cyclique $k(P)(\chi)/k(P)$ définie par $\chi \in H^1(k(P), \mathbf{Z}/r)$ ([17], Prop. 2.2.1, p. 226).

Pour un tel couple v, il existe un O_v-point N_v^0 du fermé $f = 0$ de $\mathcal{D}$ qui s'envoie isomorphiquement sur un O_v-point M_v^0 de $\mathcal{C}$.

Soit $N_v \in \mathcal{D}(O_v)$ tel que $f(N_v) \neq 0$ et soit $M_v \in \mathcal{C}(O_v) \subset C(k_v)$ son image. On a

$$\alpha(M_v) = \alpha(N_v) = \beta(N_v) + \big(f(N_v), \chi(N_v)\big) \in \mathrm{Br}\,(k_v) \simeq \mathbf{Q}/\mathbf{Z}.$$

On a $\beta(N_v) \in \mathrm{Br}\,(O_v) = 0$. Par ailleurs, comme w est inerte dans l'extension cyclique $k(P)(\chi)/k(P)$, si N_v est suffisamment proche de N_v^0 pour la topologie v-adique sur $\mathcal{D}(O_v)$, la classe $\chi(N_v) \in H^1(k(P)_w, \mathbf{Z}/r) = H^1(k_v, \mathbf{Z}/r)$ est d'ordre r. La formule standard pour le symbole modéré montre alors que $\alpha(M_v) \in \mathbf{Z}/r \subset \mathbf{Q}/\mathbf{Z}$ est égal à la classe de la valuation $v\big(f(N_v)\big)$ modulo r. Comme le fermé $f = 0$ de $\mathcal{D} \times_O O_v$ contient la O_v-section de $\mathcal{D} \times_O O_v \to \mathrm{Spec}\,(O_v)$ définie par N_v^0, on peut trouver $N_v \in \mathcal{D}(O_v)$, proche de N_v^0, tel que $v\big(f(N_v)\big) \equiv 1 \bmod r$, et donc d'image $M_v \in \mathcal{C}(O_v) \subset C(k_v)$ satisfaisant $\alpha(M_v) \neq 0$.

Remarques.

(i) Le cas le plus simple est le suivant, que le lecteur devrait établir lui-même avant de lire la démonstration ci-dessus. On considère $a \in k^*$, a non carré dans k, on prend $X = \mathbf{A}^1_k = \mathrm{Spec}\,\big(k[t]\big)$, puis U l'ouvert défini par $t \neq 0$, enfin $\alpha \in \mathrm{Br}\,\big(k(t)\big)$ la classe de l'algèbre de quaternions (a, t). Il existe une infinité de places v pour lequelles il existe $t_v \in k_v^*$ avec $(a, t_v) \neq 0 \in \mathrm{Br}\,(k_v)$.

(ii) Sur cet exemple, il existe aussi une infinité de places v telle que α s'annule identiquement sur $U(k_v)$. Ce sera plus généralement le cas pour tout X de dimension 1. Par contre, si X est géométriquement intègre de dimension au moins 2 et qu'il existe un point de codimension 1 de X où A a un résidu $\partial(A) \in H^1\big(k(Z), \mathbf{Q}/\mathbf{Z}\big)$ définissant une extension cyclique $L/k(Z)$ telle que k soit algébriquement clos dans L (possible seulement si la dimension de X est au moins 2), alors pour presque toute place v de k la classe α prend sur $U(k_v)$ au moins une valeur autre que 0.

Des propositions précédentes, par un argument essentiellement combinatoire, on établit l'énoncé fort utile suivant, qui est une variante, énoncée dans [12], du "lemme formel" de D. Harari (Corollaire 2.6.1, p. 233 de [17]).

Théorème 1.4. *Soient k un corps de nombres et X une k-variété lisse et géométriquement connexe. Soit $U \subset X$ un ouvert non vide, et soit $B \subset \mathrm{Br}\,(U)$ un sous-groupe fini. Soit $\{P_v\} \in U(\mathbf{A}_k)$. Supposons que pour tout α dans le groupe fini $B \cap \mathrm{Br}\,(X)$, on ait*

$$\sum_{v \in \Omega} \alpha(P_v) = 0.$$

Alors pour tout ensemble fini S de places de k il existe une adèle $\{M_v\} \in U(\mathbf{A}_k)$ telle que $M_v = P_v$ pour $v \in S$ et tel que pour tout $\beta \in B$ on ait

$$\sum_{v\in\Omega} \beta(M_v) = 0.$$

Démonstration. Comme le groupe B est fini, quitte à remplacer S par un ensemble fini de places plus gros, ce qui est licite, on peut supposer que l'inclusion $U \subset X$ admet un modèle entier $\mathcal{U} \subset \mathcal{X}$ au-dessus de l'anneau O des S-entiers de k tel que $P_v \in \mathcal{U}(O_v) \subset \mathcal{X}(O_v)$ pour $v \notin S$, que $B \subset \mathrm{Br}(\mathcal{U})$ et $B \cap \mathrm{Br}(X) \subset \mathrm{Br}(\mathcal{X})$. Pour tout $\beta \in B$, et pour $v \notin S$, on a $\beta(P_v) = 0$. Pour tout $\beta \in B \cap \mathrm{Br}(X)$, tout $v \notin S$ et tout $M_v \in \mathcal{X}(O_v) \subset X(k_v)$, on a $\beta(M_v) = 0$.

Soit $\alpha \in B, \alpha \notin \mathrm{Br}(X)$. D'après le théorème 1.3, il existe un ensemble infini T_α de places $v \notin S$ et une famille $\{N_v\} \in \prod_{v\in T_\alpha} U(k_v) \cap \mathcal{X}(O_v)$ telle que $\alpha(N_v) \neq 0$ pour tout $v \in T_\alpha$. Le groupe $B/\big(B \cap \mathrm{Br}(X)\big)$ est fini, donc aussi son dual $\mathrm{Hom}\,\big(B/\big(B\cap \mathrm{Br}(X)\big), \mathbf{Q}/\mathbf{Z}\big)$. Il existe donc un sous-ensemble infini de T_α tel que les applications linéaires $B/\big(B \cap \mathrm{Br}(X)\big) \to \mathbf{Q}/\mathbf{Z}$ données par $\gamma \to \gamma(N_v)$ pour v dans ce sous-ensemble coïncident. Quitte à remplacer T_α par ce sous-ensemble, on peut donc supposer qu'il existe une application linéaire $\varphi_\alpha : B/\big(B \cap \mathrm{Br}(X)\big) \to \mathbf{Q}/\mathbf{Z}$ satisfaisant $\varphi_\alpha(\alpha) \neq 0$, telle que pour tout $\beta \in B/\big(B \cap \mathrm{Br}(X)\big)$ et tout $v \in T_\alpha$ on ait

$$\varphi_\alpha(\beta) = \beta(N_v) \in \mathbf{Q}/\mathbf{Z},$$

avec $N_v \in U(k_v) \cap \mathcal{X}(O_v)$ comme ci-dessus.

Dans le groupe dual du groupe fini $B/\big(B \cap \mathrm{Br}(X)\big)$, les sommes de telles applications φ_α pour α variant dans $B/\big(B \cap \mathrm{Br}(X)\big)$ (les répétitions étant autorisées) forment un sous-groupe, que nous notons C ; cela résulte simplement du fait que le groupe $B/\big(B\cap \mathrm{Br}(X)\big)$ est, comme B, un groupe de torsion. Considérons alors l'accouplement naturel bilinéaire

$$B/\big(B \cap \mathrm{Br}(X)\big) \times C \to \mathbf{Q}/\mathbf{Z}.$$

D'après ce qu'on a vu ci-dessus, cet accouplement de groupes abéliens finis est non dégénéré à gauche. Ainsi le groupe $B/\big(B \cap \mathrm{Br}(X)\big)$ s'injecte dans le dual de C. Par comptage, ceci n'est possible que si C est égal au dual de $B/\big(B \cap \mathrm{Br}(X)\big)$ tout entier.

L'hypothèse du théorème assure que la famille $\{P_v\}_{v\in S}$ définit une application linéaire $B/\big(B \cap \mathrm{Br}(X)\big) \to \mathbf{Q}/\mathbf{Z}$, donnée par $\beta \mapsto -\sum_{v\in S} \beta(P_v)$.

D'après ce qu'on vient de voir, cette application linéaire peut s'écrire comme une somme d'applications φ_α (avec répétitions). On peut écrire chacun des termes de cette dernière somme sous la forme $\beta \mapsto \beta(N_v)$, cette fois-ci sans répétition sur les places v (puisqu'à chaque fois on a une infinité de places à disposition). On a alors trouvé un ensemble T fini de places $v \notin S$ et des points $N_v \in U(k_v) \cap \mathcal{X}(O_v)$ pour $v \in T$ tels que

$$\sum_{v \in S} \beta(P_v) + \sum_{v \in T} \beta(N_v) = 0$$

pour tout $\beta \in B/\big(B \cap \mathrm{Br}\,(X)\big)$, soit encore

$$\sum_{v \in S} \beta(P_v) + \sum_{v \in T} \beta(N_v) = 0$$

pour tout $\beta \in B$. On a alors

$$\sum_{v \in S} \beta(P_v) + \sum_{v \in T} \beta(N_v) + \sum_{v \notin S \cup T} \beta(P_v) = 0$$

pour tout $\beta \in B$. Ce qui établit l'énoncé, en choisissant $M_v = N_v$ pour $v \in T$ et $M_v = P_v$ pour $v \notin S \cup T$.

Liens entre torseurs et groupe de Brauer ([10], [12], [33], [34])

On donne ici une version résumée de la théorie de la descente. Jusqu'à une date récente, on se limitait à appliquer ladite méthode aux variétés projectives. C'était le cadre adopté dans [10] (chapitre III) et [33]. L'observation que l'on peut aussi faire des descentes utiles sur des variétés "ouvertes" est récente ([12]).

Soient k un corps et G un k-groupe algébrique lisse. Soit X une k-variété. Par définition, un torseur (espace homogène principal) sur X sous G est une k-variété Y équipée d'un k-morphisme $p : Y \to X$, d'une action $G \times_k Y \to Y$, notée $(g, y) \mapsto g.y$; on a donc $g_1.(g_2.y) = (g_1.g_2).y$, et $e.y = y$, où e désigne l'élément neutre de G. On demande que l'action $G \times_k Y \to Y$ respecte la projection $p : Y \to X$, et on demande que l'action soit fidèle et transitive dans les fibres de p. En d'autres termes, on veut que le morphisme naturel $G \times_k Y \to Y \times_X Y$ défini par $(g, y) \to (gy, y)$ soit un isomorphisme.

On s'intéressera ici seulement au cas d'un k-groupe G commutatif. Sous des hypothèses raisonnables (par exemple G groupe linéaire), le groupe (abélien) de cohomologie étale $H^1(X, G)$ classifie les G-torseurs sur X (à isomorphisme non unique près).

Dans la suite on s'intéresse au cas où le groupe G est un k-groupe de type multiplicatif, c'est-à-dire un k-groupe algébrique qui sur une clôture séparable $\overline{k}$ de k se plonge dans un produit de groupes multiplicatifs $\mathbf{G}_m$. On note $\hat{G} = \mathrm{Hom}_{k-\mathrm{groupes}}(G, \mathbf{G}_m)$ le groupe des caractères d'un tel groupe. Le groupe (des $\overline{k}$-points) de $\hat{G}$ est un groupe abélien de type fini muni d'une action continue de $\Gamma = \mathrm{Gal}\,(\overline{k}/k)$, pour la topologie discrète sur ce groupe abélien de type fini : l'action de Γ se factorise par un quotient fini. On sait que le foncteur contravariant défini par $G \mapsto \hat{G}$ est une antidualité, qui tranforme suites exactes en suites exactes.

Soient X une k-variété et G un k-groupe de type multiplicatif. Soit Y un G-torseur sur X. A tout caractère $\chi : \overline{G} \to \mathbf{G}_{m,\overline{k}}$, on associe le produit contracté $\overline{Y} \overset{\overline{G}}{\times} \mathbf{G}_{m,\overline{k}}$, qui est un $\mathbf{G}_{m,\overline{k}}$-torseur sur $\overline{X}$. Comme tout torseur pour la topologie étale sous $\mathbf{G}_m$ est localement trivial pour la topologie de Zariski (théorème 90 à la Grothendieck), on obtient ainsi un élément de $\mathrm{Pic}\,(\overline{X})$. En termes de groupes de cohomologie, on a un homomorphisme naturel

$$\chi : H^1(X, G) \to \mathrm{Hom}_\Gamma\left(\hat{G}, \mathrm{Pic}\,(\overline{X})\right).$$

L'image $\chi\big([Y]\big)$ de la classe $[Y]$ du torseur Y par cette application est appelée le *type* du torseur Y, elle est souvent simplement notée $\chi(Y)$.

L'énoncé suivant est le théorème principal de la théorie de la descente.

Théorème 1.5. *Soient k un corps de nombres et X une k-variété géométriquement intègre. Soit $Y \to X$ un torseur sur X sous un k-groupe de type multiplicatif G. Soit $\theta : \hat{G} \to \mathrm{Pic}\,(\overline{X})$ le type du torseur Y et $\tilde{\theta} : H^1(\Gamma, \hat{G}) \to H^1\big(\Gamma, \mathrm{Pic}\,(\overline{X})\big)$ l'application induite. Soit $B_\theta \subset \mathrm{Br}_1(X)$ l'image réciproque de $\tilde{\theta}\big(H^1(\Gamma, \hat{G})\big) \subset H^1\big(\Gamma, \mathrm{Pic}\,(\overline{X})\big)$ par l'application naturelle $\mathrm{Br}_1(X) \to H^1\big(\Gamma, \mathrm{Pic}\,(\overline{X})\big)$. Soit $\{P_v\} \in X(\mathbf{A}_k)$ tel que pour tout $\alpha \in B_\theta$ on ait $\sum_{v\in\Omega} \alpha(P_v) = 0$. Alors il existe $\xi \in H^1(k, G)$ tel que le torseur Y^ξ possède une adèle $\{M_v\}$ dont l'image par la projection naturelle $Y^\xi \to X$ soit l'adèle $\{P_v\}$.*

Démonstration. La famille des fibres $[Y](P_v) \in H^1(k_v, T)$ définit une classe dans le produit restreint $P^1(k, G) \subset \prod_v H^1(k_v, G)$. On dispose de la suite exacte de Poitou–Tate

$$H^1(k, G) \to P^1(k, G) \to H^1(k, \hat{G})^o,$$

où l'application $P^1(k, G) \to H^1(k, \hat{G})^o$ est induite par la somme des applications composées

$$i_v : H^1(k_v, G) \to H^1(k_v, G)^o \to H^1(k, G)^o,$$

la notation A^o désignant le dual $\mathrm{Hom}\,(A, \mathbf{Q}/\mathbf{Z})$ d'un groupe abélien A. La flèche $H^1(k_v, G) \to H^1(k_v, G)^o$, obtenue par cup-produit et théorie du corps de classes local, est un isomorphisme de groupes finis ([31], II.5.8, Thm. 6 p. 112; [27], I.2.4 p. 35). La flèche $H^1(k_v, G)^o \to H^1(k, G)^o$ est la duale de la restriction $H^1(k, G) \to H^1(k_v, G)$. La suite exacte ci-dessus est une partie d'une suite exacte longue ([27], I.4.20 (b) p. 80, où le produit devrait être remplacé par un produit restreint.)

Une vérification homologique non triviale montre que l'hypothèse d'annulation de la somme $\sum_{v \in \Omega} \alpha(P_v)$ pour tout $\alpha \in B_\theta$ entraîne que la somme $\sum i_v([Y](P_v)) \in H^1(k, \hat{G})^o$ est nulle. De l'exactitude de la suite de Poitou–Tate on tire l'existence d'une classe $-\xi \in H^1(k, G)$ d'image $[Y](P_v) \in P^1(k, G)$. Le torseur tordu Y^ξ (de classe $[Y] + \xi$) a donc une fibre triviale en chacun des P_v. Comme on peut écrire un modèle de G et du torseur $Y \to X$ au-dessus d'un ouvert du spectre de l'anneau des entiers de k, on voit aisément qu'il existe une adèle de Y^ξ d'image $\{P_v\} \in X(\mathbf{A}_k)$.

Note. Comme le groupe G est commutatif, on peut se permettre l'abus de langage consistant à noter Y^ξ un torseur défini par torsion du torseur Y par un 1-cocycle (quelconque) de Γ à valeurs dans G et de classe $\xi \in H^1(k, G)$. Le torseur Y^ξ est défini à isomorphisme non unique près.

Le théorème ci-dessus fut d'abord énoncé sous l'hypothèse que T est un tore et que X est projective et lisse ([10]). Il fut établi dans le cas général par Skorobogatov [33]. Il est surtout intéressant (= applicable numériquement) lorsque la k-variété X est projective (ou propre), car alors pour tout élément $A \in \mathrm{Br}\,(X)$ il existe seulement un nombre fini de places v pour lesquelles l'application d'évaluation $ev_A : X(k_v) \to \mathbf{Q}/\mathbf{Z}$ a une image non réduite à zéro. On peut donc espérer calculer les valeurs prises par $\sum_{v \in \Omega} A(P_v)$ lorsque $\{P_v\}_{v \in \Omega}$ parcourt les adèles de X. Pour les variétés ouvertes, l'énoncé efficace est le suivant ("descente sur une variété ouverte" [12]).

Théorème 1.6. *Soient k un corps de nombres et X une k-variété lisse géométriquement connexe satisfaisant $\overline{k}^* \simeq \overline{k}[X]^*$. Soit $X \subset X_c$ une k-compactification lisse. Soit $Y \to X$ un torseur sur X sous un k-tore T. Soient $\theta : \hat{T} \to \mathrm{Pic}\,(\overline{X})$ le type du torseur Y et $\tilde{\theta} : H^1(\Gamma, \hat{T}) \to H^1(\Gamma, \mathrm{Pic}\,(\overline{X}))$ l'application induite. Soit $B_\theta \subset \mathrm{Br}_1(X)$ l'image réciproque* *de $\tilde{\theta}\big(H^1(\Gamma,\hat{T})\big) \subset H^1\big(\Gamma,\mathrm{Pic}(\overline{X})\big)$ par l'application naturelle $\mathrm{Br}_1(X) \to H^1\big(\Gamma,\mathrm{Pic}(\overline{X})\big)$. Soit $\{P_v\} \in X(\mathbf{A}_k)$ telle que pour tout $\alpha \in B_\theta \cap \mathrm{Br}(X_c)$ on ait $\sum_{v\in\Omega} \alpha(P_v) = 0$. Soit S un ensemble fini de places de k. Alors il existe $\xi \in H^1(k,T)$ tel que le torseur Y^ξ possède une adèle $\{M_v\}$, et que pour chaque $v \in S$ l'image de M_v par la projection structurale $Y^\xi \to X$ soit le point P_v.*

Démonstration. Comme T est un k-tore, son groupe des caractères $\hat{T}$ est un groupe abélien libre de type fini, donc le groupe $H^1(\Gamma,\hat{T})$ est fini. Par ailleurs, l'hypothèse $\overline{k}^* \simeq \overline{k}[X]^*$ donne la suite exacte

$$\mathrm{Br}(k) \to \mathrm{Br}_1(X) \to H^1\big(\Gamma,\mathrm{Pic}(\overline{X})\big).$$

Le groupe $B_\theta/\mathrm{Br}(k)$ est donc un groupe fini. Soit $B \subset B_\theta$ un groupe fini s'envoyant surjectivement sur $B_\theta/\mathrm{Br}(k)$. L'hypothèse et le théorème 1.4 montrent l'existence d'une adèle $\{R_v\} \in X(\mathbf{A}_k)$ avec $M_v = P_v$ pour $v \in S$, telle que $\sum_{v\in\Omega} \alpha(R_v) = 0$ pour tout $\alpha \in B$, donc aussi pour tout $\alpha \in B_\theta$. Le théorème 1.5 permet de conclure.

Un théorème de Heath-Brown et Skorobogatov

Dans ce paragraphe, je donne une démonstration légèrement simplifiée d'un résultat récent de Heath-Brown et Skorobogatov [24]. Comme chez ces auteurs, la démonstration repose sur la descente sur une variété ouverte.

Théorème 1.7. *Soit K/k une extension finie de corps de nombres, et soit $\omega_1,\dots,\omega_n$ une base de K/k. Soient a, b des entiers strictement positifs, premiers entre eux, et soit $\varepsilon \in k^*$. Soit U l'ouvert de lissité de la k-variété affine définie par l'équation*

$$t^a(1-t)^b = \varepsilon.N_{K/k}(x_1\omega_1 + \cdots + x_n\omega_n).$$

Soit X une k-compactification lisse de U. Supposons que le corps k est le corps $\mathbf{Q}$ des rationnels. Alors l'ensemble $X(k)$ est dense dans $X(\mathbf{A}_k)^{\mathrm{Br}(X)}$, le fermé de $X(\mathbf{A}_k)$ formé des adèles sur lesquelles s'annule le groupe de Brauer de X. L'adhérence de $X(k)$ dans $X(\mathbf{A}_k)$ est ouverte. L'approximation faible-faible vaut pour X.

Démonstration. On vérifie aisément que le lieu singulier de l'hypersurface affine de $\mathbf{A}^{n+1}$ définie par

$$t^a(1-t)^b = \varepsilon.N_{K/k}(x_1\omega_1 + \cdots + x_n\omega_n)$$

est de codimension au moins 2, puis que la flèche naturelle $\overline{k}^* \to \overline{k}[U]^*$ est un isomorphisme : les seules fonctions inversibles sur $\overline{U}$ sont les constantes. Soit $U_1 \subset U$ l'ouvert défini par $t(1-t) \neq 0$. Soit $T = R^1_{K/k}\mathbf{G}_m$ le k-tore des éléments de norme 1 dans K^*. Sur la variété U, le diviseur de la fonction t, et le diviseur de la fonction $1-t$, sont des normes de diviseurs sur U_K : comme U est lisse, il suffit de considérer les diviseurs au sens de Weil, la vérification est immédiate. Ceci implique ([10], Lemme 2.7.6 p. 446 et Exemple 2.7.8 (b)) que pour $r, s \in \mathbf{Z}$, le torseur sous T sur U_1 défini par $t^s(1-t)^r = N_{K/k}(y_1\omega_1 + \cdots + y_n\omega_n) \neq 0$ est la restriction d'un torseur Y sous T sur U. Le type du torseur $Y \to U$ est un Γ-homomorphisme $\theta : \hat{T} \to \operatorname{Pic}(\overline{U})$. Soit $B_\theta \subset \operatorname{Br}_1(U)$ le groupe défini au théorème 1.6.

Soit $\{P_v\} \in X(\mathbf{A}_k)^{\operatorname{Br}(X)}$. On a donc $\{P_v\} \in X(\mathbf{A}_k)^{B_\theta}$. Pour établir le théorème, il suffit de montrer que pour tout ensemble fini S de places et toute famille d'ouverts $W_v \subset X(k_v)$, $P_v \in W_v$, il existe un point $M \in X(k)$ tel que $M \in W_v$ pour $v \in S$. Comme $B_\theta/\operatorname{Br}(k)$ est fini, on voit aisément que l'on peut remplacer $\{P_v\}$ par une adèle, encore notée $\{P_v\}$, de l'ouvert $U_1 \subset X$.

L'hypothèse et le théorème 1.6 assurent alors l'existence d'un torseur Y^ξ sur U sous T (de même type que Y), avec $\xi \in k^*/N_{K/k}(K^*) = H^1(k, T)$ convenable, qui possède une adèle $\{M_v\}$ telle que pour tout $v \in S$ l'image par la projection $Y^\xi \to U$ de M_v est P_v.

En appliquant le théorème 2.3.1, p. 421 de [10] au couple (U, U_1), la suite (2.3.1) de [10] étant ici la suite $1 \to T \to R_{K/k}\mathbf{G}_m \to \mathbf{G}_m \to 1$ définie par la norme de K à k, on voit que la restriction du torseur Y^ξ au-dessus de U_1 est donnée par une équation

$$t^s(1-t)^r = \gamma.N_{K/k}(y_1\omega_1 + \cdots + y_n\omega_n) \neq 0,$$

où $\gamma \in k^*$ est un relèvement de $\xi \in k^*/N_{K/k}(K^*)$.

Puisque a et b sont premiers entre eux, il existe des entiers $r, s \in \mathbf{Z}$ tels que $ra - sb = 1$. Appliquons les considérations ci-dessus à la paire (r, s).

La restriction de Y^ξ à U_1 est donnée, comme k-variété, par le système d'équations :

$$t^s(1-t)^r = \gamma.N_{K/k}(y_1\omega_1 + \cdots + y_n\omega_n) \neq 0,$$

$$t^a(1-t)^b = \varepsilon.N_{K/k}(x_1\omega_1 + \cdots + x_n\omega_n) \neq 0.$$

Cette k-variété est k-isomorphe à la k-variété Z définie par

$$t = \varepsilon^r.\gamma^{-b}.N_{K/k}(u_1\omega_1 + \cdots + u_n\omega_n) \neq 0,$$

$$1-t = \gamma^a.\varepsilon^{-s}.N_{K/k}(v_1\omega_1 + \cdots + v_n\omega_n) \neq 0,$$

comme on voit par le changement de variables inversible :

$$u_1\omega_1 + \cdots + u_n\omega_n = (x_1\omega_1 + \cdots + x_n\omega_n)^r.(y_1\omega_1 + \cdots + y_n\omega_n)^{-b},$$

$$v_1\omega_1 + \cdots + v_n\omega_n = (y_1\omega_1 + \cdots + y_n\omega_n)^a.(x_1\omega_1 + \cdots + x_n\omega_n)^{-s}.$$

La k-variété Z est un ouvert lisse de l'hypersurface de l'espace affine $\mathbf{A}^{2n}$ d'équation

$$1 = \varepsilon^r.\gamma^{-b}.N_{K/k}(u_1\omega_1 + \cdots + u_n\omega_n) + \gamma^a.\varepsilon^{-s}.N_{K/k}(v_1\omega_1 + \cdots + v_n\omega_n).$$

La k-variété Z possède des points P'_v pour $v \in S$ correspondant par l'isomorphisme ci-dessus aux points P_v. Par ailleurs, comme la k-variété lisse Y^ξ possède des k_v-points pour toute place v, il en est de même de son ouvert défini par $t(1-t) \neq 0$, ce qui implique que la k-variété (lisse) Z a elle aussi des points dans tous les complétés de k.

Les arguments utilisés jusqu'ici valent sur tout corps de nombres. Pour conclure on utilise le théorème suivant, établi par la méthode du cercle par Heath-Brown et Skorobogatov (sur le modèle d'un travail beaucoup antérieur de Birch, Davenport et Lewis) : *Pour* $k = \mathbf{Q}$, $\alpha, \beta, \gamma \in k^*$ *et* K/k *une extension finie de corps de degré* n, *le principe de Hasse et l'approximation faible valent pour l'ouvert de lissité de l'hypersurface projective d'équation*

$$\alpha.w^n + \beta.N_{K/k}(u_1\omega_1 + \cdots + u_n\omega_n) + \gamma.N_{K/k}(v_1\omega_1 + \cdots + v_n\omega_n) = 0.$$

(Une telle démonstration, par la méthode du cercle, établit un résultat plus fort, elle donne une estimation du nombre de points de hauteur au plus B, pour B tendant vers l'infini.)

Dans le cas particulier $a = b = 1$ et K/k de degré 3, étudié dans [9], l'idée de combiner descente et méthode du cercle avait été suggérée par P. Salberger.

Le résultat obtenu ci-dessus suppose a et b premiers entre eux. Voici un cas où cette hypothèse n'est pas requise.

Théorème 1.8. *Soit K/k une extension finie de corps de nombres, et soit $\omega_1, \ldots, \omega_n$ une base de K/k. Soient a, b des entiers strictement positifs, non nécessairement premiers entre eux, et soit $\varepsilon \in k^*$. Soit U l'ouvert de lissité de la k-variété affine définie par l'équation*

$$t^a(1-t)^b = \varepsilon.N_{K/k}(x_1\omega_1 + \cdots + x_n\omega_n).$$

Soit X une k-compactification lisse de U.

Supposons que le corps k est le corps $\mathbf{Q}$ des rationnels, et que les espaces homogènes principaux sous le tore $R^1_{K/k}\mathbf{G}_m$ satisfont le principe de Hasse et l'approximation faible (c'est le cas par exemple si K/k est une extension cyclique ou si K/k est de degré premier).

Alors l'ensemble $X(k)$ est dense dans $X(\mathbf{A}_k)^{\mathrm{Br}(X)}$. L'adhérence de $X(k)$ dans $X(\mathbf{A}_k)$ est ouverte. L'approximation faible-faible vaut pour X.

Démonstration. Comme on l'a observé ci-dessus, les fonctions t et $1-t$ ont un diviseur qui est une norme de diviseur sur U_K. Soit ici T le k-tore $(R^1_{K/k}\mathbf{G}_m)^2$. Comme ci-dessus, le torseur sous ce k-tore T défini sur l'ouvert $t(1-t) \neq 0$ de U par l'équation

$$t = N_{K/k}(y_1\omega_1 + \cdots + y_n\omega_n), \quad 1-t = N_{K/k}(z_1\omega_1 + \cdots + z_n\omega_n)$$

s'étend en un torseur sur U sous T. On applique alors le théorème 1.6.

Un changement de variables simple montre que l'on obtient le produit d'une k-variété qui est un espace homogène principal sous $R^1_{K/k}\mathbf{G}_m$ par une hypersurface d'équation

$$\alpha + \beta N_{K/k}(u_1\omega_1 + \cdots + u_n\omega_n) + \gamma N_{K/k}(v_1\omega_1 + \cdots + v_n\omega_n) = 0.$$

Les hypothèses permettent de conclure.

Remarques

(i) Pour K/k cyclique, dans la situation des théorèmes 1.7 et 1.8, il existe des contre-exemples à l'approximation faible (D. Coray, voir [9], §8).

(ii) Pour K/k de degré 3, le théorème 1.8 vaut sur tout corps de nombres ([9]).

(iii) Je renvoie à [24] pour l'observation suivante : si a ou b ou $a+b$ est premier avec n, alors tout modèle projectif de la variété définie par $t^a(1-t)^b = \varepsilon.N_{K/k}(x_1\omega_1 + \cdots + \omega_n x_n)$ possède un point k-rationnel, ce qui sous les hypothèses du théorème assure que les points k-rationnels sont (Zariski-)denses. L'existence d'un k-point n'est pas garantie si par exemple $a=2$, $b=3$, $n=30$.

(iv) La validité du principe de Hasse et de l'approximation faible pour l'ouvert de lissité de l'hypersurface projective d'équation

$$\alpha.w^n + \beta.N_{K/k}(u_1\omega_1 + \cdots + u_n\omega_n) + \gamma.N_{K/k}(v_1\omega_1 + \cdots + v_n\omega_n) = 0,$$

pour $k = \mathbf{Q}$, incite à penser que sur tout corps k, pour tout modèle projectif et lisse Z/k de cette hypersurface, le quotient $\mathrm{Br}(Z)/\mathrm{Br}(k)$ est trivial.

(v) Dans certains cas, on n'a pas besoin de recourir à la "descente ouverte". Ainsi pour la variété d'équation affine

$$t(1-t) = \varepsilon.N_{K/k}(x_1\omega_1 + \cdots + x_n\omega_n),$$

lorsque le degré de K/k est impair, le diviseur de t sur tout modèle projectif et lisse est une norme de l'extension K/k. On peut donc partir d'un torseur sous $R^1_{K/k}\mathbf{G}_m$ sur un modèle projectif et lisse.

(vi) Dans la situation des théorèmes 1.7 et 1.8, on peut prendre pour X une k-variété projective et lisse équipée d'un k-morphisme dominant $p : X \to \mathbf{P}^1_k$ étendant la projection $U \to \mathbf{A}^1_k = \mathrm{Spec}\,(k[t])$ définie par $(t, x_1, \ldots, x_n) \mapsto t$. L'invariant δ de cette fibration, comme défini dans l'introduction est en général 3 (fibres en $t = 0, 1, \infty$). En outre, pour le théorème 1.7, on ne suppose pas la validité du principe de Hasse dans les fibres (mais l'on sait que l'obstruction de Brauer–Manin est la seule pour les fibres). On comparera ce résultat avec ceux rappelés à la fin de l'introduction.

(vii) Une discussion lors du colloque de Budapest a permis, au prix d'un travail plus conséquent, de montrer que le théorème 1.7 vaut sans supposer que a et b sont premiers entre eux. Ceci fera l'objet d'une publication séparée ([6]). Une idée-clef est de faire une descente ouverte sur une k-variété lisse ouverte contenant strictement U, et contrôlable.

Chapitre 2: Pinceaux d'espaces homogènes de variétés abéliennes

Revenons à la situation générale évoquée à la fin de l'introduction. Soient donc k un corps de nombres et $p : X \to \mathbf{P}^1_k$ un morphisme dominant de k-variétés lisses géométriquement intègres. Notons $K = k(\mathbf{P}^1_k)$. Soit X_η/K la fibre générique (lisse) de ce morphisme, supposée géométriquement connexe. Supposons que pour tout point fermé $M \in \mathbf{P}^1_k$ la fibre $X_M/k(M)$ possède une composante Y de multiplicité 1 telle que la clôture intégrale de $k(M)$ dans le corps des fonctions de Y est une extension abélienne de $k(M)$. Sous l'hypothèse de Schinzel, la condition $X(\mathbf{A}_k)^{\mathrm{Br}_{\mathrm{vert}}} \neq \emptyset$ suffit à assurer l'existence de k-points $P \in \mathbf{P}^1(k)$ tels que la fibre X_P satisfasse $X_P(\mathbf{A}_k) \neq \emptyset$ (théorème 0.2). Lorsque le principe de Hasse vaut pour les fibres, ceci permet de conclure à l'existence de points k-rationnels sur X.

Dans ce chapitre, nous allons considérer le cas où la fibre générique X_η est un espace homogène principal sous une K-variété abélienne A_η, par exemple le cas où X_η est une courbe de genre 1. Pour presque tout $P \in \mathbf{P}^1(k)$, la fibre X_P est un espace homogène principal sous la k-variété abélienne spécialisée A_P. Il est bien connu que le principe de Hasse ne vaut en général pas pour les espaces homogènes principaux de variétés abéliennes : sauf *heureux accident,* il n'y a pas de raison qu'une fibre X_P, qui possède des points dans tous les complétés de k, possède un point k-rationnel (on peut même donner des contre-exemples en famille [8]).

Depuis 1994, Swinnerton-Dyer et ses collaborateurs ont développé une technique qui permet précisément de forcer d'*heureux accidents* ([35], [14], [37], [2], [5], [39]).

Avant de décrire cette technique, il convient de rappeler ce qu'est le groupe de Tate–Shafarevich et ce qu'est le calcul de Selmer.

Rappels sur les groupes de Selmer

On se limite ici au cas de la multiplication par un nombre premier l sur une courbe elliptique E. Pour tout groupe abélien A et tout entier $n > 0$ on note ${}_nA$ le sous-groupe de A formé des éléments annulés par n. Cette notation sera aussi utilisée pour les schémas en groupes abéliens. De la suite de Kummer

$$0 \to {}_lE \to E \xrightarrow{x \mapsto lx} E \to 0,$$

on tire la suite exacte courte

$$0 \to E(k)/l \to H^1(k, {}_lE) \to {}_lH^1(k, E) \to 0.$$

Le groupe $H^1(k, E)$ classifie les espaces homogènes principaux sous E. Le groupe $H^1(k, {}_lE)$ classifie les l-revêtements de E, c'est-à-dire les torseurs $C \to E$, de groupe structural ${}_lE$, qui sur la clôture algébrique de k sont isomorphes au torseur $E \to E$ donné par la multiplication par l. L'espace total C d'un l-revêtement possède un k-point si et seulement s'il a une image triviale dans $H^1(k, E)$, si et seulement si sa classe dans $H^1(k, {}_lE)$ est dans l'image de $E(k)/l$.

On envoie ensuite la suite exacte

$$0 \to E(k)/l \to H^1(k, {}_lE) \to {}_lH^1(k, E) \to 0$$

dans le produit des suites exactes correspondantes sur chacun des complétés k_v :

$$0 \to \prod_v E(k_v)/l \to \prod_v H^1(k_v, {}_lE) \to \prod_v {}_lH^1(k_v, E) \to 0.$$

Le groupe de Tate–Shafarevich d'une k-variété abélienne A est par définition le noyau de l'application diagonale $H^1(k, A) \to \prod_v H^1(k_v, A)$. Ce groupe classifie les espaces homogènes principaux sous A qui ont des points dans chaque complété k_v. On a la conjecture bien connue :

Conjecture (Tate–Shafarevich). *Pour toute variété abélienne A sur un corps de nombres k, le groupe de Tate–Shafarevich de A est un groupe fini.*

On a la suite exacte naturelle

$$0 \to E(k)/l \to \mathrm{Sel}_l(E) \to {}_l\text{Ш}^1(k, E) \to 0$$

obtenue à partir de la suite de Kummer par image réciproque via l'inclusion ${}_l\text{Ш}^1(k, E) \subset {}_lH^1(k, E)$. Le groupe de Selmer $\mathrm{Sel}_l(E)$ classifie les l-revêtements $C \to E$ dont l'espace total C possède des points dans tous les complétés k_v. L'image de $E(k)/l$ dans $\mathrm{Sel}_l(E)$ correspond exactement à ceux de ces revêtements $C \to E$ tels que $C(k) \neq \emptyset$. Ainsi la validité du principe de Hasse pour toutes les courbes C qui sont des l-revêtements de E équivaut à ${}_l\text{Ш}^1(k, E) = 0$.

Le groupe de Selmer est fini (alors que le groupe $H^1(k, {}_lE)$ est infini). Plus précisément, soit S un ensemble fini de places contenant les places archimédiennes, les places l-adiques et les places non archimédiennes où la courbe E n'a pas bonne réduction. Soit O_S l'anneau des S-entiers de k, et soit $\mathcal{E}/O_S$ le O_S-schéma abélien de fibre générique E/k. Si v est une place hors de S, l'image de la flèche $E(k_v)/l \to H^1(k_v, {}_lE)$ est le sous-groupe $H^1_{\text{ét}}(O_v, {}_l\mathcal{E}) \subset H^1(k_v, {}_lE)$. Ainsi $\text{Sel}_l(E)$ est contenu dans le sous-groupe de $H^1(k, {}_lE)$ formé des éléments non ramifiés en dehors de S, et ce groupe est fini ([31], II.6.2 Thm. 7 p. 117). Ledit sous-groupe s'identifie de fait au sous-groupe fini $H^1_{\text{ét}}(O_S, {}_l\mathcal{E}) \subset H^1(k, {}_lE)$ (c'est le principe de la démonstration du théorème de Mordell–Weil faible).

Pour une place v quelconque, on peut donner une caractérisation de l'image de $E(k_v)$ dans $H^1(k_v, {}_lE)$. On dispose de l'accouplement, symplectique et non dégénéré, de Weil

$${}_lE \times {}_lE \to \mu_l.$$

(Pour une variété abélienne, il faudrait ici faire intervenir la variété duale.) Pour tout place v de k, cet accouplement induit un accouplement, symplectique et non dégénéré, de groupes finis

$$H^1(k_v, {}_lE) \times H^1(k_v, {}_lE) \to H^2(k_v, \mu_l) \subset \mathbf{Z}/l.$$

(La dernière inclusion est fournie par le corps de classes local ; pour v non archimédienne, c'est une égalité.)

Un résultat de Cassels, généralisé aux variétés abéliennes par Tate, assure que l'image de $E(k_v)/l$ dans $H^1(k_v, {}_lE)$ est un sous-espace isotrope maximal pour cet accouplement.

Soit $V_S = \oplus_{v \in S} H^1(k_v, {}_lE)$. Définissons sur V_S un accouplement symplectique et non dégénéré par somme orthogonale des accouplements précédents. On dispose du sous-groupe isotrope maximal $\oplus_{v \in S} E(k_v)/l \subset V_S$. On a par ailleurs la flèche composée $H^1(O_S, {}_l\mathcal{E}) \to H^1(k, {}_lE) \to V_S$, la dernière flèche étant la flèche diagonale. La loi de réciprocité assure que l'image de $H^1(O_S, {}_l\mathcal{E})$ est isotrope dans V_S.

Fixons un ensemble fini S_0 de places du corps de nombres k contenant toutes les places archimédiennes, toutes les places l-adiques, et tel que le groupe des classes de l'anneau O_{S_0} soit trivial. Des arguments classiques de théorie des nombres donnent le résultat suivant ([14], Prop. 1.2.1).

Proposition 2.1. *Supposons que tous les points d'ordre l de la courbe elliptique E sont rationnels sur k (ce qui implique que le corps k contient toutes les racines l-ièmes de l'unité). Pour tout ensemble fini S de places contenant S_0 et les places de mauvaise réduction de E, l'application de $H^1(O_S, {}_l\mathcal{E})$*

dans V_S est injective, son image I^S est un sous-groupe isotrope maximal, et le groupe de Selmer $\mathrm{Sel}_l(E)$ *s'identifie au sous-groupe de $H^1(O_S, {}_l\mathcal{E}) \subset V_S$ orthogonal au sous-groupe isotrope maximal $\oplus_{v\in S} E(k_v)/l \subset V_S$.*

On voit donc que dans ce cas le groupe de Selmer est le noyau à droite de l'accouplement

$$\bigoplus_{v\in S} E(k_v)/l \times H^1(O_S, {}_l\mathcal{E}) \to \mathbf{Z}/l$$

obtenu par restriction de l'accouplement $V_S \times V_S \to \mathbf{Z}/l$. Le rang du groupe de Selmer $\mathrm{Sel}_l(E)$ (comme $\mathbf{F}_l$-vectoriel) peut donc être calculé comme le corang d'une matrice *carrée* sur le corps $\mathbf{F}_l$.

Notons que, comme E a tous ses points d'ordre l rationnels sur k, on dispose d'éléments évidents dans $E(k_v)/l$, à savoir l'image du sous-groupe ${}_lE(k_v)$ (isomorphe à $(\mathbf{Z}/l)^2$) dans $E(k_v)/l$. Ces éléments sont fort utiles pour montrer que certaines classes dans $H^1(O_S, {}_lE)$ ne sont pas dans le groupe de Selmer.

La symétrisation du calcul de Selmer

Conservons l'hypothèse que le groupe ${}_lE$ est isomorphe au groupe constant $(\mathbf{Z}/l)^2$, ce qui pour S comme ci-dessus implique l'existence d'un isomorphisme de ${}_l\mathcal{E}$ avec le O_S-groupe constant $(\mathbf{Z}/l)^2$. Dans [35] sous une forme primitive, puis dans [14] de façon systématique, on a décrit un processus plus élaboré pour calculer le groupe de Selmer, au moyen d'une matrice carrée plus petite que celle évoquée ci-dessus, et qui de plus possède la propriété étonnante d'être symétrique.

Le point de départ est un lemme général sur les sous-espaces maximaux isotropes d'une somme directe d'espaces symplectiques ([14], Lemma 1.1.3). Ce lemme montre qu'il existe des espaces isotropes maximaux $K_v \subset H^1(k_v, {}_lE)$ tels que $K_S = \oplus_{v\in S} K_v$ soit un supplémentaire de $H^1(O_S, {}_l\mathcal{E})$ dans V_S. De plus, pour $v \notin S_0$, on peut prendre, et on prendra $K_v = H^1\big(O_v, (\mathbf{Z}/l)^2\big)$.

On projette alors V_S vers $I^S \simeq H^1(O_S, {}_l\mathcal{E})$ le long de $\oplus_{v\in S} K_v$. L'accouplement ci-dessus induit un accouplement entre les $\mathbf{F}_l$-vectoriels

$$\mathbf{I}^S = I^S \cap (W^S + K_S) \quad \text{et} \quad \mathbf{W}_S = W_S/(W_S \cap K_S) = \bigoplus_{v\in S} W_v/(W_v \cap K_v).$$

On montre :

(i) La projection $V_S \to I^S$ induit par restriction à W_S un isomorphisme $\tau : \mathbf{W}_S \simeq \mathbf{I}^S$.

(ii) Le noyau de l'accouplement $\mathbf{I}^S \times \mathbf{W}_S \to \mathbf{Z}/l$ (à gauche ou à droite) est isomorphe à $\mathrm{Sel}_l(E)$.

(iii) Via l'isomorphisme τ, cet accouplement définit une forme bilinéaire symétrique sur $\mathbf{W}_S$, et aussi sur $\mathbf{I}^S$.

Un principe de Hasse pour certains espaces homogènes principaux de variétés abéliennes

Comme remarqué par Manin ([26]), la finitude des groupes de Tate–Shafarevich implique que l'obstruction de Brauer–Manin au principe de Hasse pour les espaces homogènes principaux de variétés abéliennes est la seule obstruction. Nous ne nous servirons ici que d'une variante particulière de ce fait.

Proposition 2.2. *Soient k un corps de nombres et E/k une courbe elliptique. Soit l un nombre premier. Supposons la partie l-primaire du groupe $Ш(E)$ finie. Si la dimension (sur $\mathbf{F}_l = \mathbf{Z}/l$) du sous-groupe de l-torsion de $Ш(E)$ est au plus égale à 1, alors le sous-groupe de torsion l-primaire de E est trivial : si un espace homogène principal sous E est d'ordre une puissance de l, et s'il a des points dans tous les complétés de k, alors il a un point dans k.*

Démonstration. Sous l'hypothèse que la torsion l-primaire $Ш(E)\{l\}$ est finie, Cassels a établi l'existence d'une forme alternée non dégénérée sur $Ш(E)\{l\}$. Ce groupe est donc une somme de groupes de type $(\mathbf{Z}/l^r)^2$. Sa l-torsion est donc une somme d'exemplaires de $(\mathbf{Z}/l)^2$, sa dimension sur $\mathbf{F}_l$ est paire. Le résultat suit (voir les rappels sur le groupe de Selmer).

Remarques

(i) On peut donner une généralisation aux variétés abéliennes principalement polarisées, en faisant attention au cas $l = 2$ (exemples de Poonen et Stoll).

(ii) On trouvera dans [2] et [5] une variante de cet énoncé dans le cas d'une isogénie de degré 2 (plutôt que la multiplication par 2), et une variante délicate dans le cas d'une isogénie de degré 3 dans [1] et [39].

(iii) En pratique, c'est le $\mathbf{F}_l$-rang d'un groupe de Selmer que l'on bornera. Par exemple, si la courbe elliptique E a tous ses points d'ordre l rationnels sur k, si l'application ${}_lE(k) \to E(k)/l$ est injective (cette condition est facile à assurer), et si le l-groupe de Selmer $\mathrm{Sel}_l(E)$ est de dimension 3 sur $\mathbf{F}_l$, alors le rang de la courbe E sur k est 1 et tout l-revêtement de E qui a des points partout localement possède une infinité de points dans k.

A la recherche des points rationnels dans les fibres

Revenons à la situation d'une fibration $p : X \to \mathbf{P}^1_k$ satisfaisant les hypothèses générales mentionnées au début du chapitre. On fait les hypothèses suivantes :

(a) La fibre générique X_η est une courbe de genre 1 sur le corps $K = k(\mathbf{P}^1)$, et X_η ne possède pas de K-point.

(b) Il existe un premier l tel que la jacobienne E_η de X_η ait toute sa l-torsion rationnelle sur K, i.e. il existe un isomorphisme $(\mathbf{Z}/l)^2 \simeq {}_lE_\eta$.

(c) L'espace homogène principal non trivial X_η a une classe annulée par l dans $H^1(K, E_\eta)$, donc peut être représenté par un l-revêtement $X_\eta \to E_\eta$.

(d) On a $X(\mathbf{A}_k)^{\mathrm{Br}_{\mathrm{vert}}} \neq \emptyset$.

Sous l'hypothèse de Schinzel, on trouve alors une infinité de points $P \in \mathbf{P}^1(k)$ tels que la fibre X_P soit lisse et ait des points dans tous les complétés de k (Théorème 0.2). Sauf cas très particulier le groupe de Selmer $\mathrm{Sel}_l(E_P)$, qui est un $\mathbf{F}_l$-vectoriel, contient un sous-espace vectoriel de dimension 3, noté U_P, engendré par l'image de ${}_lE_P(k) \simeq (\mathbf{Z}/l)^2$ (l'application ${}_lE_P(k) \to E_P(k)/lE_P(k)$ étant injective si $E_P(k)$ n'a pas de point de l^2-torsion non de l-torsion) et par la classe du l-revêtement $X_P \to E_P$.

Si la dimension de $\mathrm{Sel}_l(E_P)$ est 3, alors la dimension de ${}_l\text{Ш}(E_P)$ est au plus 1. La proposition 2.2 permet de conclure que ${}_l\text{Ш}(E_P) = 0$, que X_P possède un k-point, donc est isomorphe à E_P. En général, ceci montre aussi que la courbe E_P a un rang de Mordell–Weil égal à 1, et donc $X_P(k)$ est infini.

De façon générale, on cherche à montrer l'existence d'une infinité de points rationnels Q, "modifications" du point P, tels que l'on ait encore $X_Q(\mathbf{A}_k) \neq \emptyset$ mais tels que la dimension de $\mathrm{Sel}_l(E_Q)$ soit 3. On peut alors conclure que $X(k)$ est dense pour la topologie de Zariski sur X.

On ne peut pas toujours garantir l'existence de tels points Q. Par exemple, si le rang de la partie libre du groupe $E_\eta(K)$ est au moins égal à 2, alors pour presque tout $Q \in \mathbf{P}^1(k)$, le rang de $\mathrm{Sel}_l(E_Q)$ sur $\mathbf{F}_l$ est au moins 4.

On cherche à donner des conditions suffisantes contrôlables sur $X/\mathbf{P}^1_k$ permettant de trouver de tels points Q.

Il faut déjà pouvoir comparer les groupes de Selmer

$$\mathrm{Sel}_l(E_P) \subset H^1(k, {}_lE_P) \quad \text{et} \quad \mathrm{Sel}_l(E_Q) \subset H^1(k, {}_lE_Q).$$

C'est ici qu'intervient de façon essentielle l'hypothèse que la l-torsion de la fibre générique E_η est "constante" (hypothèse (b)). On a une identification naturelle de $(\mathbf{Z}/l)^2$ avec ${}_lE_P$ et ${}_lE_Q$. On peut donc voir les deux groupes de Selmer comme plongés dans le même espace $H^1\big(k, (\mathbf{Z}/l)^2\big)$.

Le groupe $\mathrm{Sel}_l(E_P)$ est un sous-groupe du groupe fini

$$H^1\big(O_{S(P)}, (\mathbf{Z}/l)^2\big) \subset H^1\big(k, (\mathbf{Z}/l)^2\big),$$

où $S(P)$ est un ensemble fini de places contenant la réunion des places de S_0 et des places où la courbe E_P a mauvaise réduction.

Commençons par préciser la définition de $S(P)$. On peut supposer la fibre à l'infini de $p : X \to \mathbf{P}^1_k$ lisse. Soit $\mathbf{A}^1_k = \mathrm{Spec}\big(k[t]\big)$ le complémentaire du point à l'infini, et soient $R_i(t) \in k[t]$ les polynômes unitaires irréductibles définissant les points fermés $M \in \mathbf{A}^1_k$ avec mauvaise fibre (pas de composante de multiplicité 1 géométriquement intègre sur $k(M)$). Soit S_1 un ensemble fini de places contenant l'ensemble S_0 (donc les places archimédiennes et les places au-dessus de l), les places où l'un des polynômes $R_i(t)$ n'est pas entier, les places telles que chaque $R_i(t)$ est entier mais le produit $\prod_i R_i(t)$ ne se réduit pas en un polynôme séparable, et les places restantes où la fibration $X \to \mathbf{P}^1_k$ a mauvaise réduction. Pour $P \in \mathbf{A}^1(k)$ donné par $t = \lambda \in k$, à fibre X_P lisse, les places $v \notin S_1$ où la jacobienne E_P a mauvaise réduction sont parmi les places v telles que $v\big(R_i(\lambda)\big) \neq 0$ pour au moins un i. Pour le calcul du groupe de Selmer de E_P, on prendra $S(P)$ égal à la réunion des places dans S_1 et des places v telles que l'un des $v\big(R_i(\lambda)\big) \neq 0$.

On cherche un nouveau point Q très proche de P pour la topologie v-adique en toutes les places de $S(P)$. Si $\lambda_0 \in k$ est le paramètre de P, on cherche $\lambda \in k$ très proche de λ_0 aux places $v \in S(P)$. La nouvelle courbe E_Q est alors isomorphe à E_P sur chaque complété k_v pour $v \in S(P)$. On peut supposer $S(P) \subset S(Q)$ donc $H^1\big(O_{S(P)}, (\mathbf{Z}/l)^2\big) \subset H^1\big(O_{S(Q)}, (\mathbf{Z}/l)^2\big)$.

Soit $Z_P \subset \mathrm{Sel}_l(E_P) \subset H^1\big(O_{S(P)}, (\mathbf{Z}/l)^2\big)$ un complémentaire du vectoriel U_P. On cherche Q (proche de P aux places dans $S(P)$) tel que l'image de Z_P dans $H^1\big(O_{S(Q)}, (\mathbf{Z}/l)^2\big)$ ait une intersection nulle avec Sel (E_Q). Pour éliminer un élément non nul $\alpha \in Z_P$ par ce procédé, on cherche une place v non dans $S(P)$ et $Q \in \mathbf{P}^1(k)$ tels que α ne soit pas dans le noyau de l'accouplement

$$\bigoplus_{v \in S(Q)\setminus S(P)} {}_lE_Q(k_v) \times Z_P \to \mathbf{Z}/l$$

induit par

$$\bigoplus_{v \in S(Q)} E_Q(k_v)/l \times H^1\big(O_{S(Q)}, (\mathbf{Z}/l)^2\big) \to \mathbf{Z}/l.$$

Une telle place v si elle existe est à chercher parmi les v tels que l'un des $R_i(\lambda)$ ne soit pas une unité en v. En ce qui concerne le paramètre λ du point Q, la condition sera satisfaite pour tout Q dans un ouvert v-adique fixe.

On ne peut pas toujours trouver de tels v et Q, mais comme expliqué plus loin, on a su donner *des conditions suffisantes (conditions (Alg) ou (Arith) décrites ci-dessous)* le garantissant. Soit alors S_2 un ensemble de places v comme ci-dessus. Un argument simple montre qu'on peut trouver un ensemble de telles places, de cardinal exactement égal à la dimension de Z_P, et tel que l'accouplement

$$\bigoplus_{v \in S_2} {}_lE_Q(k_v) \times Z_P \to \mathbf{Z}/l$$

soit non dégénéré à droite.

Pour pouvoir contrôler le groupe de Selmer de E_Q, il ne faut pas que dans son calcul trop de nouvelles places apparaissent. Par ailleurs il faut aussi assurer $X_Q(\mathbf{A}_k) \neq \emptyset$.

Ici l'hypothèse de Schinzel joue à nouveau un rôle (rappelons qu'elle permet déjà d'établir l'existence d'un point P tel que $X_P(\mathbf{A}_k) \neq \emptyset$). Elle permet maintenant de trouver Q de paramètre λ proche de λ_0 aux places de $S(P) \cup S_2$, et tel que dans la décomposition de chaque $R_i(\lambda)$ il n'intervient outre les places précédentes qu'une seule nouvelle place v_i, où $R_i(\lambda)$ est une uniformisante. Soit S_3 la réunion des v_i. On peut alors calculer le groupe de Selmer de E_Q en se limitant aux places de $S(P) \cup S_2 \cup S_3$.

Tout ceci ne permet pas encore de trouver un groupe de Selmer de rang 3, en particulier parce que l'on ne voit pas comment contrôler l'accouplement

$$\bigoplus_{v \in S(Q)} E_Q(k_v)/l \times H^1\big(O_{S(Q)}, (\mathbf{Z}/l)^2\big) \to \mathbf{Z}/l$$

sur les éléments de $H^1(O_{S(Q)}, (\mathbf{Z}/l)^2)$ qui ne proviennent pas de

$$H^1(O_{S(P)}, (\mathbf{Z}/l)^2).$$

C'est ici qu'intervient la *symétrisation* du calcul de Selmer. Grâce à cette symétrisation, on peut "faire comme si" l'accouplement

$$\bigoplus_{v \in S} E_Q(k_v)/l \times H^1(O_{S(Q)}, (\mathbf{Z}/l)^2) \to \mathbf{Z}/l$$

était donné par une forme bilinéaire symétrique $B_P : V_Q \times V_Q \to \mathbf{Z}/l$ (dans ce processus, on diminue la dimension des espaces concernés).

On décompose l'espace vectoriel V_P sous la forme $V_P = V_0 \oplus V_1 \oplus V_2$, où $V_0 \oplus V_1$ est le noyau de la forme bilinéaire B_P. La restriction de B_P à V_2 est non dégénérée. Le sous-vectoriel V_0, de dimension 3, est celui engendré par les points d'ordre l de E_P et la classe de X_P. En utilisant ce qui précède (et en particulier en utilisant l'hypothèse de Schinzel) on trouve un point Q tel que $V_Q = V_P \oplus V_3$, que la forme bilinéaire symétrique B_Q se restreigne sur V_P à la forme bilinéaire symétrique B_P, que la dimension de V_3 soit égale à la dimension de V_1, et que l'accouplement $B_Q : V_1 \times V_3$ soit non dégénéré. La *symétrie* de la forme bilinéaire B_Q implique alors que le noyau de la forme bilinéaire B_Q est réduit à V_0, donc le rang du groupe de Selmer de E_Q est 3.

Il nous reste à préciser les conditions suffisantes, mentionnées ci-dessus, qui permettent de réaliser ce programme. Elles sont en pratique performantes, mais sont difficiles à énoncer de façon formelle.

Le premier type de condition est de nature algébrique. On parlera ici d'une condition de type (Alg). C'est la "Condition (D)" apparaissant sous des formes diverses dans [35], [14], [37], [2], [5]. On demande qu'un certain sous-groupe d'un l-groupe de Selmer de nature algébrico-géométrique attaché à la fibre générique E_η soit aussi petit que possible, plus précisément qu'il soit de dimension 3 sur $\mathbf{F}_l$ (la condition impose en particulier automatiquement que le rang de la partie libre de $E_\eta(K)$ est au plus 1) (pour cette interprétation de la "Condition (D)" en ces termes, voir [14], §4). Le l-groupe de Selmer classique (ci-dessus) est défini à partir d'une courbe elliptique E et des complétés k_v du corps de nombres k, ou encore de ses hensélisés. Le groupe de Selmer algébrico-géométrique défini dans [14] est associé à une courbe elliptique définie sur le corps $K = k(\mathbf{P}^1)$ et à l'ensemble des hensélisés stricts aux points fermés de $\mathbf{P}^1_k$ (la définition de ce groupe ne fait pas intervenir l'arithmétique du corps k). Le sous-groupe est défini au moyen d'applications associées aux fibres de mauvaise réduction de la famille de courbes sur $\mathbf{P}^1_k$. Si cette condition de type (Alg) est satisfaite, alors on arrive à éliminer le groupe Z_P. (Il y a là une analogie, a priori purement formelle, avec le théorème d'irréductibilité de Hilbert.)

Le second type de condition est arithmétique. On parlera ici d'une condition de type (Arith). C'est le type de condition imposé dans [39]. On demande a priori que pour chacune des classes non nulles (elles sont en nombre fini) dans Z_P il existe une place v (nécessairement non dans $S(P)$) telle que l'accouplement ${}_lE(k_v) \times H^1(k_v, (\mathbf{Z}/l)^2) \to \mathbf{Z}/l$ permette de l'éliminer. Cette description n'est qu'une approximation. La situation dans [39] est plus subtile : le nombre fini de classes “résistantes” dans $H^1\big(k, (\mathbf{Z}/l)^2\big)$ à prendre en compte ne dépend pas du point P trouvé initialement, il dépend seulement de la mauvaise réduction de la variété X. La condition imposée est qu'un certain nombre de variétés auxiliaires n'aient pas de points locaux.

Remarque. Pour obtenir l'existence de points rationnels sur X, outre les conjectures de Schinzel et de Tate–Shafarevich, les hypothèses ont été :

(i) $X(\mathbf{A}_k)^{\mathrm{Br}_{\mathrm{vert}}} \neq \emptyset$.

(ii) La condition (Alg) ; ou la condition (Arith).

On peut être surpris du résultat, qui semble ignorer la possibilité que le sous-ensemble $X(\mathbf{A}_k)^{\mathrm{Br}(X)} \subset X(\mathbf{A}_k)^{\mathrm{Br}_{\mathrm{vert}}}$ soit vide.

En fait, dans le cas considéré dans [14], on montre (algébriquement, sans utiliser de conjectures) que la condition (Alg) est proche d'une autre condition : le sous-groupe l-primaire du groupe de Brauer de X coïncide avec celui de $\mathrm{Br}_{\mathrm{vert}}(X/\mathbf{P}^1_k)$.

Dans la situation considérée dans [39], le groupe de Brauer n'est pas entièrement vertical, et la partie non verticale peut donner naissance à une obstruction au principe de Hasse. On peut vérifier (sans utiliser de conjectures) que la condition (Arith) imposée dans la preuve implique $X(\mathbf{A}_k)^{\mathrm{Br}(X)} \neq \emptyset$.

En bref, pour chaque type de surface X discuté ci-après, les conditions qu'on est amené à imposer pour obtenir l'existence de points rationnels sont plus fortes que la simple condition $X(\mathbf{A}_k)^{\mathrm{Br}(X)} \neq \emptyset$, hypothèse que nous ne savons pas utiliser directement, du moins en ce qui concerne la partie non verticale du groupe de Brauer.

Quelques résultats obtenus par cette méthode

Théorème 2.3 ([35]; [14] Prop. 3.2.1). *Soient k un corps de nombres et a_i, b_i, $i = 0, \cdots, 4$ des éléments de k^*. Soit $X \subset \mathbf{P}^4_k$ l'intersection de deux quadriques définie par*

$$\sum_{i=0}^{4} a_i T_i^2 = 0, \quad \sum_{i=0}^{4} b_i T_i^2 = 0.$$

Supposons :

(i) *la torsion 2-primaire des groupes de Tate–Shafarevich des courbes elliptiques définies sur k est finie ;*

(ii) *l'hypothèse de Schinzel est satisfaite.*

Si les a_i, b_i sont suffisamment généraux, le principe de Hasse vaut pour X : si X a des points rationnels dans tous les complétés de k, alors X a des points k-rationnels.

(La condition précise sur les a_i, b_i est une condition d'indépendance sur les divers $d_{ij} = a_i b_j - a_j b_i$ ($i \neq j$) vus comme éléments de k^*/k^{*2}. Elle est satisfaite pour $a_0, \ldots, a_4, b_0, \ldots, b_4$ dans un ensemble de Hilbert de k^{10}.)

La démonstration de ce théorème utilise une fibration en courbes de genre 1 dont la fibre générique est un 2-revêtement d'une courbe elliptique dont tous les points d'ordre 2 sont rationnels. L'hypothèse que les a_i, b_i sont suffisamment généraux garantit d'une part que la condition (Alg) est satisfaite, d'autre part que la 2-torsion du groupe de Brauer vertical est triviale.

On trouvera dans [2] et [5] un résultat analogue pour des couples de formes quadratiques plus généraux (mais encore partiellement simultanément diagonales). On a une fibration analogue à celle évoquée ci-dessus, mais ici la courbe elliptique générique possède seulement un point de 2-torsion non trivial rationnel sur K.

Le dernier paragraphe de [2] laisse espérer la validité du théorème ci-dessus pour une intersection complète lisse de deux quadriques dans $\mathbf{P}^4_k$ sans aucune condition de diagonalisation simultanée (mais bien sûr avec des coefficients suffisamment généraux).

Dans [14], nous décrivons de nombreuses classes de surfaces fibrées en courbes de genre 1 pour lesquelles la méthode s'applique (le théorème ci-dessus apparaissant comme un cas très particulier). Une variante de [14] permet d'obtenir l'élégant résultat suivant.

Théorème 2.4 [37]. *Soient $k = \mathbf{Q}$ et a_i, $i = 0, \ldots, 3$ des éléments de k^* dont le produit $a_0.a_1.a_2.a_3$ est un carré. Soit $X \subset \mathbf{P}^3_k$ la surface quartique définie par l'équation*

$$\sum_{i=0}^{3} a_i T_i^4 = 0.$$

Supposons :

(i) *La torsion 2-primaire des groupes de Tate–Shafarevich des courbes elliptiques définies sur k est finie ;*

(ii) *L'hypothèse de Schinzel est satisfaite.*

(iii) *Le produit des a_i n'est pas dans k^{*4} et aucun des $\pm a_i a_j$ pour $i \neq j$ n'est un carré.*

(iv) *L'ensemble $X(\mathbf{A}_k)^{\mathrm{Br}_1(X)}$ est non vide.*

Alors X possède des points k-rationnels.

L'énoncé donné dans [37] n'est pas exactement formulé comme ci-dessus, mais un petit travail analogue à celui fait dans [5] devrait permettre de faire la traduction et d'étendre le résultat à tout corps de nombres.

La démonstration de ce théorème utilise une fibration en courbes de genre 1 dont la fibre générique est un 2-revêtement d'une courbe elliptique dont tous les points d'ordre 2 sont rationnels. L'hypothèse que les a_i sont suffisamment généraux garantit que la condition (Alg) est satisfaite.

La situation est donc tout à fait analogue à celle du théorème 2.3 et des résultats généraux de [14]. La différence dans l'exécution tient au fait suivant. Dans l'énoncé précis de la condition (Alg), la structure des fibres dégénérées (du modèle de Néron) de la fibration elliptique joue un rôle important. Ces fibres sont différentes dans les théorèmes 2.3 et 2.4.

Le théorème suivant a l'immense avantage de ne pas recourir à l'hypothèse de Schinzel.

Théorème 2.5 [39]. *Soient $a_i \in \mathbf{Z}$, $i = 0, \ldots, 3$ des entiers non nuls, sans facteur commun, chacun d'entre eux non divisible par un cube. Soit $X \subset \mathbf{P}^3_{\mathbf{Q}}$ la surface cubique diagonale donnée par*

$$\sum_{i=0}^{3} a_i T_i^3 = 0.$$

Supposons que sur toute extension quadratique k de $\mathbf{Q}$, la partie 3-primaire des groupes de Tate–Shafarevich des courbes elliptiques d'équation $X^3+Y^3 = aT^3$ ($a \in k^*$) est finie. Supposons satisfaite l'une des hypothèses :

(i) *Il existe un premier $p \neq 3$ divisant a_0 mais aucun des autres a_i et il existe un premier $q \neq 3$ divisant a_1 mais aucun des autres a_i.*

(ii) *Il existe un premier $p \neq 3$ divisant a_0 mais aucun des autres a_i, et les classes de a_1, a_2, a_3 dans $\mathbf{F}_p^*/\mathbf{F}_p^{*3}$ ne sont pas toutes égales.*

Alors le principe de Hasse vaut pour X : Si X a des points rationnels dans tous les complétés de $\mathbf{Q}$, alors X a des points $\mathbf{Q}$-rationnels.

Un argument relativement standard de fibration permet de déduire de ce théorème le corollaire spectaculaire suivant :

Théorème 2.6 [39]. *Supposons que sur toute extension quadratique k de $\mathbf{Q}$, la partie 3-primaire des groupes de Tate–Shafarevich des courbes elliptiques d'équation $X^3 + Y^3 = aT^3$ ($a \in k^*$) est finie. Alors, sur le corps $\mathbf{Q}$ des rationnels, pour tout $n \geq 4$, le principe de Hasse vaut pour toute hypersurface cubique diagonale*

$$\sum_{i=0}^{n} a_i T_i^3 = 0.$$

Rappelons qu'un tel énoncé est conjecturé pour toute hypersurface cubique lisse dans $\mathbf{P}^n_{\mathbf{Q}}$, $n \geq 4$, et que la méthode du cercle a permis de l'établir pour les hypersurfaces cubiques diagonales avec $n \geq 6$ (Davenport, 1959, pour $n \geq 8$; R. C. Baker, 1989, pour $n \geq 6$; pour $n \geq 6$, il existe toujours des solutions locales.)

On pourrait essayer de démontrer le théorème 2.5 en prenant une droite rationnelle assez générale dans $\mathbf{P}^3_{\mathbf{Q}}$, et en considérant le pinceau de cubiques découpé sur la surface X par les plans passant par cette droite. Si le pinceau est assez général, toutes les courbes du pinceau sont géométriquement intègres – et on n'aurait pas alors besoin de recourir à l'hypothèse de Schinzel. Mais pour un pinceau général on n'espère pas trouver de sous-groupe "constant" $\mathbf{Z}/l \subset J_\eta$ dans la jacobienne de la courbe générique du pinceau.

Le point de départ de Swinnerton-Dyer est, après un changement de variable évident, une réécriture de l'équation considérée sous la forme

$$a_0T_0^3 + a_1T_1^3 = a_2T_2^3 + a_3T_3^3.$$

(C'est aussi le point de départ de la démonstration par Hasse de son principe pour les formes quadratiques en 4 variables.) Plus précisément, on considère le cône affine Y dans $\mathbf{A}^4_{\mathbf{Q}}$ sur la surface X, défini par l'équation ci-dessus. Il suffit de trouver un $\mathbf{Q}$-point sur un modèle lisse de ce cône. Ce dernier est le produit fibré de deux courbes de genre 1 au-dessus de la droite $\mathbf{A}^1_{\mathbf{Q}} = \mathrm{Spec}\,(\mathbf{Q}[t])$:

$$a_0T_0^3 + a_1T_1^3 = t, \quad a_2T_2^3 + a_3T_3^3 = t.$$

Il existe un modèle projectif et lisse Z de Y équipé d'une fibration $p : Z \to \mathbf{P}^1_{\mathbf{Q}}$ étendant la projection $Y \to \mathbf{A}^1_{\mathbf{Q}}$ donnée par t. Pour tout point fermé M de $\mathbf{P}^1_{\mathbf{Q}}$ différent de $t = 0, \infty$, la fibre X_M possède une composante de multiplicité 1 géométriquement intègre. Ainsi $\delta(p) \leq 2$ et on peut appliquer le théorème 0.2 (a), qui recourt au théorème de Dirichlet mais pas à l'hypothèse de Schinzel générale.

L'autre aspect agréable de la fibration p est que la jacobienne de la fibre générique est le produit fibré des courbes elliptiques $T_0^3 + T_1^3 = a_0a_1tU^3$ et $T_2^3 + T_3^3 = a_2a_3tV^3$. Chacune de ces courbes possède un sous-groupe μ_3 constant. On peut donc espérer utiliser la méthode générale décrite plus haut.

Il ne semble pas y avoir dans le cas présent d'analogue de la condition (Alg) (comme les fibres de $Z \to \mathbf{P}^1_{\mathbf{Q}}$ sont de dimension plus grande que 1, on ne peut ici suivre l'analyse du §4 de [14]).

De façon un peu surprenante, le groupe de Brauer vertical associé à la situation est trivial (proposition 3.5 ci-après). Sous l'hypothèse que X, et donc Z, possède partout localement des points, le théorème 0.2 (a) permet de trouver des points $P \in \mathbf{P}^1(\mathbf{Q})$ tels que Z_P, qui est le produit de deux courbes de genre 1, ait des points dans chacun des complétés. A chacune de ces deux courbes on associe un élément d'un groupe de Selmer convenable (associé à une isogénie de degré 3). On essaye de remplacer le point P par un autre point Q tel que la somme des dimensions (sur $\mathbf{F}_3$) des deux groupes de Selmer en ce point soit strictement plus petite. On a ici un analogue de la "symétrisation" du groupe de Selmer, mais le processus pour décroître l'ordre des deux groupes est extrêmement subtil. Ce processus bute à un certain moment. C'est ce qui amène à imposer la condition (Arith) mentionnée plus haut, qui est satisfaite sous les hypothèses (i) ou (ii) du théorème 2.5. On rencontre ici une difficulté supplémentaire lorsqu'on cherche à établir, pour les isogénies concernées, l'analogue de la proposition 2.2. Cette difficulté tient à l'existence de multiplication complexe par $\mathbf{Q}(\mu_3)$ sur les courbes considérées. Pour obtenir le substitut convenable, on utilise le fait que $\mathbf{Q}$ ne contient pas les racines cubiques de l'unité ! L'analogue des théorèmes 2.5 et 2.6 vaut d'ailleurs sur tout corps de nombres k tel que l'extension $k(\mu_3)/k$ est non triviale.

Remarque. Soient $g_1(x)$ et $g_2(x)$ des polynômes séparables de degré 4 sur un corps de nombres k. Considérons la k-variété affine définie par les équations

$$y_1^2 g_1(x_1) = t = y_2^2 g_2(x_2)$$

dans l'espace affine $\mathbf{A}_k^5$ avec coordonnées t, x_1, x_2, y_1, y_2. Cette variété est k-birationnelle au produit fibré, au-dessus de $\mathbf{A}_k^1 = \mathrm{Spec}\,(k[t])$, des courbes de genre 1 d'équation affine $y_1^2 = tg_1(x_1)$ et $y_2^2 = tg_2(x_2)$. Chacune de ces courbes a bonne réduction en dehors de $t = 0, \infty$. La jacobienne de chacune de ces courbes est une courbe elliptique sur le corps $k(t)$ dont la 2-torsion est un $k(t)$-groupe qui provient d'un k-groupe. Si les résolvantes de g_1 et g_2 ont toutes leurs solutions dans k, ce k-groupe est $(\mathbf{Z}/2)^2$. On est donc dans une situation analogue à celle étudiée dans [39]. En outre, sauf pour des polynômes g_1 et g_2 très particuliers, on n'a pas la difficulté due à la multiplication complexe. On peut espérer dans une telle situation obtenir des résultats parallèles à ceux de [39].

Chapitre 3: Surfaces et hypersurfaces cubiques diagonales sur un corps de fonctions

Dans ce chapitre, je montre que les résultats de l'article [39] de Swinnerton-Dyer, que l'on vient de discuter, et qui dépendent de la finitude de groupes de Tate–Shafarevich, admettent des analogues inconditionnels lorsque l'on remplace les corps de nombres par des corps globaux de caractéristique positive.

Quand on parle d'un corps de fonctions k sur un corps fini $\mathbf{F}_q$, on sous-entend que le corps $\mathbf{F}_q$ est algébriquement clos dans k. Ainsi la condition $q \equiv 2$ mod. 3 dans les énoncés suivants équivaut au fait que la caractéristique de k est différente de 3 et que le corps k ne contient pas les racines cubiques primitives de l'unité.

Théorème 3.1. *Soit k un corps de fonctions d'une variable sur un corps fini $\mathbf{F}_q$ avec q impair, $q \equiv 2$ mod. 3. Soient $a, b, c, d \in k^*$. Si la surface cubique V d'équation*

$$ax^3 + by^3 + cz^3 + dt^3 = 0$$

possède des points dans tous les complétés k_v de k, alors, sous chacune des hypothèses suivantes, elle possède un point dans k.

(i) *Il existe des places v, w de k satisfaisant : $v(b)$, $v(c)$, $v(d)$, $w(a)$, $w(c)$, $w(d)$ sont tous divisibles par 3, mais ni $v(a)$ ni $w(b)$ ne le sont.*

(ii) *Il existe une place v de k telle que $v(a)$ n'est pas divisible par 3 alors que $v(b)$, $v(c)$, $v(d)$ le sont, et l'un des quotients $b/c, c/d$ n'est pas un cube dans le complété k_v.*

(iii) *Il existe une place v de k telle que $v(c)$ et $v(d)$ sont divisibles par 3 alors que ni $v(a)$ ni $v(b)$ ne le sont, et la surface cubique $V \times_k k_v$ n'est pas birationnelle à un plan sur le corps k_v.*

(Cette dernière condition peut s'exprimer de façon explicite en terme des coefficients.)

Théorème 3.2. *Soit k un corps de fonctions d'une variable sur un corps fini $\mathbf{F}_q$ avec q impair et $q \equiv 2$ mod. 3. Si une hypersurface cubique diagonale*

$$ax^3 + by^3 + cz^3 + dt^3 + eu^3 = 0$$

a des points rationnels dans tous les complétés de k, alors elle a des points dans k.

Le travail de Swinnerton-Dyer utilise un certain nombre de résultats qui pourraient a priori paraître spécifiques aux corps de nombres. Dans ce chapitre, qui renvoie constamment à l'article [39], on indique simplement les substituts nécessaires pour couvrir le cas des corps globaux de caractéristique positive. Certains des commentaires ont aussi leur intérêt dans le cas des corps de nombres. Le lecteur est donc supposé lire ce chapitre avec l'article [39] sous la main. Sauf mention contraire, les références dans ce chapitre sont aux lemmes, propositions, paragraphes etc. de [39].

La théorie du corps de classes pour les corps de fonctions d'une variable sur un corps fini est due à Hasse, F. K. Schmidt et Witt. Elle fut développée dans un cadre plus géométrique par Lang (voir le livre [29] de Serre). Le théorème de Tchebotarev dans le cas fonctionnel est établi par Lang [25] et Serre [30]. Voir en particulier le théorème 1, chap. VI, §16, p. 135 de [29].

On note $\mathbf{F}$ un corps fini de cardinal congru à 2 modulo 3, c'est-à-dire un corps fini qui ne contient pas de racine cubique non triviale de 1. On note $\mathbf{F}'$ le corps $\mathbf{F}[X]/(X^2+X+1)$. On considère une $\mathbf{F}$-courbe C, projective, lisse, géométriquement connexe. On note $k=\mathbf{F}(C)$ son corps des fonctions. On note $C'=C\times_{\mathbf{F}}\mathbf{F}'$ et $K=\mathbf{F}'(C')$ le corps des fonctions de C', corps qui est donc obtenu à partir de k par ajout des racines cubiques de 1 : c'est une extension "constante" du corps de fonctions d'une variable k. On note $G=\mathrm{Gal}\,(\mathbf{F}'/\mathbf{F})=\mathrm{Gal}\,(K/k)$, et on note σ ($\sigma^2=1$) l'élément non trivial de G.

Etant donnés $a,b,c,d\in k^*$, on considère la surface cubique non singulière V/k définie par

$$ax^3+by^3+cz^3+dt^3=0.$$

On fixe un ensemble fini S_0 de places de K contenant un ensemble de générateurs du groupe de Picard $\mathrm{Pic}\,(C')$ (on rappelle que le sous-groupe $\mathrm{Pic}^0(C')\subset\mathrm{Pic}\,(C')$ des classes de degré zéro est le groupe des points rationnels d'une variété abélienne sur un corps fini, donc est fini), et on suppose S_0 stable par G. On note S_1 un ensemble fini de places de K, stable par G, contenant S_0 et contenant toutes les places v de K pour lesquelles l'un de $v(a)$, $v(b)$, $v(c)$, $v(d)$ n'est pas divisible par 3.

Il n'y a rien à modifier à la description de la "première descente" sur une courbe $X^3+Y^3=AZ^3$ avec $A\in k^*$ au début du paragraphe 2.

Dans le cas fonctionnel, le paragraphe précédant le lemme 1 (voir formules (8)) est à remplacer par les considérations (plus simples) suivantes. On se donne un ensemble fini S de places (points fermés) de C', stable par σ. Soit n son ordre. On note $U'\subset C'$ l'ouvert complémentaire de S. On considère l'application naturelle de $\mathbf{Z}/3$-vectoriels

$$\varphi:\mathbf{F}'[U']^*/\mathbf{F}'[U']^{*3}\to\bigoplus_{v\in S}K_v^*/K_v^{*3}$$

induite par les inclusions $\mathbf{F}'[U']\subset K\subset K_v$.

Fixons un isomorphisme $\mathbf{Z}/3 = \mu_3$ au-dessus de K. Le second espace vectoriel, qui par la théorie de Kummer s'identifie à $\oplus_{v \in S} H^1(K_v, \mu_3)$, est alors muni d'une structure naturelle d'espace alterné non dégénéré, donnée par la somme des cup-produits (symboles de Hilbert)

$$H^1(K_v, \mu_3) \times H^1(K_v, \mu_3) \to H^2(K_v, \mu_3^{\otimes 2}) = H^2(K_v, \mu_3) = \mathbf{Z}/3,$$

dont chacun est alterné non dégénéré ([31], II.5.8 Thm. 6 p. 112).

L'analogue du lemme de départ de [39] (remarques précédant le lemme 1) est ici :

Lemme 3.3. *Avec les notations ci-dessus,*

a) *la dimension du* $\mathbf{Z}/3$*-vectoriel* $\oplus_{v \in S} K_v^*/K_v^{*3}$ *est* $2n$ *;*

b) *l'application* φ *a une image totalement isotrope;*

c) *si de plus* $\mathrm{Pic}\,(U') = 0$, *alors l'application* φ *est injective, et son image est un sous-espace totalement isotrope maximal de* $\oplus_{v \in S} K_v^*/K_v^{*3}$.

Démonstration. Pour tout point fermé v de C', le complété K_v est isomorphe à $\mathbf{F}_v((t))$, où le corps résiduel $\mathbf{F}_v$ est un corps de caractéristique différente de 3 et contenant les racines cubiques de 1. Le quotient $\mathbf{F}_v^*/\mathbf{F}_v^{*3}$ est donc d'ordre 3, et le quotient K_v^*/K_v^{*3} isomorphe à $(\mathbf{Z}/3)^2$, ce qui établit le point a). Le point b) résulte par exemple de la loi de réciprocité pour le groupe de Brauer de $\mathbf{F}'(C)$. Pour le point c), la dualité de Poitou–Tate (Milne [27], I, Thm. 4.10 p. 70) montre que le noyau de φ s'identifie au dual du noyau de l'application

$$H^2_{\text{ét}}(U', \mathbf{Z}/3) \to \bigoplus_{v \in S} H^2(K_v, \mathbf{Z}/3).$$

Sous l'hypothèse $\mathrm{Pic}\,(U') = 0$, la suite de Kummer donne un isomorphisme $H^2_{\text{ét}}(U', \mu_3) \simeq {}_3\mathrm{Br}\,(U')$. Comme $\mathbf{Z}/3 \simeq \mu_3$ sur C', le noyau ci-dessus s'identifie donc au noyau de

$${}_3\mathrm{Br}\,(U') \to \prod_{v \in S} {}_3\mathrm{Br}\,(K_v),$$

c'est-à-dire, la courbe C' étant lisse, à un sous-groupe de ${}_3\mathrm{Br}\,(C')$. Et l'on sait ([16] III (2.15)) qu'une courbe projective lisse connexe sur un corps fini a un groupe de Brauer nul.

Pour établir que l'image de φ est un sous-espace totalement isotrope maximal, il suffit maintenant de montrer que la dimension du $\mathbf{Z}/3$-vectoriel

$\mathbf{F}'[U']^*/\mathbf{F}'[U']^{*3}$ est n. Ceci s'établit simplement en considérant la multiplication par 3 sur la suite exacte

$$0 \to \mathbf{F}'[U']^*/\mathbf{F}'^* \to \bigoplus_{v \in S} \mathbf{Z} \to \mathrm{Pic}\,(C') \to 0$$

(l'exactitude à droite provient de l'hypothèse $\mathrm{Pic}\,(U') = 0$), en utilisant la finitude du noyau $\mathrm{Pic}^0(C')$ de l'application degré $\mathrm{Pic}\,(C') \to \mathbf{Z}$, et en tenant compte du fait déjà mentionné $F'^*/F'^{*3} \simeq \mathbf{Z}/3$.

Remarque. Dans le cas fonctionnel, pour *toute* place v de K, notant O_v l'anneau des entiers du complété K_v, on voit que $T_v = O_v^*/O_v^{*3} \simeq \mathbf{Z}/3$ est un sous-espace isotrope maximal de $K_v^*/K_v^{*3} \simeq (\mathbf{Z}/3)^2$ (alors que pour les corps de nombres on doit faire attention aux places 3-adiques).

Le lemme 1 de [39] est un lemme d'algèbre linéaire abstrait, il est délicat mais ne requiert aucune modification dans notre contexte.

Les considérations suivant ce lemme valent encore, avec quelques simplifications. Pour E/k définie par une équation $X^3 + Y^3 = AT^3$ et v place de K, l'image W_v de l'application de Kummer $E(K_v) \to K_v^*/K_v^{*3} \simeq (\mathbf{Z}/3)^2$ est $T_v = O_v^*/O_v^{*3} \simeq \mathbf{Z}/3$ si $v(A)$ est divisible par 3, sinon c'est le groupe $\mathbf{Z}/3$ engendré par la classe de A dans K_v^*/K_v^{*3}.

Discutons maintenant le paragraphe précédant le lemme 2, et le groupe noté G_0 par Swinnerton-Dyer. Pour la commodité du lecteur, rappelons que la "norme absolue" n'est autre que le cardinal du corps résiduel. A toute place $v \in C$ on associe la classe de l'élément de Frobenius dans le groupe $\pi_1^{ab}(C)$, groupe de Galois de l'extension abélienne maximale non ramifiée de $k = \mathbf{F}(C)$. On obtient ainsi une application du groupe des diviseurs de C dans $\pi_1^{ab}(C)$, et cette application passe au quotient par l'équivalence rationnelle, définissant un homomorphisme $\mathrm{Pic}\,(C) \to \pi_1^{ab}(C)$. La flèche $\mathrm{Pic}\,(C) \to \mathbf{Z}$ définie par le degré sur $\mathbf{F}$ et la flèche $\pi_1^{ab}(C) \to \hat{\mathbf{Z}}$, où $\hat{\mathbf{Z}}$ est le groupe de Galois absolu de $\mathbf{F}$ et la seconde flèche est induite par la flèche structurale $C \to \mathrm{Spec}\,(\mathbf{F})$, sont compatibles. Le groupe $G_0 \subset \mathrm{Pic}\,(C)$ est le sous-groupe noyau de l'application composée $\mathrm{Pic}\,(C) \to \mathbf{Z} \to \mathbf{Z}/2$, une place $v \in C$ est dans le noyau de cette flèche si et seulement si le Frobenius F_v correspondant induit $1 \in \mathrm{Gal}\,(\mathbf{F}'/\mathbf{F})$.

Le début du lemme 2 est inchangé. On pourrait remplacer le quatrième paragraphe (commençant par "In this case I claim ...") par les considérations suivantes, aussi valables dans le cas des corps de nombres, dès que S contient les places au-dessus de 3, considérations qui n'utilisent pas l'hypothèse que le générateur σ de $\mathrm{Gal}\,(K/k)$ agit sans point fixe sur S. Soient $U = C \setminus S$ et $U' = U \times_{\mathbf{F}} \mathbf{F}'$. Le revêtement U'/U est galoisien, fini et étale de groupe G. La suite spectrale de Hochschild–Serre $E_2^{pq} = H^p(G, H^q_{\text{ét}}(U', \mu_3)) \Longrightarrow H^n_{\text{ét}}(U, \mu_3)$ a tous ses termes avec $p > 0$ nuls, puisqu'annulés par 2 (ordre de G) et 3 (exposant de μ_3). Ainsi $H^1_{\text{ét}}(U, \mu_3) = H^1_{\text{ét}}(U', \mu_3)^G$. L'hypothèse que $\mathrm{Pic}\,(U') = 0$ implique que le groupe $\mathrm{Pic}\,(U) = H^1_{\text{ét}}(U, \mathbf{G}_m)$ est annulé par 2, en particulier n'a pas de 3-torsion. Les suites de cohomologie déduites de la suite de Kummer

$$1 \to \mu_3 \to \mathbf{G}_m \overset{x \mapsto x^3}{\longrightarrow} \mathbf{G}_m \to 1$$

pour U, resp. U', donnent alors

$$\mathbf{F}[U]^*/\mathbf{F}[U]^{*3} = H^1_{\text{ét}}(U, \mu_3) \quad \text{et} \quad \mathbf{F}'[U]^*/\mathbf{F}'[U]^{*3} = H^1_{\text{ét}}(U', \mu_3),$$

isomorphismes compatibles aux flèches de restriction. On a donc

$$\mathbf{F}[U]^*/\mathbf{F}[U]^{*3} = \left(\mathbf{F}'[U]^*/\mathbf{F}'[U]^{*3}\right)^G.$$

Remarque. On pourrait donner une démonstration plus élémentaire en considérant la cohomologie de la suite exacte de G-modules

$$1 \to \mathbf{F}'[U]^* \to \mathbf{F}'(C)^* \to \mathrm{Div}\,(U') \to 0.$$

Un calcul un peu plus subtil, spécialement dans le cas des corps de nombres, est celui apparaissant au paragraphe suivant ("Denote the order ..."). Montrons que ce calcul, qui suppose que σ agit sans point fixe sur S, vaut aussi dans le cas des corps de fonctions. Soit $n = 2m$ l'ordre de S. On considère la suite exacte de G-modules

$$1 \to \mathbf{F}'[U]^*/\mathbf{F}'^* \to P \to \mathrm{Pic}\,(C') \to 0,$$

où $P = \oplus_{v \in S} \mathbf{Z}$ désigne le groupe des diviseurs de C' engendré par les places de S. Ce G-module est un module de permutation, en particulier $H^1(G, P) = 0$. C'est un groupe abélien libre de rang $n = 2m$, et l'hypothèse que σ agit sans point fixe sur S assure que le groupe P^G est libre de rang m. La suite exacte ci-dessus donne naissance à la suite exacte

$$1 \to \left(\mathbf{F}'[U]^*/\mathbf{F}'^*\right)^G \to P^G \to \left(\mathrm{Pic}\,(C')\right)^G \to H^1\left(G, \mathbf{F}'[U]^*/\mathbf{F}'^*\right) \to 0.$$

Comme on a $H^1(G, \mathbf{F}'^*) = 0$ (théorème 90 de Hilbert) et $H^2(G, \mathbf{F}'^*) = 0$ (trivialité du groupe de Brauer du corps fini $\mathbf{F}$), on peut réécrire cette suite exacte sous la forme

$$1 \to \mathbf{F}[U]^*/\mathbf{F}^* \to P^G \to \mathrm{Pic}\,(C) \to H^1\big(G, \mathbf{F}'[U]^*/\mathbf{F}'^*\big) \to 0.$$

En utilisant le fait que $H^1\big(G, \mathbf{F}'[U]^*/\mathbf{F}'^*\big)$ est annulé par 2, la considération de la multiplication par 3 sur cette suite et une chasse au diagramme facile donnent une suite exacte

$$0 \to {}_3\mathrm{Pic}\,(C) \to \mathbf{F}[U]^*/\mathbf{F}^*.\mathbf{F}[U]^{*3} \to P^G/3 \to \mathrm{Pic}\,(C)/3 \to 0.$$

Le groupe Pic (C) est une extension de $\mathbf{Z}$ par un groupe fini. De la suite ci-dessus on tire que la dimension du $\mathbf{Z}/3$-vectoriel $\mathbf{F}[U]^*/\mathbf{F}^*.\mathbf{F}[U]^{*3}$ est $m-1$. Comme $\mathbf{F}^*/\mathbf{F}^{*3}$ est d'ordre 1, on conclut que la dimension du $\mathbf{Z}/3$-vectoriel $\mathbf{F}[U]^*/\mathbf{F}[U]^{*3}$ est $m-1$.

Par ailleurs de la suite exacte

$$1 \to \mathbf{F}'[U]^*/\mathbf{F}'^* \to P \to \mathrm{Pic}\,(C') \to 0$$

on déduit que $\mathbf{F}'[U]^*/\mathbf{F}'^*$ est un groupe abélien libre de rang $2m-1$. Comme $\mathbf{F}'^*/\mathbf{F}'^{*3}$ est d'ordre 3, on voit que le $\mathbf{Z}/3$-vectoriel $\mathbf{F}'[U]^*/\mathbf{F}'[U]^{*3}$ a pour dimension $n = 2m$.

La situation est donc identique à celle décrite dans [39] (avec

$$X_S = \mathbf{F}'[U]^*/\mathbf{F}'[U]^{*3} \quad \text{et} \quad X_\Sigma = \mathbf{F}[U]^*/\mathbf{F}[U]^{*3}).$$

Il n'y a aucune modification à apporter à la fin du paragraphe 2 de [39].

Passons au paragraphe 3. L'énoncé à établir est le suivant.

Proposition 3.4. *Soit Σ un ensemble de places de k contenant toutes les places v telles que l'un de $v(a), v(b), v(c), v(d)$ ne soit pas divisible par 3. Soit, pour chaque place v dans Σ, un élément $\gamma_v \in k_v^*$ tel que le système*

$$ax^3 + by^3 = \gamma_v = cz^3 + dt^3$$

admette une solution dans k_v. Il existe alors $\gamma \in k^*$ tel que, pour toute place v de k, le système

$$ax^3 + by^3 = \gamma = cz^3 + dt^3$$

admette une solution dans k_v. On peut choisir γ tel que $\gamma/\gamma_v \in k_v^*$ soit un cube pour chaque place $v \in \Sigma$, et que le support du diviseur de γ soit formé *de places de Σ et de places inertes dans K. Si de plus σ agit sans point fixe sur Σ, on peut faire en sorte que la partie première à Σ du diviseur de γ soit de la forme $2p+q+r$ avec p, q, r places de k (i.e. points fermés de la courbe C) inertes dans K, i.e. telles que le corps résiduel ait un ordre congru à 2 modulo 3.*

Démonstration. Par la théorie du corps de classes géométrique, le groupe I_k/k^* des classes d'idèles de k s'identifie au sous-groupe du groupe $\mathrm{Gal}\,(k^{ab}/k)^0$ du groupe de Galois $\mathrm{Gal}\,(k^{ab}/k)$ formé des classes qui induisent une puissance entière de l'élément de Frobenius dans le groupe de Galois absolu de $\mathbf{F}$. A un sous-groupe ouvert H_L de $\mathrm{Gal}\,(k^{ab}/k)^0$ correspond une extension abélienne finie L de k.

Par approximation faible, on trouve $\gamma_1 \in k^*$ tel que $\gamma_1/\gamma_v \in k_v^{*3}$ pour chaque $v \in \Sigma$. Ecrivons le diviseur de γ_1 sur C comme la somme d'un diviseur Δ à support dans Σ et d'un diviseur Δ_1 étranger à Σ :

$$\mathrm{div}_C(\gamma_1) = \Delta + \Delta_1.$$

Considérons le sous-groupe ouvert $W = \prod_{v\in\Sigma} k_v^{*6} \times \prod_{v\notin\Sigma} O_v^*$ de I_k, où $O_v^* \subset k_v^*$ désigne le sous-groupe des unités. Le sous-groupe $k^*.W \subset I_k$ est ouvert d'indice fini, la théorie du corps de classes définit un isomorphisme entre $I_k/k^*.W$ et le groupe de Galois $\mathrm{Gal}\,(L/k)$ d'une extension finie abélienne L de k contenant K/k (ce dernier point est assuré par le fait que l'on a pris des puissances sixièmes dans la définition de W). Au diviseur Δ_1 est associée une classe bien définie dans $I_k/k^*.W$. Soit τ son image dans $\mathrm{Gal}\,(L/k)$. L'image de τ dans $\mathrm{Gal}\,(K/k) \simeq \mathbf{Z}/2$ est la classe, dans $\mathbf{Z}/2$, du degré du diviseur Δ_1 de C (par rapport au corps $\mathbf{F}$).

Supposons d'abord que τ induise l'élément non trivial de $\mathrm{Gal}\,(K/k)$. Le théorème de Tchebotarev assure l'existence d'une place $v_1 \notin \Sigma$ dont l'élément de Frobenius associé dans $\mathrm{Gal}\,(L/k)$ est τ. La place v_1 est donc inerte dans l'extension K/k. Il existe un élément $\delta \in k^*$ et un diviseur Δ_2 à support dans Σ tels que

$$\mathrm{div}_C(\delta) = \Delta_1 - v_1 + \Delta_2$$

et que de plus $\delta \in k_v^{*6}$ pour $v \in \Sigma$. Soit $\gamma = \gamma_1/\delta$. On a donc

$$\mathrm{div}_C(\gamma_1/\delta) = v_1 + \Delta_3$$

avec Δ_3 à support dans Σ, et $v_1 \notin \Sigma$, v_1 inerte dans K. Par ailleurs $\gamma/\gamma_v = (\gamma_1/\gamma_v).(1/\delta) \in k_v^{*3}$ pour chaque place $v \in \Sigma$.

Si τ induit l'élément trivial de Gal (K/k), on choisit, ce qui est loisible par le théorème de Tchebotarev, une place $v_0 \notin \Sigma$ dont l'élément de Frobenius associé dans Gal (L/k) induise l'élément non trivial de Gal (K/k). Au diviseur $\Delta_1 + v_0$ est associée une classe bien définie dans $I_k/k^*.W$, dont l'image dans Gal (L/k) induit l'élément non trivial de Gal (K/k). Utilisant le théorème de Tchebotarev comme ci-dessus, on trouve une place $v_1 \notin \Sigma$, inerte dans K/k et distincte de v_0, un élément $\delta \in k^*$ et un diviseur Δ_2 à support dans Σ tels que

$$\mathrm{div}_C(\delta) = \Delta_1 - v_1 - v_0 + \Delta_2$$

et que de plus $\delta \in k_v^{*6}$ pour $v \in \Sigma$. Soit $\gamma = \gamma_1/\delta$. On a alors

$$\mathrm{div}_C(\gamma_1/\delta) = v_1 + v_0 + \Delta_3$$

avec Δ_3 à support dans Σ, et $v_0, v_1 \notin \Sigma$, $v_0 \neq v_1$ inertes dans K. Par ailleurs, pour chaque place $v \in \Sigma$, on a $\gamma/\gamma_v = (\gamma_1/\gamma_v).(1/\delta) \in k_v^{*3}$.

Dans l'un comme dans l'autre cas, l'équation

$$ax^3 + by^3 = \gamma = cz^3 + dt^3$$

a des solutions dans tous les complétés k_v. C'est clair pour les places $v \notin \Sigma \cup \{v_1\} \cup \{v_0\}$, tous les coefficients étant alors des unités (à des cubes près). Pour $v \in \Sigma$, c'est aussi clair d'après l'hypothèse, puisqu'on a $\gamma/\gamma_v \in k_v^{*3}$. Les places v_0 et v_1 sont inertes dans K/k, le corps résiduel en une telle place est un corps fini qui ne contient pas de racine cubique non triviale de l'unité, donc l'élévation au cube dans ce corps résiduel est une bijection, et toute unité dans k_v^* est un cube. Pour une telle place v, chacune des deux courbes cubiques lisses

$$ax^3 + by^3 = \gamma.u^3, \quad cz^3 + dt^3 = \gamma.u^3$$

admet un point rationnel dans k_v avec $u = 0$, elle admet donc d'autres points avec $u \neq 0$, et le système

$$ax^3 + by^3 = \gamma = cz^3 + dt^3$$

admet une solution dans k_{v_0} et dans k_{v_1}.

La démonstration ci-dessus établit plus généralement le résultat suivant. Soient $v_1, \ldots, v_n$ des places de k non dans Σ, et soient $r_1, \ldots, r_n$ des entiers. Partant de γ_1 comme ci-dessus, on peut trouver $\gamma \in k^*$ tel que $\gamma/\gamma_v \in k_v^{*3}$

pour $v \in \Sigma$ et tel que la partie étrangère à Σ du diviseur de γ sur C soit de la forme $\sum_i r_i v_i + w_0 + \varepsilon w_1$ avec w_0 et w_1 inertes dans K, et $\varepsilon = 0$, resp. $\varepsilon = 1$, si le degré (sur $\mathbf{F}$) du diviseur $\Delta_1 + \sum_i r_i v_i$ est impair, resp. pair.

Le degré de $\mathrm{div}_C(\gamma_1) = \Delta + \Delta_1$ est nul. Si l'action de G sur S est sans point fixe, chaque place v dans Σ est décomposée dans K, et donc est de degré pair sur $\mathbf{F}$. Le degré de Δ est pair. Celui de Δ_1 est donc pair. Le dernier énoncé de la proposition est donc établi.

La proposition ci-dessus, et sa démonstration, peuvent paraître étonnants. On va montrer que ce résultat arithmétique est la conséquence d'un résultat purement algébrique.

Soit k un corps de caractéristique différente de 3. Le système d'équations

$$a_1x_1^3 + a_2x_2^3 = t = a_3x_3^3 + a_4x_4^3$$

définit une variété affine de dimension 3, lisse en dehors du point dont toutes les coordonnées sont nulles. Soit V le complémentaire de ce point, et soit $p : V \to \mathbf{A}^1_k$ la flèche définie par t. Soit X une k-compactification lisse de V, équipée d'un k-morphisme $p : X \to \mathbf{P}^1_k$ étendant p. On peut choisir X telle que, pour tout point M de $\mathbf{P}^1_k \setminus \{0, \infty\}$, la fibre X_M soit simplement le produit des deux cubiques projectives lisses $a_1x_1^3+a_2x_2^3 = tz^3$ et $a_3x_3^3+a_4x_4^3 = tw^3$, en particulier soit géométriquement connexe (et lisse). Au-dessus du point zéro, la fibre X_0 est un diviseur de X dont une union de composantes, soit Δ, est la fermeture de la surface d'équation affine

$$a_1x_1^3 + a_2x_2^3 = 0 = a_3x_3^3 + a_4x_4^3$$

(les x_i n'étant pas tous nuls).

Proposition 3.5. *Le groupe de Brauer vertical* $\mathrm{Br}_{\mathrm{vert}}(X) \subset \mathrm{Br}\,(X)$ *relativement à p est réduit à l'image de* $\mathrm{Br}\,(k)$.

(Note : en l'absence d'un modèle projectif et lisse X, on pourrait établir l'énoncé en remplaçant le groupe $\mathrm{Br}\,(X)$ par le groupe de Brauer non ramifié $\mathrm{Br}_{nr}\,(k(X)/k)$.)

Démonstration. Supposons d'abord que k contient une racine primitive cubique de 1, soit ζ, laquelle définit une identification $\mathbf{Z}/3 = \mu_3$. Les fibres du morphisme $p : X \to \mathbf{P}^1_k$ au-dessus de tout point de $\mathbf{P}^1_k \setminus \{0, \infty\}$ possèdent une composante géométriquement intègre (on peut en fait choisir X tel que ces fibres soient simplement le produit de deux cubiques lisses).

Soit $A \in \mathrm{Br}\left(k(t)\right)$, d'exposant premier à la caractéristique de k, et dont l'image réciproque par p est non ramifiée sur X. Ses résidus aux points génériques des composantes des fibres X_M sont triviaux. Pour $M \notin \{0, \infty\}$, ceci implique que le résidu de A en M est trivial.

Soit $K = k\left((a_1/a_2)^{1/3}, (a_3/a_4)^{1/3}\right)$. C'est une extension galoisienne de k de groupe de Galois G un quotient de $\mathbf{Z}/3 \times \mathbf{Z}/3$. Les composantes géométriques de Δ sont définies sur K. En $t = 0$, le résidu de A, qui appartient à $H^1(k, \mathbf{Q}/\mathbf{Z})$ doit s'annuler dans $H^1(K, \mathbf{Q}/\mathbf{Z})$, donc appartenir au groupe $H^1(G, \mathbf{Q}/\mathbf{Z})$, groupe annulé par 3. Ainsi ce résidu appartient au noyau de $H^1(k, \mathbf{Z}/3) \to H^1(K, \mathbf{Z}/3)$, i.e. au noyau de $k^*/k^{*3} \to K^*/K^{*3}$. Ce noyau est le groupe engendré par les classes a_1/a_2 et a_3/a_4. Soit donc

$$(a_1/a_2)^i.(a_3/a_4)^j \in k^*/k^{*3} = H^1(k, \mathbf{Z}/3)$$

le résidu de A en $t = 0$. Le résidu en $t = 0$ de l'algèbre $\left(t, (a_1/a_2)^i.(a_3/a_4)^j\right)_\zeta$ est précisément

$$(a_1/a_2)^i.(a_3/a_4)^j \in k^*/k^{*3} = H^1(k, \mathbf{Z}/3).$$

L'algèbre $\left(t, (a_1/a_2)^i.(a_3/a_4)^j\right)_\zeta$ n'a pas de résidu en dehors de $t = 0$ et $t = \infty$. Ainsi $A - \left(t, (a_1/a_2)^i.(a_3/a_4)^j\right)_\zeta \in \mathrm{Br}\left(k(t)\right)$ a ses résidus triviaux en tout point de $\mathbf{P}^1$ sauf peut-être en l'infini. Un résultat classique (Faddeev, cf. [15], §1.2) montre alors que $A - \left(t, (a_1/a_2)^i.(a_3/a_4)^j\right)_\zeta$ appartient à $\mathrm{Br}\,(k) \subset \mathrm{Br}\left(k(t)\right)$.

Je dis que l'image de $\left(t, (a_1/a_2)\right)_\zeta$ dans $\mathrm{Br}\left(k(X)\right)$ appartient à l'image de $\mathrm{Br}\,(k)$. En effet, dans le corps des fonctions de X, on a l'égalité $t = a_1x_1^3 + a_2x_2^3$. Donc

$$\begin{aligned}\left(t, (a_1/a_2)\right)_\zeta &= (a_1x_1^3 + a_2x_2^3, a_1/a_2)_\zeta \\ &= (a_2, a_1/a_2)_\zeta + \left((a_1/a_2)x_1^3 + x_2^3, a_1/a_2\right)_\zeta\end{aligned}$$

et le symbole $\left((a_1/a_2)x_1^3 + x_2^3, a_1/a_2\right)_\zeta$ est nul (tout symbole $(\alpha, 1-\alpha)_\zeta$ avec $\alpha \neq 0, 1$ est nul). Le même argument vaut pour $\left(t, (a_3/a_4)\right)_\zeta$. Ainsi l'image de A dans $\mathrm{Br}\left(k(X)\right)$ vient de $\mathrm{Br}\,(k)$.

Si k ne contient pas de racine primitive cubique de 1, on passe au corps $k(\zeta)$, extension quadratique de k. Un argument un tout petit peu astucieux de corestriction-restriction permet d'établir le résultat à partir de celui déjà établi.

Nous pouvons maintenant déduire de la proposition algébrique précédente la première partie de la proposition arithmétique qui la précède.

Soit Σ un ensemble de places de k contenant toutes les places v telles que l'un de $v(a)$, $v(b)$, $v(c)$, $v(d)$ ne soit pas divisible par 3. Soit pour chaque place $v \in \Sigma$, un élément $\gamma_v \in k_v^*$ tel que le système

$$ax^3 + by^3 = \gamma_v = cz^3 + dt^3$$

admette une solution dans k_v. Il existe alors $\gamma \in k^*$ tel que, pour toute place v de k, le système

$$ax^3 + by^3 = \gamma = cz^3 + dt^3$$

admette une solution dans k_v.

Démonstration. La fibration $X/\mathbf{P}^1_k$ a seulement deux fibres non lisses ($\lambda = 0$ et $\lambda = \infty$). La combinaison du théorème 0.2 (a) (Theorem 2.2.1 de [13]) et de la proposition 3.5 ci-dessus donne le résultat.

Remarque. La variété X est k-birationnelle au produit de $\mathbf{P}^1_k$ par la surface cubique lisse V d'équation $ax^3 + by^3 = cz^3 + dt^3$. Pour a, b, c, d généraux, on a $\mathrm{Br}\,(V)/\,\mathrm{Br}\,(k) = \mathbf{Z}/3$. On a donc $\mathrm{Br}\,(X)/\,\mathrm{Br}\,(k) = \mathbf{Z}/3$. Il existe donc un élément non trivial dans $\mathrm{Br}\,(X)$, mais cet élément n'est pas "vertical" par rapport à la fibration p. Il ne semble pas y avoir ici d'analogue de la condition (Alg) (condition (D) de [14]). On observera que l'analyse du §4 de [14] est spécifique aux fibrations $p : X \to \mathbf{P}^1_k$ dont la fibre est une courbe.

Remarque. Il est intéressant de noter que les mêmes arguments donnent :

Soient $p \geq 2$ un nombre premier et k un corps global de caractéristique différente de p. Soient $a, b, c, d \in k^*$. Si pour chaque place v il existe $\gamma_v \in k_v^*$ tel que l'équation

$$ax^p + by^p = \gamma_v = cz^p + dt^p$$

ait une solution dans k_v, alors il existe $\gamma \in k^*$ tel que l'équation

$$ax^p + by^p = \gamma = cz^p + dt^p$$

ait une solution dans chaque k_v.

Sous des hypothèses raisonnables sur les coefficients a, b, c, d, peut-on trouver γ tel que de plus la p-torsion du groupe de Tate–Shafarevich de la jacobienne de la courbe $ax^p + by^p = \gamma$ et celui de la courbe $cz^p + dt^p = \gamma$ soit trivial ? Ceci suffirait à assurer l'existence d'un zéro-cycle de degré 1 sur la surface $ax^p + by^p = cz^p + dt^p$.

Le paragraphe 4 de [39] (Lemme de Fisher), où l'hypothèse que k ne contient pas les racines de l'unité est cruciale, ne requiert ici aucune modification.

Au paragraphe 5 de [39], on trouve au lemme 7 une utilisation du théorème de densité de Tchebotarev. La seule modification à apporter dans notre contexte est qu'il convient de remplacer, dans l'énoncé de ce lemme, les "premiers de degré (absolu) 1" de K par les points fermés de C' qui sont de degré 1 par rapport à C, i.e. sont au-dessus de places de k (= points fermés de C) inertes dans l'extension K/k. Les points fermés de C' qui sont de degré 2 par rapport à C ont une densité de Dirichlet nulle. Le théorème 7 de [30] assure ici aussi l'existence de places de K du type requis. On pourrait d'ailleurs, dans [39] comme dans la situation analogue ici, raisonner directement avec l'extension non abélienne $F = K\big((a_1/a_2)^{1/3}, (a_3/a_4)^{1/3}, \xi^{1/3}\big)/k$, en appliquant le théorème 7 de [30] à la classe de conjugaison de $\mathrm{Gal}(F/k)$ formée des éléments non triviaux induisant l'identité sur $K\big((a_1/a_2)^{1/3}, (a_3/a_4)^{1/3}\big)$.

Immédiatement après la démonstration du lemme 7 de [39] on trouve une autre version du théorème de Tchebotarev (sous la forme du théorème de Dirichlet). Ici il convient d'une part de remplacer les "premiers de degré (absolu) 1" de K par les points fermés de C' qui sont de degré 1 par rapport à C, d'autre part de choisir un sous-groupe ouvert convenable du groupe des idèles de K comme il a été fait (pour k) dans la discussion du paragraphe 3 (proposition établie ci-dessus).

L'argument extrêmement élaboré qui suit l'énoncé de la Condition 5 vaut tout autant dans le cas fonctionnel. On y trouve une utilisation du théorème de Tchebotarev (Dirichlet) deux paragraphes avant l'énoncé du théorème 3. On peut ici utiliser la forme suivante du théorème de Tchebotarev, qui m'a été communiquée par Serre :

Théorème 3.6. *Soit K un corps global. On se donne un ensemble fini S de places, contenant les places archimédiennes s'il y en a. On se donne pour chaque $v \in S$ un sous-groupe ouvert U_v du groupe multiplicatif local K_v^* ; on suppose qu'il existe au moins une place $v \in S$ pour laquelle U_v est d'indice fini dans K_v^* (c'est automatique si v est une place archimédienne). Soit, pour tout v dans S, un élément x_v de K_v^*. Alors il existe un élément x de K^* et une place w non dans S tels que l'on ait :*

(i) *$x/x_v \in U_v$ pour $v \in S$;*

(ii) *x est une unité en toute place non dans S distincte de w ;*

(iii) *x est une uniformisante en w.*

Lors de la démonstration du théorème 3, à la fin du paragraphe 5, Swinnerton-Dyer doit faire une extension quadratique $k = k_0(\sqrt{\alpha})$. Ceci nous amène ici à nous limiter au cas où q est impair.

En retranscrivant les toutes dernières lignes avant le paragraphe 6, on a ici besoin de montrer : soit $C/\mathbf{F}$ une courbe projective, lisse, géométriquement connexe, soient $\mathbf{F}'/\mathbf{F}$ une extension quadratique et $C' = C \times_{\mathbf{F}} \mathbf{F}'$. Alors le groupe Pic (C') est engendré par des points fermés de degré 1 au-dessus de C. Pour l'établir, notons d'abord que le théorème de Tchebotarev garantit l'existence de points fermés de C décomposés dans l'extension C'/C. Soit $P \in C$ un tel point, et $Q_1, Q_2 \in C'$ les points de C' au-dessus de P. Le groupe Pic $\big(C' \setminus \{Q_1, Q_2\}\big)$ est un quotient fini de Pic (C') qui correspond à un revêtement abélien connexe C''/C'. Le théorème de Tchebotarev assure alors que ce groupe est engendré par les éléments de Frobenius associés à un nombre fini de points fermés de C' de degré 1 sur C. La réunion de ces points fermés et de $\{Q_1, Q_2\}$ engendre Pic (C').

Nous avons maintenant vérifié que la démonstration de Swinnerton-Dyer s'étend au cas où le corps k est un corps de fonctions d'une variable sur un corps fini $\mathbf{F}$, de caractéristique différente de 2 et 3, ne contenant pas de racine cubique non triviale de 1. Comme indiqué dans l'introduction, elle s'étend de façon inconditionnelle : nous allons en effet rappeler comment les résultats de Tate [42] impliquent la finitude des groupes de Tate–Shafarevich qui nous intéressent ici.

Théorème 3.7. *Soit $K = \mathbf{F}(C)$ le corps des fonctions d'une courbe C sur un corps fini $\mathbf{F}$, de caractéristique impaire première à 3. Soit $a \in K^*$. Le sous-groupe de torsion 3-primaire du groupe de Tate–Shafarevich de la K-courbe elliptique $x^3 + y^3 + at^3 = 0$ est un groupe fini.*

Démonstration. Soit E_K la courbe elliptique définie sur K par l'"équation $x^3+y^3+at^3 = 0$. Soit $L = K(a^{1/3})$. C'est une extension finie de K, de degré 1 ou 3. Le noyau de l'application de restriction $Ш(E_K) \to Ш(E_K \times_K L)$ est un sous-groupe de la 3-torsion de $Ш(E_K)$, et ce dernier groupe est fini. En effet, si $U \subset C$ est un ouvert de Zariski non vide au-dessus duquel la courbe E admet pour modèle un U-schéma abélien $\mathcal{E}/U$, la 3-torsion de $Ш(E_K)$ est un sous-quotient $H^1_{\text{ét}}(U, {}_3\mathcal{E})$, et ce groupe est clairement fini.

Pour établir le théorème, quitte à changer les notations, on peut supposer que a est un cube dans le corps de fonctions $K = \mathbf{F}(C)$, et donc que la courbe E_K n'est autre que la courbe “constante” $x^3+y^3+t^3 = 0$, vue comme courbe sur K. La courbe E_K admet alors pour modèle au-dessus de C le produit $X = E \times_{\mathbf{F}} C$, où l'on note $E/\mathbf{F}$ la courbe d'équation $x^3 + y^3 + t^3 = 0$ sur $\mathbf{F}$.

Suivant (dans un cas très particulier) des arguments d'Artin et Tate ([41], Theorem 3.1 ; la démonstration est donnée au paragraphe 4 de l'article [16] de Grothendieck), on voit que $Ш(E_K)$ s'identifie au groupe de Brauer $\mathrm{Br}\,(X)$ de $X = E \times_{\mathbf{F}} C$.

Soit l premier différent de la caractéristique de $\mathbf{F}$, par exemple $l = 3$. Etant donnée une $\mathbf{F}$-variété projective et lisse W, la torsion l-primaire de $\mathrm{Br}\,(W)$ est finie si et seulement si la conjecture de Tate $T_l^1(W)$ vaut pour les diviseurs sur W ([41], Thm. 5.2, où l'énoncé est limité au cas des surfaces, qui nous suffirait ici ; le résultat est général, voir [43], Prop. 4.3). Une des versions de cette conjecture dit que l'application naturelle

$$\mathrm{Pic}\,(W) \otimes_{\mathbf{Q}} \mathbf{Q}_l \to H^2_{\text{ét}}\big((W \times_{\mathbf{F}} \overline{\mathbf{F}}, \mathbf{Q}_l)(1)\big)^{\mathrm{Gal}\,(\overline{\mathbf{F}}/\mathbf{F})}$$

est un isomorphisme. La conjecture de Tate T_l^1 pour une courbe est triviale. Si l'on connaît la conjecture T_l^1 pour deux variétés W_1 et W_2, on la connaît pour leur produit. Cet énoncé difficile résulte, comme expliqué par Tate dans [40] et [42], de son théorème [42] selon lequel pour deux variétés abéliennes A et B sur un corps fini $\mathbf{F}$, la flèche

$$\mathbf{Q}_l \otimes \mathrm{Hom}_{\mathbf{F}}(A, B) \to \mathrm{Hom}_{\mathrm{Gal}\,(\overline{\mathbf{F}}/\mathbf{F})}\big(H^1(B), H^1(A)\big)$$

est un isomorphisme. Ainsi la conjecture de Tate T_l^1 vaut pour $X = E \times_{\mathbf{F}} C$, produit de deux courbes (les deux variétés abéliennes A et B sont la courbe elliptique E d'une part, la jacobienne de C d'autre part.) Par l'équivalence mentionnée plus haut, on conclut que la torsion l-primaire de $\mathrm{Br}\,(X)$ est finie.

Remarque. On peut en fait montrer que le groupe $\mathrm{Br}\,(X)$ tout entier est fini, voir [41], Thm. 5.2, pour la torsion première à la caractéristique, et les références (à un travail de Milne) après la proposition 4.3 de [43]. On peut ainsi voir que sous les hypothèses du théorème, la torsion première à la caractéristique de $\mathbf{F}$ du groupe de Tate–Shafarevich de la K-courbe elliptique

$$x^3 + y^3 + at^3 = 0$$

est un groupe fini. Ceci vaut plus généralement pour le groupe de Tate–Shafarevich de la jacobienne d'une $\mathbf{F}(C)$-courbe qui devient constante après extension finie de son corps de base.

Ceci achève d'établir le théorème 3.1 énoncé au début de ce chapitre. Le théorème 3.2 en résulte par une méthode de sections hyperplanes classique ([39], p. 911).

BIBLIOGRAPHIE

[1] C. L. Basile et T. A. Fisher, Diagonal cubic equations in four variables with prime coefficients, *Rational points on algebraic varieties* (E. Peyre, Yu. Tschinkel, éd.), Progress in Mathematics, vol. **199**, Birkhäuser (2001), pp. 1–12.

[2] A. O. Bender et Sir Peter Swinnerton-Dyer, Solubility of certain pencils of curves of genus 1, and of the intersection of two quadrics in $\mathbf{P}^4$, *Proc. London Math. Soc. (3)*, **83** (2001), 299–329.

[3] M. Borovoi et B. È. Kunyavskiĭ, Formulas for the unramified Brauer group of a principal homogeneous space of a linear algebraic group, *J. Algebra*, **225** (2000), 804–821.

[4] J.-L. Colliot-Thélène, The Hasse principle in a pencil of algebraic varieties, *Number theory (Tiruchirapalli, 1996)*, Contemp. Math. vol. **210**, Amer. Math. Soc. (Providence, RI, 1998), pp. 19–39.

[5] J.-L. Colliot-Thélène, Hasse principle for pencils of curves of genus one whose jacobians have a rational 2-division point, in : *Rational points on algebraic varieties* (E. Peyre, Yu. Tschinkel, éd.), Progress in Mathematics, vol. **199**, Birkhäuser, (2001), pp. 117–161.

[6] J.-L. Colliot-Thélène, D. Harari et A. N. Skorobogatov, Valeurs d'un polynôme à une variable représentées par une norme, prépublication (distribuée antérieurement sous le titre "Sur l'équation $P(t) = \mathrm{Norme}_{K/k}(\mathbf{z})$").

[7] J.-L. Colliot-Thélène et B. È. Kunyavskiĭ, Groupe de Brauer non ramifié des espaces principaux homogènes de groupes linéaires, *J. Ramanujan Math. Soc.*, **13** (1998), 37–49.

[8] J.-L. Colliot-Thélène et B. Poonen, Algebraic families of nonzero elements of Shafarevich-Tate groups, *J. Amer. Math. Soc.*, **13** (1) (1999), 83–99.

[9] J.-L. Colliot-Thélène et P. Salberger, Arithmetic on some singular cubic surfaces, *Proc. London Math. Soc. (3)*, **58** (1989), 519–549.

[10] J.-L. Colliot-Thélène et J.-J. Sansuc, La descente sur les variétés rationnelles II, *Duke Math. J.*, **54** (1987), 375–492.

[11] J.-L. Colliot-Thélène, J.-J. Sansuc et Sir Peter Swinnerton-Dyer, Intersections of two quadrics and Châtelet surfaces, I., *J. reine angew. Math.*, **373** (1987) 37–107; II., *J. reine angew. Math.*, **374** (1987), 72–168.

[12] J.-L. Colliot-Thélène et A. N. Skorobogatov, Descent on fibrations over $\mathbf{P}^1_k$ revisited, *Math. Proc. Camb. Phil. Soc.*, **128** (2000), 383–393.

[13] J.-L. Colliot-Thélène, A. N. Skorobogatov et Sir Peter Swinnerton-Dyer, Rational points and zero-cycles on fibred varieties : Schinzel's hypothesis and Salberger's device., *J. reine angew. Math.*, **495** (1998), 1–28.

[14] J.-L. Colliot-Thélène, A. N. Skorobogatov et Sir Peter Swinnerton-Dyer, Hasse principle for pencils of curves of genus one whose Jacobians have rational 2-division points, *Invent. math.*, **134** (1998), 579–650.

[15] J.-L. Colliot-Thélène et Sir Peter Swinnerton-Dyer, Hasse principle and weak approximation for pencils of Severi–Brauer and similar varieties, *J. reine angew. Math.*, **453** (1994), 49–112.

[16] A. Grothendieck, Le groupe de Brauer I, II, III. Exemples et compléments, *Dix exposés sur la cohomologie des schémas*, North-Holland (Amsterdam) ; Masson (Paris, 1968), pp. 67–188.

[17] D. Harari, Méthode des fibrations et obstruction de Manin, *Duke Math. J.*, **75** (1994), 221–260.

[18] D. Harari, Flèches de spécialisations en cohomologie étale et applications arithmétiques, *Bull. Soc. Math. France,* **125** (1997), 143–166.

[19] D. Harari, Weak approximation and fundamental groups, *Ann. Sci. Éc. Norm. Sup.*, **33** (2000), 467–484.

[20] D. Harari, Groupes algébriques et points rationnels, *Math. Ann.*, **322** (2002), 811–826.

[21] D. Harari et A. N. Skorobogatov, Non-abelian cohomology and rational points, *Compositio Math.*, **130** (2002), 241–273.

[22] D. Harari et A. N. Skorobogatov, The Brauer group of torsors and its arithmetic applications, Max Planck Institut Preprint **33** (2002).

[23] D. R. Heath-Brown, The solubility of diagonal cubic diophantine equations, *Proc. London Math. Soc. (3),* **79** (1999), 241–259.

[24] D. R. Heath-Brown et A. N. Skorobogatov, On certain equations involving norms, *Acta Math.*, **189** (2002), 161–177.

[25] S. Lang, Sur les séries L d'une variété algébrique, *Bull. Soc. Math. France,* **84** (1956), 385–407.

[26] Yu. I. Manin, Le groupe de Brauer–Grothendieck en géométrie diophantienne, *Actes du congrès Intern. Math. (Nice, 1970)*, Gauthier-Villars (Paris, 1971), vol. **1**, pp. 401–411.

[27] J. S. Milne, *Arithmetic Duality Theorems*, Perspectives in Mathematics, vol. **1**, Academic Press (Boston, 1986).

[28] B. Poonen, The Hasse principle for complete intersections in projective space, in : *Rational points on algebraic varieties* (E. Peyre, Yu. Tschinkel, éd.), Progress in Mathematics, vol. **199**, Birkhäuser (2001), pp. 307–311.

[29] J-P. Serre, *Groupes algébriques et corps de classes*, Actualités scientifiques et industrielles **1264**, Hermann (Paris, 1959).

[30] J-P. Serre, Zeta and L functions, *Arithmetical Algebraic Geometry (Proc. Conf. Purdue Univ., 1963)*, (O.F.G. Schilling, éd.), Harper and Row (New York, 1965), pp. 82–92.

[31] J-P. Serre, *Cohomologie galoisienne,* Cinquième édition, révisée et complétée, LNM 5, Springer (1994).

[32] A. N. Skorobogatov, Descent on fibrations over the projective line, *Amer. J. of Math.*, **118** (1996), 905–923.

[33] A. N. Skorobogatov, Beyond the Manin obstruction, *Invent. math.*, **135** (1999), 399–424.

[34] A. N. Skorobogatov, *Torsors and rational points,* Cambridge Tracts in Mathematics, vol. **144**, Cambridge University Press (2001).

[35] Sir Peter Swinnerton-Dyer, Rational points on certain intersections of two quadrics, *Abelian varieties* (W. Barth, K. Hulek and H. Lange, éd.), Walter de Gruyter (Berlin, 1995), pp. 273–292.

[36] Sir Peter Swinnerton-Dyer, Rational points on some pencils of conics with 6 singular fibres, *Ann. Fac. Sci. Toulouse Math. (6),* **8** (1999), 331–341.

[37] Sir Peter Swinnerton-Dyer, Arithmetic on diagonal quartic surfaces, II, *Proc. London Math. Soc. (3),* **80** (2000), 513–544. Corrigenda : *Proc. London Math. Soc. (3),* **85** (2002), 564.

[38] Sir Peter Swinnerton-Dyer, Weak approximation and R-equivalence on cubic surfaces, in : *Rational points on algebraic varieties* (E. Peyre, Yu. Tschinkel, éd.), Progress in Mathematics, vol. **199**, Birkhäuser (2001), pp. 359–406.

[39] Sir Peter Swinnerton-Dyer, The solubility of diagonal cubic surfaces, *Ann. Sci. Éc. Norm. Sup.*, **34** (2001), 891–912.

[40] J. Tate, Algebraic cycles and poles of zeta functions, in *Arithmetical Algebraic Geometry (Proc. Conf. Purdue Univ., 1963)* (O.F.G. Schilling, éd.), Harper and Row (New York, 1965), pp. 93–110.

[41] J. Tate, On the conjectures of Birch and Swinnerton-Dyer and a geometric analog, Séminaire Bourbaki, Exp. 306, réimpression : *Séminaire Bourbaki, Années 1964/65-1965/66, Exposés 277-312.* Soc. Math. France (Paris, 1995).

[42] J. Tate, Endomorphisms of abelian varieties over finite fields, *Invent. math.*, **2** (1966), 134–144.

[43] J. Tate, Conjectures on algebraic cycles in l-adic cohomology, in : *Motives* (U. Jannsen et al. éd.), Proc. Sympos. Pure Math., vol. **55.1**, Amer. Math. Soc. (Providence, RI, 1994), pp. 71–83.

Jean-Louis Colliot-Thélène

CNRS, UMR 8628, Mathématiques
Université de Paris-Sud
Bâtiment 425
F-91405 Orsay
France

`Jean-Louis.Colliot-Thelene@math.u-psud.fr`

BOLYAI SOCIETY
MATHEMATICAL STUDIES, 12

Higher Dimensional Varieties and Rational Points, pp. 223–282.

Potential Density of Rational Points on Algebraic Varieties

BRENDAN HASSETT

1. Introduction

Let K be a number field and consider a collection of equations

$$f_1(x_1,\dots,x_n) = f_2(x_1,\dots,x_n) = \dots = f_r(x_1,\dots,x_n) = 0$$

where the f_j are polynomials with coefficients in K. What are the solutions to these equations over K? Sometimes, every solution $(a_1,\dots,a_n) \in K^n$ necessarily satisfies further equations which are not algebraic consequences of our original collection. For instance, every solution over $\mathbb{Q}$ to $x_1^3 + x_2^3 = 1$ satisfies the additional equation $x_1x_2 = 0$. Of course, solutions over extensions L/K may fail to satisfy these further equations, e.g., the solution $\left(-1, \sqrt[3]{2}\right)$ to the Fermat equation.

This paper is dedicated to studying collections of equations with the following desideratum: All the equations satisfied by the solutions over K are algebraic consequences of the equations we start with. For instance, the solutions to $x_1^2 + x_2^2 = 1$ over $\mathbb{Q}$ satisfy no equations that are not multiples of the original equation.

Our approach is geometric. The desired property may be restated in the language of algebraic geometry: We seek classes of algebraic varieties whose K-rational points are dense in the Zariski topology. Here 'classes' of algebraic varieties are distinguished by invariants, like the geometric genus or plurigenera, or by geometric properties, like the existence of fibrations by rational or elliptic curves. One of the most profound questions of higher dimensional geometry is the logical relationship between the values of the invariants and the presence of fibrations.

However, there is a price to be paid for working geometrically: It is exceedingly rare for all the varieties with common geometric properties to have dense rational points. Even varieties which are geometrically $\mathbb{P}^n$ may or may not have rational points over a given number field, so we allow ourselves finite extensions of the given field. Then more uniform statements on rational points, depending only on the underlying geometry, are possible.

The paper is organized as follows. Basic definitions and notation are laid out in §2. General properties of density are discussed systematically in §3. The remainder of the paper is devoted to classes of examples. Abelian varieties and fibrations are addressed in §4. Fano varieties, especially those of small dimension, are discussed in §5. K3 surfaces, the simplest class of varieties where density of rational points remains a matter of controversy, are studied in §6. Several sections are devoted to the twisting method of Bogomolov and Tschinkel, as applied to elliptic K3 surfaces (Theorem 6.4.) Basic properties of the Tate–Shafarevich group are reviewed in §7. The simplest application of the method, to nonisotrivial elliptic K3 surfaces, is given in §8. This approach yields an infinite number of irreducible, nodal, rational curves in the K3 surface (see Corollaries 8.12 and 8.13). A more refined approach covering the isotrivial cases, and closer in spirit to the original paper [5] of Bogomolov and Tschinkel, can be found in §9. A third, independent approach, originating from a letter of J. Kollár, is given in §10. Finally, §11 contains the strongest results now known for general K3 surfaces, as well as some statements for higher dimensional varieties. The Appendix is a short resumé of Galois cohomology and principal homogeneous spaces.

Acknowledgments. The author is very grateful to F. Bogomolov and Y. Tschinkel for their generosity in explaining the ideas in their work, both through conversations and correspondence. He would also like to thank J. L. Colliot-Thélène for his encouragement and for sharing unpublished expository accounts of some of the results covered here. The argument of §10 is due to J. Kollár, who provided numerous other constructive suggestions. The author also benefitted from conversations with D. Abramovich and A. Kresch, who proposed Remark 6.5 and Proposition 8.2 respectively. The manuscript was greatly improved by insightful comments from T. Szamuely.

This paper is based on lectures given at the European Summer School *Higher Dimensional Varieties and Rational Points*, held at the Alfréd Rényi

Institute of Mathematics of the Hungarian Academy of Sciences. The author was partially supported by National Science Foundation Grant 0196187.

2. The Notion of Potential Density

We fix some notation. Throughout this paper, we work over a number field K. A variety X/K is a geometrically integral scheme of finite type over K. Its K-rational points are denoted $X(K)$.

Definition 2.1. Let X be a variety over K. The K-rational points of X are *dense* if $X(K)$ is not contained in any Zariski closed subset of X. Rational points of X are *potentially* dense if there exists a finite field extension L/K over which rational points are dense.

In the sequel, when we say that 'potential density holds for X' we mean that rational points are potentially dense on X.

Our motivation for considering *potential* density is to isolate properties of rational points that follow from the *geometric properties* of X, rather than those that depend on the ground field.

Example 2.2. Rational points of $\mathbb{P}^n$ are dense over $\mathbb{Q}$.

Example 2.3. Consider the curve

$$X = \{x^2 + y^2 = -z^2\} \subset \mathbb{P}^2_{\mathbb{Q}}.$$

This has no rational points whatsoever, but rational points are potentially dense. Indeed, over $\mathbb{Q}(i)$ we have an isomorphism

$$\begin{aligned} X &\xrightarrow{\simeq} \{x^2 + y^2 = z^2\} \\ (x, y, z) &\rightarrow (x, y, iz). \end{aligned}$$

However, the curve $\{x^2 + y^2 = z^2\}$ is isomorphic to $\mathbb{P}^1$

$$\begin{aligned} \mathbb{P}^1 &\xrightarrow{\simeq} \{x^2 + y^2 = z^2\} \\ (s, t) &\rightarrow (2st, s^2 - t^2, s^2 + t^2), \end{aligned}$$

which has dense rational points.

Example 2.4. The example above admits generalizations to higher dimensions. A *Brauer–Severi* variety X/K is a variety such that $X_{K^a} \simeq \mathbb{P}^N_{K^a}$, i.e., a variety geometrically isomorphic to $\mathbb{P}^n$. Of course, such an isomorphism may be realized over some finite extension L/K, and L-rational points of X are dense. The conic curve of Example 2.3 is the simplest example of a Brauer–Severi variety.

From an arithmetic standpoint, the precise nature of the field extensions L/K over which rational points become dense is an extremely interesting topic. Indeed, Brauer–Severi varieties first arose in the study of 'generic splitting fields' for central simple algebras [1] [33]. However, our primary interest lies in the interaction between density of rational points and geometric properties of algebraic varieties. Consequently, we will not keep track of the field extension over which points become dense. Indeed, in the sequel we will take finite extensions of the ground field without explicit comment, and even suppress the notation for the ground field.

3. Basic Properties of Potential Density

3.1. Behavior under morphisms

Proposition 3.1. *Let $g : X \dashrightarrow Y$ be a dominant rational map of projective varieties over a number field. Assume that rational points of X are potentially dense. Then rational points of Y are potentially dense.*

In particular, potential density is a birational property.

Proof. Choose a number field L over which X, Y, and g are defined, and $X(L)$ is Zariski dense. Let $U \subset X$ be an open subset over which g is a morphism. We have $g(U(L)) \subset Y(L)$, and the image of a dense set under a dominant rational map remains dense. ∎

Definition 3.2. A variety Y defined over a number field is *unirational* if, over some finite field extension, there exists a dominant rational map $g : \mathbb{P}^N \dashrightarrow Y$.

Combining Proposition 3.1 with Example 2.2, we obtain

Corollary 3.3. *Let Y be a variety over a number field and assume that Y is unirational. Then rational points of Y are potentially dense.*

Proposition 3.4 (Chevalley–Weil Theorem [43]). *Let $\phi : X \to Y$ be an étale morphism of proper varieties over a number field. Assume rational points of Y are potentially dense. Then rational points of X are also potentially dense.*

Sketch Proof. A good modern account can be found in [36]. Let L be a number field over which X, Y, and g are defined, and over which rational points of Y are dense. Choose a ring of integers $\mathcal{O} = \mathcal{O}_{L,S}$ (with S a finite set of places, including the infinite ones) over which we have the following:

1. X and Y have models $\mathcal{X}$ and $\mathcal{Y}$ over $\mathcal{O}$, i.e., $\mathcal{X} \to \operatorname{Spec}(\mathcal{O})$ and $\mathcal{Y} \to \operatorname{Spec}(\mathcal{O})$ are flat and projective, with generic fibers X_L and Y_L respectively;
2. g extends to an étale morphism $g : \mathcal{X} \to \mathcal{Y}$.

We choose S to exclude primes over which g has ramification or indeterminacy. Since $\mathcal{Y}$ is proper, the valuative criterion implies each L-rational point of $\mathcal{Y}$ extends to a $\mathcal{O}$-integral point. The $\mathcal{O}$-integral points of $\mathcal{Y}$ yield points of X defined over extensions of L of degree $\leq \deg(\phi)$, with discriminant contained in S. It is a classical theorem of Hermite that there are a finite number of such extensions $M_1, \ldots, M_r$. Over one of them rational points of X are Zariski dense. ■

Question 3.5. What classes of morphisms $X \to Y$ share the lifting property of Proposition 3.4?

One other class of such morphisms is étale projective bundles (cf. [19]).

3.2. Negative results

Any dense subset of a curve is infinite, so we have the following restatement of the Mordell Conjecture:

Proposition 3.6 (Faltings Theorem [11]). *Let X be a curve of genus ≥ 2 over a number field. Then rational points are not potentially dense.*

Example 3.7. Colliot-Thélène, Skorobogatov, and Swinnerton-Dyer [8] give an example of a variety for which rational points are not potentially dense, and the proof of nondensity requires Propositions 3.1, 3.4, and 3.6. They describe a smooth projective surface Y, which itself does not dominate any curve of genus ≥ 2 (or, indeed, any variety of general type), but admits an étale double cover X that does dominate a curve of genus two:

$$\begin{array}{rcl} & X & \rightarrow C \\ \text{étale double cover} & \downarrow & \\ & Y & \end{array}$$

If rational points were potentially dense on Y they would be potentially dense on X by Proposition 3.4, and thus potentially dense on C as well, contradicting Proposition 3.6.

Let X be smooth and projective with canonical bundle ω_X. We say X is of *general type* if ω_X is *big*, i.e.,

$$h^0(\omega_X^{\otimes n}) \sim Cn^{\dim(X)}, \quad C > 0,$$

for sufficiently large n. The following conjecture would preclude potential density of rational points for such varieties:

Conjecture 3.8 (Lang–Bombieri Conjecture). Let X be a projective variety of general type defined over a number field. Then rational points on X are not potentially dense.

This is known for subvarieties of abelian varieties which are of general type [12].

Remark 3.9. What is the largest class of projective varieties for which potential density might hold? In light of Conjecture 3.8 and Propositions 3.1 and 3.4, a variety with dense rational points should admit no étale covers that dominate varieties of general type. It is not known whether there are any further constraints. We refer the reader to [19] and [7] for further discussion of this question.

4. Abelian Fibrations

4.1. Abelian varieties

We start with some definitions:

Definition 4.1. Let $(A, 0)$ be an abelian variety defined over a field F of characteristic zero. A point $p \in A(F)$ is *nontorsion* if the set

$$\mathbb{Z}p = \{np \,:\, n \in \mathbb{Z}\}$$

is infinite and *nondegenerate* if this set is dense.

Proposition 4.2. *Let A be an abelian variety over a number field K. After passage to a finite extension L/K, $A(L)$ contains a nondegenerate point and L-rational points are dense.*

Sketch Proof. (cf. Proposition 3.1 of [18], where a second argument can be found.) The Mordell–Weil Theorem (see [36]) says that $A(K)$ is a finitely generated abelian group for any number field K; the *rank* of $A(K)$ is just the rank of this group. The main ingredient in the proof is a result of Jarden and Frey [13], Theorem 10.1: After passage to a finite extension L/K we have

$$\operatorname{rank} A(L) > \operatorname{rank} A(K).$$

The argument of [13] uses $\mathfrak{p}$-adic techniques: Consider primes $\mathfrak{p}$ totally ramified over a fixed prime of K. One creates a point, over a suitable extension of K, with prescribed reduction mod $\mathfrak{p}^n$. The condition on the reduction mod $\mathfrak{p}^n$ is used to show that this point is not in the span of $A(K)$.

This result proves the proposition when A is geometrically simple: For any nontorsion point $p \in A(L)$, the Zariski closure of $\mathbb{Z}p \subset A(L)$ contains a positive-dimensional abelian subvariety, and hence is equal to A. In general, the argument proceeds by induction on the number of simple components. We assume $A = A_1 \times A_2$ where A_2 is simple and $p_1 \in A_1(K), p_2 \in A_2(K)$ are nondegenerate. If $(p_1, p_2) \in A_1 \times A_2$ is degenerate then it is contained in a proper abelian subvariety $B \subset A_1 \times A_2$, which corresponds to a rational homomorphism $\beta \in \operatorname{Hom}(A_1, A_2) \otimes \mathbb{Q}$. It follows that $d\beta(p_1) = dp_2$ for some positive d. However, the group $\operatorname{Hom}(A_1, A_2)$ is finitely generated. Again applying the result of [13], we obtain a finite extension L/K and a point $q \in A_2(L)$ so that $q \notin \beta(p_1)$ for any $\beta \in \operatorname{Hom}(A_1, A_2) \otimes \mathbb{Q}$. ∎

4.2. Abelian fibrations over a field and principal homogeneous spaces

Throughout this section, F is a field of characteristic zero with algebraic closure F^a.

Definition 4.3. A variety X_F over F is an *abelian fibration* if it is geometrically an abelian variety, i.e., $X_{F^a} := X_F \times_F F^a$ is isomorphic to an abelian variety over F^a.

Remark 4.4. (Warning) We do *not* assume that X_F has a point over F. In particular, there may not be a group law on X_F defined over F (cf. Definition 4.12.)

There are a number of auxiliary abelian fibrations associated to X_F. Let $J^0(X_F)$ denote the *Albanese* of X_F, an abelian variety over F satisfying the following ([25] II §3):

1. The formation of the Albanese commutes with field extensions E/F, i.e., $J^0(X_E) = J^0(X_F) \times_F E$.
2. There is a morphism over F
$$s : X_F \times_F X_F \to J^0(X_F).$$
If E/F is an extension over which there is a point $x \in X_F(E)$ then $s \mid (X_F \times \{x\})$ induces an isomorphism $i_x : X_E \simeq J^0(X_E)$ so that $s(x_1, x_2) = i_x(x_1) - i_x(x_2)$.
3. Suppose we have another morphism to an abelian variety
$$s' : X_F \times_F X_F \to A,$$
such that $s'(x_1, x_2) = i(x_1) - i(x_2)$, for some $i : X_E \to A_E$ defined over an extension E/F. Then factors s' through s.

Of course, X_F always has a point over its function field $F(X)$ (the generic point x), and we may consider the composition

$$J^0(X_F) \times_F F(X) = J^0\big(X_{F(X)}\big) \xrightarrow{i_x^{-1}} X_{F(X)} = X_F \times_F F(X) \xrightarrow{\pi_1} X_F$$

as a rational map $J^0(X_F) \times_F X_F \dashrightarrow X_F$. Any such rational map extends uniquely to a morphism $a : J^0(X_F) \times_F X_F \to X_F$ [25] II §1. Over any field extension over which $X_E \neq \emptyset$ we can write $a(j, x) = j + x$. Thus X_F has the structure of a *principal homogeneous space* for $J^0(X_F)$, classified by a cocycle $[X_F] \in H^1(\Gamma, J^0(X_F))$ (see the Appendix).

For each $m \in \mathbb{Z}$, let $J^m(X_F)$ denote the principal homogeneous space corresponding to $m[X_F]$, which by descent is also an abelian fibration over F [26] §2, Prop. 4. Descent also gives the following morphisms over F:

1. the cycle-class morphism
$$\underbrace{X_F \times_F \cdots \times_F X_F}_{m \text{ times}} \to J^m(X_F)$$
$$(x_1, \ldots, x_m) \to x_1 + \cdots + x_m;$$

2. the addition morphism
$$J^n(X_F) \times_F J^m(X_F) \to J^{m+n}(X_F),$$
compatible with the cycle class morphism;
3. the multiplication by N morphism
$$\mu_N : J^m(X_F) \to J^{mN}(X_F).$$

Any zero-cycle on X_F of degree m, defined over F, yields an F-rational point of $J^m(X_F)$; each point of X_F defined over an extension E/F of degree m gives such a point. The existence of such a cycle implies $[X_F] = \left[J^1(X_F)\right] \in H^1(\Gamma, J^0(X_F))$ has order dividing m [26], §2, Prop. 5.

Definition 4.5. A *multisection* of an abelian fibration X_F is a point defined over a finite extension of F. The *degree* of the multisection is the smallest degree of an extension E/F over which the point can be defined. We shall identify multisections of degree d with collections of d conjugate points of $X(F^a)$
$$M = \{m_1, \ldots, m_d\}.$$

The zero-cycle $m_1 + \cdots + m_d$ is defined over F and thus yields an F-rational point of $J^d(X_F)$. If E/F is a degree-d extension over which some m_i is defined then we may consider the point
$$\tau_M = (m_1 + m_2 + \cdots + m_d) - dm_i \in J^0(X_F)(E).$$

This may be zero even when $d > 1$.

Definition 4.6. A multisection M of an abelian fibration X_F is *torsion* (resp. *degenerate*) if τ_M is torsion (resp. degenerate.) The *order* of a multisection is the smallest positive integer N such that $N(m_i - m_j) = 0$ for each $i, j = 1, \ldots, d$. The order is *infinite* when no such integer exists.

Proposition 4.7. *A multisection is torsion if and only if it has finite order.*

Proof. Let L/F be a Galois extension containing E, and let $\tau_{M,i} \in J^0(X_F)(L)$ be associated to the various conjugate points m_i, $i = 1, \ldots, d$. If one $\tau_{M,i}$ is torsion then all are, and we have

$$d(m_i - m_j) = \tau_{M,j} - \tau_{M,i}.$$

so each $m_i - m_j$ must also be torsion. Conversely, if each $m_i - m_j$ is torsion then

$$\tau_{M,i} = \sum_{j=1}^{d} (m_j - m_i)$$

is also torsion. ∎

Proposition 4.8. *Let $M \subset X_F$ be a torsion multisection of order N. Then M is contained in a principal homogeneous space for $J^0(X_F)[N]$, the N-torsion subgroup scheme.*

Proof. Suppose that the multisection M has order N, so that $Nm_i = Nm_j$ for each i, j. Then multiplication by N

$$\mu_N : X = J^1(X_F) \to J^N(X_F)$$

takes M to a single point $\Sigma \in J^N(X_F)$, defined over F. It follows that $M \subset \mu_N^{-1}(\Sigma)$, which is a principal homogeneous space for $J^0(X_F)[N]$. ∎

Remark 4.9. A torsion multisection M in an abelian variety A need not be contained in the torsion of A, only in a principal homogeneous space for the torsion!

Remark 4.10. Given a multisection $M' = \{m_1, \ldots, m_d\}$ of an abelian fibration X_F, it may happen that $m_i - m_j$, $i \neq j$, is torsion for certain pairs (i, j) but not all such pairs. Let N denote the smallest positive integer so that $N(m_i - m_j) = 0$ for each difference that is torsion. Then μ_N maps M' onto multisection $M \subset J^N(X_F)$, which is nontorsion if and only if M' is nontorsion. Furthermore, M' is contained in a $J^0(X_E)[N]$-principal homogeneous space over M, where E is the field of definition of M.

4.3. Potential density argument

Let B be a variety over a field of characteristic zero.

Definition 4.11. An *abelian fibration* $\eta : \overline{A} \to B$ is a projective morphism of varieties with generic fiber an abelian fibration over the function field of B. An abelian fibration of relative dimension one is an *elliptic fibration.*

A *multisection* is an irreducible closed subvariety $M \hookrightarrow \overline{A}$ so that the induced $M \to B$ is generically finite, i.e., the closure of a multisection of the generic fiber. A multisection of η is *nontorsion* or *nondegenerate* if the induced multisection of the generic fiber is.

Additional assumptions are necessary if one wants a good group law:

Definition 4.12. An abelian fibration $\eta : \overline{A} \to B$ is *Jacobian* if it satisfies the following conditions

1. η is a flat morphism;
2. the locus $A \subset \overline{A}$ where η is smooth admits a group law

$$A \times_B A \to A$$

with identity $0 : B \to A$.

For any abelian fibration admitting a rational section $0 : B \dashrightarrow \overline{A}$, there is a nonempty open subset $B' \subset B$ so that $\overline{A}_{B'} := \overline{A} \times_B B' \to B'$ is Jacobian, e.g., the open subset over which η is smooth. Given an arbitrary fibration $\overline{A} \to B$ with multisection M, after basechange to M there exists a rational section: the image of the diagonal in $M \times_B M \subset \overline{A} \times_B M$.

Proposition 4.13. *Let $\eta : \overline{A} \to B$ be an abelian fibration with multisection M, defined over a number field K. Assume that*

1. *M is nondegenerate;*
2. *K-rational points of M are dense.*

Then K-rational points of $\overline{A}$ are also dense.

In the case of an elliptic fibration, nondegeneracy and nontorsion are equivalent conditions.

Proof. After basechange to an open subset of M, we obtain a Jacobian abelian fibration $\overline{A}_M$ with a nondegenerate section τ_M. Thus $\mathbb{Z}\tau_M$ is dense in A_M and each multiple $n\tau_M$ has dense rational points. It follows that rational points in A_M are dense. Since A_M dominates $\overline{A}$, rational points in $\overline{A}$ are also dense. ■

5. Fano Varieties

In the remainder of this paper, we identify classes of algebraic varieties for which rational points are potentially dense.

Definition 5.1. A smooth projective variety X is Fano if ω_X^{-1} is ample.

Fano varieties admit no nontrivial étale covers [6] [10] and cannot dominate varieties of general type.

Here are the Fano varieties known to have potentially dense rational points:

Example 5.2 (Del Pezzo surfaces). Fano varieties of dimension two are called *Del Pezzo surfaces.* Classically, it was known that any Del Pezzo surface X is birational to $\mathbb{P}^2$, and thus has potentially dense rational points.

Example 5.3 (Cubic hypersurfaces). Nonsingular cubic hypersurfaces of dimension ≥ 2 are unirational and therefore have potentially dense rational points (Cor. 3.3).

Example 5.4 (Fano threefolds). Smooth Fano threefolds are known to be unirational, except in three cases [20] (see also [17]):

1. quartic hypersurfaces in $\mathbb{P}^4$;
2. weighted hypersurfaces of degree six in $\mathbb{P}(1,1,1,2,3)$;
3. double covers of $\mathbb{P}^3$ totally branched over a sextic.

The first two cases admit elliptic fibrations over $\mathbb{P}^2$. Arguments similar to those in the second half of this survey prove potential density (see [17] and [4]). The third case remains completely open.

Example 5.5 ($\mathbb{Q}$-Fano threefolds). There are many examples remaining where potential density has not yet been studied. For example, there are altogether 95 different families of $\mathbb{Q}$-Fano hypersurfaces in weighted projective spaces. See [9] for a concrete account of their geometry.

6. K3 Surfaces

Definition 6.1. A smooth projective surface X is a *K3 surface* if

1. $h^0(\Omega^1_X) = 0$;
2. $\omega_X \simeq \mathcal{O}_X$, i.e., the canonical bundle is trivial.

Such surfaces have been classified in great detail (see, for example, [2] VIII and [27]). In particular, the underlying complex manifolds $X(\mathbb{C})$ are all deformation equivalent and simply connected. K3 surfaces do not dominate varieties of general type, so they could very well have potentially dense rational points.

A *polarization* f of X is a primitive, ample class in the Néron–Severi group $\mathrm{NS}\,(X)$. The degree of (X, f) is the self-intersection $f.f$, a positive even integer. The K3 surfaces of degree d admit common realizations as projective varieties.

Example 6.2.
$d = 2$ a surface X admitting a degree-two cover $X \to \mathbb{P}^2$ branched over a smooth plane sextic curve.
$d = 4$ a nonsingular quartic hypersurface $X \subset \mathbb{P}^3$, e.g., $x_0^4 + x_1^4 = x_2^4 + x_3^4$.
$d = 6$ a complete intersection of quadric and cubic hypersurfaces in $\mathbb{P}^4$.

We summarize the known results for K3 surfaces:

Theorem 6.3 (Infinite automorphism group) [5]. *Let X be a K3 surface defined over a number field. Assume that the complex manifold $X(\mathbb{C})$ admits an infinite automorphism group. Then rational points on X are potentially dense.*

A proof is sketched in §6.1.

Theorem 6.4 (Elliptic K3 surfaces) [5]. *Let X be a K3 surface defined over a number field. Suppose that X admits an elliptic fibration $\eta : X \to \mathbb{P}^1$. Then rational points on X are potentially dense.*

More precise results on quartic surfaces containing a line (see Theorem 6.19) can be found in [17].

We give several approaches to Theorem 6.4; in each the key technical tool is the Tate–Shafarevich group, discussed in §7. The basic insight — that twisting an elliptic fibrations should make it easier to find rational points — is due to Bogomolov and Tschinkel. The first approach in §8 works primarily in the nonisotrivial case. It relies on the irreducibility of the p-torsion points for large p (see Theorem 8.3). One interesting by-product is the existence of rational multisections of unbounded degree (see Theorem 8.7 and its corollaries.) The second approach in §9 is perhaps the most natural, although logically it depends on the first. Essentially, one shows that torsion multisections have large genera so that rational multisections must be nontorsion. Our genus estimates for the p-torsion and the associated principal homogeneous spaces closely follow [5]. However, the presentation of the intermediate technical results and the analysis of the isotrivial cases differ to some extent. (Lemma 3.25 of [5] is not quite correct as stated: The Kummer surface associated to a product of general elliptic curves is a counterexample. This necessitates further *ad hoc* analysis in the isotrivial case.) The third approach in §10 is based on a letter of J. Kollár. Most of the detailed computation of the previous approaches is replaced by deformation-theoretic properties of rational and elliptic curves in K3 surfaces. The only classification results needed are the multiplicities of the components of degenerate fibers (see Figure 2 of §10).

Remark 6.5 (Conditional potential density for elliptic surfaces). Let $\iota : \overline{J} \to B$ be a nonisotrivial Jacobian elliptic fibration over a curve of genus zero or one. In particular, $\overline{J}$ is smooth and projective and ι admits a section (cf. §7.2). If $\overline{J}$ is defined over a number field we expect its rational points to be potentially dense. Indeed, Grant and Manduchi [14] [15] prove this conditionally, assuming a strong version of the Birch/Swinnerton-Dyer Conjecture formulated by Deligne and Gross. However, there are *isotrivial* elliptic fibrations $X \to \mathbb{P}^1$ for which rational points are known not to be potentially dense [8] (see Example 3.7).

Remark 6.6 (Enriques surfaces). Potential density results for K3 surfaces have application to other classes of surfaces dominated by them. By definition, an *Enriques surface* Y is a quotient of a K3 surface X by a fixed-point free involution. Propositions 3.1 and 3.4 imply that potential density of rational points for X and Y are equivalent. For results on Enriques surfaces and their K3 double covers, we refer the reader to [3].

6.1. K3 surfaces with automorphisms

We sketch the proof of Theorem 6.3, following [5].

Lemma 6.7. *Let X be a K3 surface. Then the automorphism group of X acts faithfully on the cohomology $H^2(X, \mathbb{Z})$. In particular, an automorphism $a : X \to X$ is uniquely determined by the homology class of its graph in $X \times X$.*

This follows from the strong version of the Torelli Theorem for K3 surfaces (see [27] for one account.)

Lemma 6.8. *Let X be a K3 surface defined over a number field K. There exists a finite extension L/K so that each automorphism of the complex manifold $X(\mathbb{C})$ is realized as an algebraic morphism defined over L.*

Proof. The automorphism group of X is finitely generated [42], so it suffices to show that any automorphism $a : X \to X$ can be defined over a finite extension of K. Choose a realization of $X \times X$ as a projective variety over K. The graph $\Gamma(a) \subset X \times X$ is a projective subvariety by the GAGA principle. Consider the connected component $\mathcal{H}$ of the Hilbert scheme of subschemes of $X \times X$ containing $\Gamma(a)$, which is defined over a finite extension L/K. The locus in $\mathcal{H}$ corresponding to graphs of automorphisms of X is clearly open, but $\Gamma(a)$ is the only graph of an automorphism of X in its homology class. It follows that $\mathcal{H} = [\Gamma(a)]$, so $\Gamma(a)$ and a are defined over L. ∎

Lemma 6.9. *Let X be a K3 surface with infinite automorphism group. Then there exists an indecomposible effective divisor class D so that the orbit of D is infinite.*

We refer the reader to §2 of [27] for a good description of the indecomposible elements in the monoid of effective divisors on a K3 surface.

Proof. The indecomposible effective divisors D with fixed (even) $d := D.D \geq -2$ divide into a finite number of orbits under the action of the automorphism group [42]. Suppose that there exists such a divisor with $d > 0$. This is always the case when the effective cone admit an irrational extremal ray; just take indecomposible divisors near this ray. The automorphisms fixing such a D admit a faithful representation in $D^{\perp} \subset \mathrm{NS}(X)$, a negative definite lattice by the Hodge index theorem. Thus each such stabilizer is finite.

Suppose X has no indecomposible effective divisors with $d > 0$. It follows that the effective cone of X is generated by divisors with $d = 0, -2$, permuted by the automorphism group. If the Néron–Severi group $\mathrm{NS}(X)$ had rank two, then the automorphism group would admit a subgroup of finite index fixing the generators of the effective cone, but all such automorphisms are trivial. If the Néron–Severi group has rank greater than two then we apply the 'alternative theorem' of Kovács [24]. Either X contains no (-2)-curves or the effective cone of X is generated by (-2)-curves. In the first case, the effective cone is 'circular' and admits an irrational extremal ray; hence the argument of the previous paragraph applies. In the second case, there must be an infinite number of (-2)-curves. If not, a finite index subgroup of the automorphism acts trivially on the effective cone, a contradiction. ■

Lemma 6.10. *Let X be a K3 surface and D an indecomposible effective divisor class. Then D contains a (possibly singular) rational curve.*

Proof. In the case where $D.D = -2$ this is clear. When $D.D = 0$, D is the class of an elliptic fibration, which admits degenerate fibers. In the case where $D.D > 0$. we use the results of [29]: A generic polarized K3 surface contains a singular rational curve in the polarization class f. However, since D is indecomposible with $D.D > 0$, (X, D) arises as the specialization of a polarized K3 surface. The rational curves of the polarized K3 surface specialize to rational curves on X in the class D. ■

We complete the proof of Theorem 6.3. Let X be a K3 surface with infinite automorphism group, defined over a number field K. Let D be a indecomposible effective divisor class with infinite orbit under the automorphism group. Then after passage to a finite extension L/K, we may assume that

1. the automorphisms of X are defined over L;
2. there is a rational curve R with class $[R] = D$, defined over L, with dense rational points $R(L) \subset R$.

The orbit $\cup_{a \in \mathrm{Aut}(X)} a(R)$ is dense in X and L-rational points are dense in the orbit, so L-rational points are dense in X. ■

Example 6.11 [41]. Consider K3 surfaces X, defined over a number field, which are realized as a complete intersection

$$X = \{F_{11} = 0\} \cap \{F_{22} = 0\} \subset \mathbb{P}^2 \times \mathbb{P}^2$$

of bihomogeneous forms of bidegree $(1, 1)$ and $(2, 2)$ respectively. The Néron–Severi lattice is

	f_1	f_2
f_1	2	4
f_2	4	2

where f_1 and f_2 are the pull-backs of the polarizations from the $\mathbb{P}^2$-factors. The projections $X \to \mathbb{P}^2$ are degree-two covers, and the corresponding pair of involutions generate an infinite group of automorphism on X. These K3 surfaces have potentially dense rational points.

More generally, we may consider K3 surfaces with Néron–Severi lattice

	f_1	f_2
f_1	2	n
f_2	n	2

with $n \geq 4$. These also admit pairs of noncommuting involutions.

6.2. Examples of elliptic K3 surfaces

Example 6.12. Harris and Tschinkel [17] consider the following special class of elliptic K3 surfaces. Let $X \subset \mathbb{P}^3$ be a quartic surface containing a line ℓ. Choose coordinates so that $\ell = \{(w, x, y, z) \;:\; w = x = 0\}$ and consider the map

$$\begin{aligned} \eta : \qquad & X \to \mathbb{P}^1 \\ & (w, x, y, z) \to (w, x). \end{aligned}$$

This is an elliptic fibration: Each hyperplane containing ℓ intersects X in the union of ℓ and a cubic plane curve

$$\{w + \alpha x = 0\} \cap X = \ell \cup E_\alpha \quad E_\alpha \quad \text{a plane cubic curve.}$$

We have $\eta^{-1}(\alpha) = E_\alpha$ and E_α is a smooth genus one curve for generic α, hence η is an elliptic fibration.

A particularly simple example is the Fermat surface $x_0^4 + x_1^4 = x_2^4 + x_3^4$, which contains the line $x_0 - x_2 = x_1 - x_3 = 0$.

Of course, there are numerous other examples of elliptic K3 surfaces. We give characterizations of elliptic K3 surfaces in Proposition 11.1.

6.3. Salient ramification and torsion multisections

In this section, we sketch some particularly elegant geometric methods for finding nontorsion multisections. We work in characteristic zero.

Definition 6.13 [3]. Consider an elliptic fibration $\eta : X \to \mathbb{P}^1$ and a multisection M. We say that M is *saliently ramified* if $M \to \mathbb{P}^1$ ramifies over a point in a smooth fiber of η.

Proposition 6.14. *Let $\eta : X \to \mathbb{P}^1$ be an elliptic fibration and M a saliently ramified multisection. Then M is nontorsion.*

Proof. Consider the fibration

$$X \times_{\mathbb{P}^1} M \to M$$

obtained after base change. This has a section (see §4.3). If the map $M \to \mathbb{P}^1$ ramifies at m then the map $M \times_{\mathbb{P}^1} M \to M$ ramifies at (m, m). Indeed, for a finite flat morphism, ramification occurs precisely where the cardinality of fibers drops; this cardinality is unchanged under base extension. (Also, being ramified is a local property in the faithfully flat topology [16] 17.7.4.) Observe that we do *not* normalize $M \times_{\mathbb{P}^1} M$, as this would destroy some of the ramification.

However, the torsion of any group scheme in characteristic zero is étale over the base, and we have a contradiction. ■

Combining this with Proposition 4.13, we obtain:

Corollary 6.15. *Let $\eta : X \to \mathbb{P}^1$ be an elliptic fibration, defined over a number field. Assume that η admits a saliently ramified multisection of genus zero or one. Then rational points of X are potentially dense.*

Remark 6.16. This approach has advantages and disadvantages. The main disadvantage is the difficulty in producing saliently ramified multisections for large classes of varieties: In general, there is no easy way to produce them out of 'thin air'.

The main advantage is that, once we are given a multisection, it is relatively easy to check whether it is saliently ramified. Furthermore, if X, η, and M are defined over K and $M(K)$ is dense, then $X(K)$ is also dense. This gives a good tool for checking density over a *given* field.

6.4. Reprieve: Quartic surfaces containing a line

We return to Example 6.12, again following [17]. Assume that the quartic surface X and the line ℓ are defined over a number field K. The line ℓ itself intersects fibers of η in three points and defines a degree-three multisection. This multisection is saliently ramified provided the following geometric condition is satisfied:

> **Tangency condition** Some *smooth* fiber of η intersects the line ℓ tangentially.

Rational points of $\ell \simeq \mathbb{P}^1$ are clearly dense over K, so the Tangency condition suffices to guarantee density of $X(K)$.

The geometric condition is satisfied for the *generic* quartic surface containing a line — but not every such surface. It fails for the Fermat surface $x_0^4 + x_1^4 = x_2^4 + x_3^4$ with $\ell = \{x_0 - x_2 = x_1 - x_3 = 0\}$. The map $\eta : X \to \mathbb{P}^1$ is given by the rational function

$$\frac{x_0 - x_2}{x_3 - x_1} = \frac{x_1^3 + x_1^2 x_3 + x_1 x_3^2 + x_3^3}{x_0^3 + x_0^2 x_2 + x_0 x_2^2 + x_2^3}.$$

Taking x_0 and x_1 as coordinates on ℓ, the rational function restricts to x_1^3/x_0^3, which ramifies (to order three) at the points $(1,0,1,0)$ and $(0,1,0,1)$. Here the fibers of η are three coincident lines.

Combining the analysis of the salient ramification with a case-by-case study of the possible torsion multisections, Harris and Tschinkel obtain the following:

Theorem 6.17 ([17] Theorem 4.1). *Let X be a smooth quartic surface containing a line ℓ and $\eta : X \to \mathbb{P}^1$ the elliptic fibration obtained by projecting from ℓ. Assume there do not exist six lines contained in X and meeting ℓ. Then ℓ is a nontorsion multisection of ℓ.*

Corollary 6.18 ([17] Theorem 1.5.a). *Retain the assumptions above, and assume that X and ℓ are defined over a number field K. Then the rational points $X(K)$ are dense.*

After further analysis of degenerate cases, Harris and Tschinkel prove the following general result, which is a special case of Theorem 6.4:

Theorem 6.19. *Let X be a smooth quartic surface defined over a number field. If X contains a line ℓ then rational points of X are potentially dense.*

7. Twisting Elliptic Fibrations

7.1. The Tate–Shafarevich group

We refer the reader to [21], [37], and [31] for full details.

Let $F = \mathbb{C}(\mathbb{P}^1)$ be the function field of $\mathbb{P}^1$, with absolute Galois group Γ. Let $(J_F, 0)$ be an elliptic curve over F. The *Tate-Shafarevich group* $\mathrm{Sh}(J_F)$ is defined as the J_F-principal homogeneous spaces which are locally trivial at each place, i.e., the kernel

$$H^1(\Gamma, J_F) \to \prod_{b \in \mathbb{P}^1} H^1(\hat{\Gamma}_b, J_{\hat{F}_b})$$

where $\hat{F}_b$ is the completion/henselization of F at b and $\hat{\Gamma}_b$ its absolute Galois group.

We recall some properties of $\mathrm{Sh}(J_F)$.

Proposition 7.1. 1) *Let X_F be a J_F-principal homogeneous space. Then $[X_F] \in H^1(\Gamma, J_F)$ has order m if and only if there is a point of X_F defined over a field extension of degree m.*
2) *There are exact sequences*

$$\begin{array}{ccccccc} 0 \to & J_{\hat{F}_b}(\hat{F}_b)/mJ_{\hat{F}_b}(\hat{F}_b) & \to & H^1(\hat{\Gamma}_b, J_{\hat{F}_b}[m]) & \to & H^1(\hat{\Gamma}_b, J_{\hat{F}_b})[m] & \to 0 \\ & \downarrow & & \downarrow & & \downarrow & \\ 0 \to & J_F(F)/mJ_F(F) & \to & H^1(\Gamma, J_F[m]) & \to & H^1(\Gamma, J_F)[m] & \to 0. \end{array}$$

3) *$\mathrm{Sh}(J_F)$ is infinitely divisible.*

Proof. We have already seen that $J^m(X_F)(F) \neq \emptyset$ whenever X_F has a point over a degree m extension of F. Conversely, suppose we have an F-rational point of $J^m(X_F)$. Such a point need not come from a line bundle L on X_F defined over F; the obstruction lies in the Brauer group of F (by the Hochschild–Serre spectral sequence

$$0 \to H^1(\mathcal{O}^*_{X_F}) \to H^0(\Gamma, H^1(\mathcal{O}^*_{X_{F^a}})) \to H^2(\Gamma, H^0(\mathcal{O}^*_{X_{F^a}})) = H^2(\Gamma, F^{a*}).)$$

The Brauer group of $F = \mathbb{C}(\mathbb{P}^1)$ is trivial because it is a C^1 field. The Riemann–Roch formula applied to L gives a degree-m cycle on X_F.

The second assertion is quite standard (see §1, 2.1 of [37]). The third statement may be found in [37] §2.5. It uses the fact that there are no nontrivial families of elliptic curves over $\mathbb{P}^1$ without degenerate fibers. ■

In light of the exact seqences above, the reduction map

$$H^1\big(\Gamma, J_F[m]\big) \to H^1\big(\hat{\Gamma}_b, J_{\hat{F}_b}[m]\big)$$

can often shed light on Tate–Shafarevich group:

Proposition 7.2 (Corollary of [37], §1.2.). *Let* $\hat{F} = \mathbb{C}\big((t)\big)$ *with absolute Galois group* $\hat{\Gamma}$, *and* $J_{\hat{F}}$ *an elliptic curve over* $\hat{F}$. *Then for each* m

$$\big|H^1\big(\hat{\Gamma}, J_{\hat{F}}[m]\big)\big| = \big|J_{\hat{F}}[m]\big|.$$

7.2. Models of elliptic surfaces

Recall that an elliptic fibration $\eta : X \to \mathbb{P}^1$ is *relatively minimal* if its fibers contain no (-1)-curves, i.e., K_X is numerically effective relative to η. Given an elliptic curve X_F over $F = \mathbb{C}(\mathbb{P}^1)$, there is a unique (smooth, projective) relatively minimal model $\eta : X \to \mathbb{P}^1$. In particular, for each elliptic fibration $\eta : X \to \mathbb{P}^1$ there is an associated Jacobian fibration $\iota : \overline{J} \to \mathbb{P}^1$, which admits a zero-section. It is obtained by taking the relative minimal model of the Jacobian $J_F := J^0(X_F)$. We also have $J^m(X)$, associated to $J^m(X_F)$. Multiplication by N,

$$\mu_N : J^m(X_F) \to J^{mN}(X_F),$$

induces a dominant rational map over $\mathbb{P}^1$

$$\mu_N : J^m(X) \dashrightarrow J^{mN}(X).$$

Definition 7.3. Let $\eta : X \to \mathbb{P}^1$ be a relatively minimal elliptic fibration. A fiber $\eta^{-1}(b)$, $b \in \mathbb{P}^1$, is *multiple* if each of its irreducible components has multiplicity > 1.

A Jacobian fibration always has a zero section, and thus has no multiple fibers. A fiber may have some nonreduced irreducible components without being multiple.

Proposition 7.4. *Let* $X \to \mathbb{P}^1$ *be a relatively minimal elliptic fibration. Then* X *has no multiple fibers if and only if* X_F *is an element of* $\mathrm{Sh}\,\big(J(X_F)\big)$.

Proof. The fiber $\eta^{-1}(b)$ is not multiple if and only if η has a section in an analytic/étale neighborhood of b. This is equivalent to $X_{\hat{F}_b} \neq \emptyset$. ■

Proposition 7.5. *Let $\eta : X \to \mathbb{P}^1$ be an elliptic surface without multiple fibers, with Jacobian J_F. Then $m[X_F] = 0$ in $\mathrm{Sh}(J_F)$ if and only if there is a multisection $M \subset X$ of relative degree m over $\mathbb{P}^1$.*

Proof. The first part of Proposition 7.1 gives this; M is obtained by taking the closure of the cycle in X_F of relative degree m. ■

Proposition 7.6. *Let $\eta : X \to \mathbb{P}^1$ be a relatively minimal elliptic fibration without multiple fibers, with Jacobian fibration $\iota : \overline{J} \to \mathbb{P}^1$. For each $b \in \mathbb{P}^1$, X and $\overline{J}$ are isomorphic over an analytic/étale neighborhood of b.*

Proof. Since X and $\overline{J}$ are relatively minimal, they remain relatively minimal after completion/henselization, and minimal models of surfaces (and Néron models of elliptic curves) are unique. ■

We remarked in Proposition 7.1 that $\mathrm{Sh}(J_F)$ is infinitely divisible for an elliptic curve J_F over $F = \mathbb{C}(\mathbb{P}^1)$. The precise structure of this group admits an elegant interpretation in terms of the transcendental cohomology of the compact Kähler manifold $\overline{J}$:

Proposition 7.7 ([38] §VII.8, Theorems 11 and 12, [39]). *Let $\overline{J} \to \mathbb{P}^1$ be a nontrivial Jacobian elliptic surface with generic fiber J_F. Then*

$$\mathrm{Sh}(J_F) = H^2(\overline{J}, \mathbb{Z})_{\mathrm{tran}} \otimes \mathbb{Q}/\mathbb{Z},$$

where

$$H^2(\overline{J}, \mathbb{Z})_{\mathrm{tran}} = H^2(\overline{J}, \mathbb{Z}) / \left(H^2(\overline{J}, \mathbb{Z}) \cap H^1(\overline{J}, \Omega^1_{\overline{J}}) \right) = H^2(\overline{J}, \mathbb{Z}) / \mathrm{NS}(\overline{J}),$$

the integral classes modulo the Néron–Severi group.

Remark 7.8. For elliptic K3 surfaces, the formula

$$\mathrm{Sh}(J_F) \simeq (\mathbb{Q}/\mathbb{Z})^e, \quad e = \mathrm{rank}\, H^2(\overline{J}, \mathbb{Z})_{\mathrm{tran}}$$

may also be deduced from the Ogg-Shafarevich formula ([37] §2.3, Theorem 3, [31] Theorem 2) and the formula for the rank of the transcendental classes quoted in §7.4[40]. This allows us to deduce e from the rank of J_F and the degenerate fibers of $\overline{J} \to \mathbb{P}^1$.

7.3. Twisting elliptic K3 surfaces

Proposition 7.9. *An elliptic K3 surface $\eta : X \to \mathbb{P}^1$ is relatively minimal and never has multiple fibers.*

Proof. The relative minimality follows because $K_X = 0$. We may then use the classification of singular fibers for minimal elliptic surfaces [2] pp. 151. The possible multiple fibers take the form mD, where D is one of the degenerate fibers enumerated in Kodaira's classification. Now $D \subset X$ is effective and numerically effective, with self-intersection $D.D = 0$. The classification theory of linear series on K3 surfaces [34] implies $|D|$ induces an elliptic fibration (cf. Proposition 11.1). ∎

Proposition 7.10. *Let X be a smooth projective variety, $\eta : X \to \mathbb{P}^1$ a relatively minimal elliptic fibration, and $\iota : \overline{J} \to \mathbb{P}^1$ its Jacobian fibration. If X is a K3 surface then $\overline{J}$ is a K3 surface. If $\overline{J}$ is a K3 surface and X has no multiple fibers then X is a K3 surface.*

Proof. Proposition 7.9 and our assumptions imply that X and $\overline{J}$ are both relatively minimal elliptic surfaces without multiple fibers. In particular, they have isomorphic fibers (cf. Prop. 7.6.) The topological Euler characteristics $\chi(X)$ and $\chi(\overline{J})$ are therefore equal. Both $\overline{J}$ and X are elliptic surfaces, so $K_{\overline{J}}^2 = K_X^2 = 0$ and Noether's formula implies $\chi(\mathcal{O}_X) = \chi(\mathcal{O}_{\overline{J}}) = 2$. The canonical bundle formula for elliptic surfaces [2] V.12.1,12.2 implies $K_X = 0$ if and only if $K_{\overline{J}} = 0$. Any surface with $K = 0$ and $\chi(\mathcal{O}) = 2$ is a K3 surface. ∎

Remark 7.11. Let B be a smooth projective curve, $X \to B$ an abelian fibration, and $\overline{J} \to B$ the Néron model of its Albanese. Then $h^i(X, \omega_X) = h^i(\overline{J}, \omega_{\overline{J}})$ for each i (by [23], Theorem 2.6 and Corollary 3.2).

One distinguishing property of elliptic K3 surfaces is that they admit *rational* multisections.

Proposition 7.12. *Let $X \to \mathbb{P}^1$ be an elliptic K3 surface with Jacobian J_F, so that $[X_F] \in \mathrm{Sh}(J_F)$ has order m. Then there exists a rational multisection $M \subset X$ of relative degree m.*

Proof. Proposition 7.5 gives a multisection $\hat{M}$ of relative degree m. There exists an indecomposible effective divisor class D so that $\hat{M} - D$ is effective and D has positive degree over the base. It must have degree exactly m because the order of $[X_F]$ equals m. Lemma 6.10 implies D contains an irreducible rational curve M. ∎

Combining Propositions 7.10, 7.5, and 7.7 with the multiplication map introduced in §7.2, we obtain

Corollary 7.13. *Let $X \to \mathbb{P}^1$ be an elliptic K3 surface, with Jacobian $\overline{J}$ and generic fiber J_F. For each integer N, there exists an elliptic K3 surface $X' \to \mathbb{P}^1$ with Jacobian $\overline{J}$ so that*

$$N[X'_F] = [X_F] \quad \text{and} \quad \text{order}\,[X'_F] = N \cdot \text{order}\,[X_F]$$

in $\text{Sh}\,(J_F)$. There is a dominant rational map $X' \dashrightarrow X$ over $\mathbb{P}^1$.

7.4. Results of Shioda–Inose

To compute transcendental cohomology and the Tate–Shafarevich group in particular examples, we use the following (see [40]):

Proposition 7.14. *Let $\eta : X \to \mathbb{P}^1$ be a elliptic K3 surface with Jacobian $\iota : \overline{J} \to \mathbb{P}^1$. Then the Néron Severi group has rank*

$$\text{rank}\,\text{NS}\,(X) = 2 + r(J_F) + \sum_b (m_b - 1)$$

and the topological Euler characteristic is

$$\chi(X) = 24 = \sum_b \varepsilon_b,$$

where $r(J_F)$ is the rank of the group of sections and m_b and ε_b are given by the following table:

Kodaira type	$I_a, a > 0$	II	III	IV	$I_a^*,\ a \geq 0$	II^*	III^*	IV^*
m_b	a	1	2	3	$a+5$	9	8	7
ε_b	a	2	3	4	$a+6$	10	9	8

Subtracting the two formulas of Proposition 7.14 we obtain

Corollary 7.15. *Let $\eta : X \to \mathbb{P}^1$ be an elliptic K3 surface. Let N_1 denote the number of degenerate fibers of type I_a, $a > 0$, and N_2 the number of degenerate fibers of other types. Then we have*

$$\text{rank}\,\text{NS}\,(X) + N_1 + 2N_2 = 26 + r(J_F).$$

7.5. A remark on Fourier–Mukai transforms

Perhaps the simplest way to construct isogenies of K3 surfaces is by taking twists of elliptic fibrations. This is not widely known, so we give a brief account here.

By definition, two K3 surfaces X and Y are *isogenous* if the Hodge structures $H^2(X)$ and $H^2(Y)$ are isomorphic over $\mathbb{Q}$. A natural problem, first studied systematically by Mukai [30], is how such isogenies are induced by correspondences between X and Y. Indeed the Hodge conjecture, applied to $X \times Y$, predicts that each isogeny between X and Y should be induced by an algebraic cycle, which ideally should admit a geometric description. Typically, Y arises as a moduli space of simple sheaves on X, with the isogeny induced by Chern classes of the universal bundle on $X \times Y$.

Assume that $\iota : \overline{J} \to \mathbb{P}^1$ is a Jacobian elliptic K3 suface and $\eta : X \to \mathbb{P}^1$ represents an element $[X_F] \in \mathrm{Sh}\,(J_F)$ of order $m > 1$. We interpret $\overline{J}$ as $J^m(X)$, the degree-m component of the Picard group relative to η. The generic point of $\overline{J}$ parametrizes a line bundle $\mathcal{L}$ of degree m supported on some fiber of η. We extract a simple sheaf $\mathcal{E}$ from the kernel of the global section map

$$0 \to \mathcal{E} \to H^0(X, \mathcal{L}) \otimes \mathcal{O}_X \to \mathcal{L} \to 0,$$

with rank m and Chern classes $c_1(\mathcal{E}) = -D$ and $c_2(\mathcal{E}) = m$, where D is the class of a fiber of η. The moduli space of such sheaves is the K3 surface $\overline{J}$.

We recall Mukai's procedure for computing the isogeny between X and $\overline{J}$. Consider

$$\Lambda := \left\{ a \in H^2(X, \mathbb{Z}) : a.D \equiv 0 \pmod{m} \right\} \subset H^2(X, \mathbb{Z}),$$

a sublattice of index m. We assumed that $[X_F] \in \mathrm{Sh}\,(J_F)$ has order m, so the algebraic classes

$$H^2(X, \mathbb{Z})_{\mathrm{alg}} := H^2(X, \mathbb{Z}) \cap H^1(X, \Omega^1_X) \subset \Lambda.$$

Set $D' = D/m$ and take Λ' to be the lattice obtained from Λ by adjoining D', so that Λ'/Λ is cyclic of order m. Observe also that

1. The intersection form on $H^2(X, \mathbb{Z})$ induces quadratic forms on Λ and Λ'; the form on Λ' is integral and unimodular.
2. $\Lambda \otimes \mathbb{C}$ and $\Lambda' \otimes \mathbb{C}$ inherit Hodge structures from $H^2(X)$, so we may define Λ_{alg} (resp. Λ'_{alg}) as the integral $(1,1)$-classes and $\Lambda_{\mathrm{tran}} = \Lambda/\Lambda_{\mathrm{alg}}$ (resp. $\Lambda'_{\mathrm{tran}} = \Lambda'/\Lambda'_{\mathrm{alg}}$).

3. $D' \in \Lambda'_{\text{alg}}$ with $D' \cdot D' = 0$, and there exists a class $\Sigma \in \Lambda'_{\text{alg}}$ with $D' \cdot \Sigma = 1$ (see Proposition 7.5).
4. $\Lambda_{\text{tran}} = \Lambda'_{\text{tran}}$ and $\Lambda_{\text{tran}} \subset H^2(X, \mathbb{Z})_{\text{tran}}$ so that the quotient is cyclic of order m.

Indeed, Mukai proves that Λ' is the Hodge structure for the moduli space of sheaves of type $\mathcal{E}$, which coincides with the Jacobian elliptic fibration $\overline{J}$. Finally, $H^2(X, \mathbb{Z})_{\text{tran}}/H^2(\overline{J}, \mathbb{Z})_{\text{tran}}$ is cyclic of order m and thus determines an element of $H^2(\overline{J}, \mathbb{Z})_{\text{tran}} \otimes (\mathbb{Q}/\mathbb{Z})$ of order m. This coincides with the element alluded to in Proposition 7.7.

8. Approach I: Irreducibility of Torsion in the Nonisotrivial Case

8.1. Group-theoretic results

We use the notation of the Appendix, in particular, the exact sequence

$$(1) \qquad 1 \to (\mathbb{Z}/p\mathbb{Z})^{\oplus 2} \to \mathrm{AffSL}_2(\mathbb{Z}/p\mathbb{Z}) \xrightarrow{q} \mathrm{SL}_2(\mathbb{Z}/p\mathbb{Z}) \to 1,$$

with splitting $\sigma : \mathrm{SL}_2(\mathbb{Z}/p\mathbb{Z}) \hookrightarrow \mathrm{AffSL}_2(\mathbb{Z}/p\mathbb{Z})$.

Proposition 8.1. *Let $p \neq 2, 3$ be prime. Let $H' \subset \mathrm{AffSL}_2(\mathbb{Z}/p\mathbb{Z})$ be a subgroup with $q(H') = \mathrm{SL}_2(\mathbb{Z}/p\mathbb{Z})$. Then one of the following is true:*

1. *there is a point in $(\mathbb{Z}/p\mathbb{Z})^{\oplus 2}$ fixed by each element in H', so this group is conjugate to $\mathrm{SL}_2(\mathbb{Z}/p\mathbb{Z})$;*
2. *$H' = \mathrm{AffSL}_2(\mathbb{Z}/p\mathbb{Z})$.*

Proof. First, assume that $q : H' \to \mathrm{SL}_2(\mathbb{Z}/p\mathbb{Z})$ has nontrivial kernel. Each element of the kernel is in $(\mathbb{Z}/p\mathbb{Z})^{\oplus 2} \subset \mathrm{AffSL}_2(\mathbb{Z}/p\mathbb{Z})$ and acts via translation by a nonzero element $v \in (\mathbb{Z}/p\mathbb{Z})^{\oplus 2}$. Any other nonzero element $w \in (\mathbb{Z}/p\mathbb{Z})^{\oplus 2}$ can be obtained by applying an element of $\mathrm{SL}_2(\mathbb{Z}/p\mathbb{Z})$, which acts by conjugation on $(\mathbb{Z}/p\mathbb{Z})^{\oplus 2} \subset \mathrm{AffSL}_2(\mathbb{Z}/p\mathbb{Z})$. The surjectivity assumption implies w can also be obtained conjugating with an element of H'. Hence every translation is contained in H' and the second alternative holds.

Now we assume $q : H' \to \mathrm{SL}_2(\mathbb{Z}/p\mathbb{Z})$ is an isomorphism. Consider the action of $H' \simeq \mathrm{SL}_2(\mathbb{Z}/p\mathbb{Z})$ on polynomials over $(\mathbb{Z}/p\mathbb{Z})^{\oplus 2}$ of degree ≤ 1

$$c_0 + c_1x_1 + c_2x_2, \quad c_0, c_1, c_2 \in \mathbb{Z}/p\mathbb{Z}.$$

We claim the resulting representation is completely reducible

$$\langle 1 \rangle \oplus \langle x_1 - v_1, x_2 - v_2 \rangle, \quad \dagger$$

where (v_1, v_2) is the fixed point of the action. We know that $\phi(-I)$, where I is the identity, is semisimple and the ± 1-eigenvalue decomposition of the linear polynomials takes the form †. It suffices to show that every other element of H' respects this decomposition. Consider

$$S = \begin{pmatrix} 0 & -1 \\ 1 & 0 \end{pmatrix} \quad T = \begin{pmatrix} 0 & -1 \\ 1 & 1 \end{pmatrix} \in \mathrm{SL}_2(\mathbb{Z}),$$

which *generate* $\mathrm{SL}_2(\mathbb{Z})$; [35] §7.1 contains a proof they generate $\mathrm{SL}_2(\mathbb{Z})/\pm I$ and they obviously yield $\pm I$. These matrices therefore generate $\mathrm{SL}_2(\mathbb{Z}/p\mathbb{Z})$ as well, so it suffices to show they respect decomposition †. These matrices are semisimple over $\mathbb{Z}/p\mathbb{Z}$ and satisfy the relations

$$S^2 = T^3 = -I,$$

i.e., they have eigenspace decompositions respecting our decomposition. ■

Proposition 8.2. *For any proper subgroup* $H \subsetneq \mathrm{SL}_2(\mathbb{Z}/p\mathbb{Z})$*, the index of* H *is at least* p*.*

Proof. We have $\mathrm{SL}_2(\mathbb{Z}/2\mathbb{Z})$ the symmetric group $\mathfrak{S}_3$ and $\mathrm{SL}_2(\mathbb{Z}/3\mathbb{Z})$ a central extension of the alternating group A_4 by a group of order two. The result holds in these special cases, so we may restrict attention to cases where $p \neq 2, 3$. Let $r = \mathrm{index}(H) = |\mathrm{SL}_2(\mathbb{Z}/p\mathbb{Z})/H|$ and consider the associated coset representation

$$\mathrm{SL}_2(\mathbb{Z}/p\mathbb{Z}) \to \mathfrak{S}_r.$$

The kernel $K \subset \mathrm{SL}_2(\mathbb{Z}/p\mathbb{Z})$ is a normal subgroup, as is its image $K' \subset \mathrm{SL}_2(\mathbb{Z}/p\mathbb{Z})/\langle \pm I \rangle$. For $p > 3$ the group $\mathrm{SL}_2(\mathbb{Z}/p\mathbb{Z})/\langle \pm I \rangle$ is simple, so either K' is trivial or the entire group, which is impossible. If K' is trivial then $|\mathrm{SL}_2(\mathbb{Z}/p\mathbb{Z})/\langle \pm I \rangle| = p(p^2-1)/2$ divides $r!$, so the index $r \geq p$. ■

8.2. Irreducibility of torsion

Throughout this section, F denotes the function field of $\mathbb{P}^1$. We consider the irreducibility of certain schemes over F:

Theorem 8.3. *Let $\overline{J} \to \mathbb{P}^1$ be a nonisotrivial Jacobian elliptic fibration, with generic fiber J_F. Let p be a sufficiently large prime number.*

1. *The nonzero p-torsion $J_F[p] - \{0\}$ is irreducible.*
2. *Any nontrivial $J_F[p]$-prinicipal homogeneous space P_F is irreducible.*

Proof. Let $U \subset \mathbb{P}^1$ be the open subset over which the fibration is smooth. Consider the monodromy representation

$$\varrho : \pi_1(U) \to \mathrm{SL}_2(\mathbb{Z}),$$

with image Γ'.

Lemma 8.4. $\Gamma' \subset \mathrm{SL}_2(\mathbb{Z})$ *has finite index.*

Proof. Let $V \to U$ be the covering space with covering group Γ', so that $U = \Gamma' \setminus V$. The pull-back of $\overline{J}$ to V has trivial monodromy, so we have

$$\begin{array}{ccc} V & \to & H \\ \downarrow & & \downarrow \\ U & \xrightarrow{j} & \mathbb{A}^1 \simeq \mathrm{SL}_2(\mathbb{Z}) \setminus H \end{array}$$

where H is the upper half plane and $\mathbb{A}^1$ is the j-line. Both vertical maps are Γ'-equivariant, so we have a factorization

$$\begin{array}{ccccc} & & \Gamma' \setminus H & & \\ & \nearrow & & \searrow & \\ U & & \xrightarrow{j} & & \mathrm{SL}_2(\mathbb{Z}) \setminus H. \end{array}$$

Since j has finite degree, Γ' has finite index. ∎

In the sequel, we take Γ to be the absolute Galois group of the function field $F = \mathbb{C}(\mathbb{P}^1)$. The profinite completion of the fundamental group $\pi_1(U)$ is a quotient of Γ, corresponding to the maximal extension of F unramified over U. The (mod p) reduction of the monodromy

$$\rho : \pi_1(U) \to \mathrm{SL}_2(\mathbb{Z}/p\mathbb{Z})$$

factors through the profinite completion and induces a representation

$$\Gamma \to \mathrm{SL}_2(\mathbb{Z}/p\mathbb{Z}).$$

We use the notation of the Appendix. The p-torsion $J_F[p]$ is a Γ-twist of $(\mathbb{Z}/p\mathbb{Z})^{\oplus 2}$ and is classified by a representation $\alpha : \Gamma \to \mathrm{SL}_2(\mathbb{Z}/p\mathbb{Z})$, which coincides with the representation $\rho(\mathrm{mod}\ p)$.

Lemma 8.5. *For sufficiently large primes p, α is surjective.*

This proves the first assertion of the Theorem.

Proof of lemma. The index of $\alpha(\Gamma)$ in $\mathrm{SL}_2(\mathbb{Z}/p\mathbb{Z})$ divides the index of Γ' in $\mathrm{SL}_2(\mathbb{Z})$. By Proposition 8.2, any proper subgroup of $\mathrm{SL}_2(\mathbb{Z}/p\mathbb{Z})$ has index at least p. It follows that $\alpha(\Gamma) = \mathrm{SL}_2(\mathbb{Z}/p\mathbb{Z})$ for p greater than the index. The irreducible components of $J_F[p] - \{0\}$ correspond to the orbits of $\alpha(\Gamma)$ on $(\mathbb{Z}/p\mathbb{Z})^{\oplus 2} - \{0\}$. ■

To each $J_F[p]$-principal homogenenous space P_F, Proposition A.2 assigns a representation $\phi : \Gamma \to \mathrm{AffSL}_2(\mathbb{Z}/p\mathbb{Z})$ with $q \circ \phi = \alpha$.

Lemma 8.6. *For sufficiently large primes p, ϕ is surjective provided P_F is nontrivial.*

Then $\phi(\Gamma)$ acts transitively on $(\mathbb{Z}/p\mathbb{Z})^{\oplus 2}$ and P_F is irreducible, and the second assertion of the Theorem is proved.

Proof of lemma. We may assume α is surjective. If $\phi(\Gamma)$ were contained in some conjugate of $\mathrm{SL}_2(\mathbb{Z}/p\mathbb{Z}) \subset \mathrm{AffSL}_2(\mathbb{Z}/p\mathbb{Z})$, then it would fix an element of $(\mathbb{Z}/p\mathbb{Z})^{\oplus 2}$, contradicting the nontriviality of P_F. Then Proposition 8.1 implies $\phi(\Gamma) = \mathrm{AffSL}_2(\mathbb{Z}/p\mathbb{Z})$. ■

8.3. Production of rational multisections

To prove potential density for an elliptic K3 surface $\eta : X \to \mathbb{P}^1$, we produce a plethora of rational multisections, with the hope that some might be nontorsion. The next result gives an infinite sequence of such multisections, with unbounded degree.

Theorem 8.7. *Let $X \to \mathbb{P}^1$ be a nonisotrivial elliptic K3 surface, with Jacobian fibration $\overline{J} \to \mathbb{P}^1$ and generic fiber J_F. Suppose that $[X_F] \in \mathrm{Sh}(J_F)$ has order m and let $p \gg m$ be prime. Then there exists a rational, nontorsion. multisection $M \subset X$ of degree mp.*

Combining this with Proposition 4.13, we easily deduce the following special case of Theorem 6.4:

Theorem 8.8. *Let $\eta : X \to \mathbb{P}^1$ be a nonisotrivial elliptic K3 surface over a number field K. Then rational points of X are potentially dense.*

Proof of Theorem 8.7. Corollary 7.13 yields an elliptic K3 surface $X' \to \mathbb{P}^1$ with Jacobian $\overline{J}$ so that

$$\mathrm{order}\,[X'_F] = p \cdot \mathrm{order}\,[X_F] \quad \text{and} \quad p[X'_F] = [X_F]$$

as well as a dominant rational map

$$\mu_p : X' \dashrightarrow J^p(X') = X.$$

Proposition 7.12 gives a rational curve $M' \subset X'$ of relative degree mp. Let M be the image $\mu_p(M')$, which is also a rational curve, and consider the induced finite morphism

$$\mu_p : M' \to M.$$

Lemma 8.9. *μ_p maps M' birationally onto M, which thus has degree mp over $\mathbb{P}^1$.*

Proof of lemma. Suppose not, so that M has degree strictly smaller than mp. Since $[X_F] \in \mathrm{Sh}\,(J_F)$ has order m, Proposition 7.5 guarantees M has relative degree equal to m. Furthermore, the induced $M' \to M$ has degree p.

Let $\{m'_1, \ldots, m'_{mp}\}$ denote the geometric points of M'_F, lying over the generic point of the base $\mathbb{P}^1$. Since $\mu_p(m'_{i_1}) = \mu_p(m'_{i_2})$ for some $i_1 \neq i_2$, we deduce that some of the $m'_{i_1} - m'_{i_2}$ are p-torsion, e.g., the points lying over a given $m_i \in M_F$. We basechange to M to obtain a elliptic fibration $X' \times_{\mathbb{P}^1} M \to M$. Using the inclusion and the map μ_p, imbed $M' \hookrightarrow X' \times_{\mathbb{P}^1} M$ so the image is a p-torsion multisection of degree p over M. Let E denote the function field of M, a degree m extension of F, and J_E the basechange of J_F to E. Then M'_E is contained in a principal homogeneous space P_E for $J_E[p]$ (see Remark 4.10).

We analyze the structure of $J_E[p]$. First, if p is chosen sufficiently large, we may assume the monodromy representation

$$\alpha : \Gamma \to \mathrm{SL}_2(\mathbb{Z}/p\mathbb{Z})$$

is surjective (see Lemma 8.5). By Proposition 8.2, the degree-m extension E/F is linearly disjoint from the $\mathrm{SL}_2(\mathbb{Z}/p\mathbb{Z})$ extension of F associated with $J_F[p]$. Indeed, the associated Galois extension of F has no subextensions with degree dividing m. Hence, even after base extension to E, the monodromy representation still surjects onto $\mathrm{SL}_2(\mathbb{Z}/p\mathbb{Z})$.

We return to the principal homogeneous space P_E. If it is trivial then Theorem 8.3 shows it has just two components, the identity and the nonidentity, neither of which has degree p. If it is nontrivial then an application of Proposition 8.1, as in the argument for Lemma 8.6, shows it is irreducible with just one component of degree p^2. In either case we deduce a contradiction, so M must be a multisection of degree mp in X. ∎

It remains to show that M is nontorsion, which is the case if M' is nontorsion:

Lemma 8.10. *$M' \subset X'$ is nontorsion.*

Proof of lemma. Suppose that M' is torsion of order N, so that $\mu_N(M') \subset J^N(X')$ is a section. Proposition 7.5 implies that $mp = \text{order}\,[X'_F]$ divides N. Write $N = np$ and consider the image

$$M'' = \mu_n(M') \subset J^n(X'),$$

a multisection of order p and degree $d > 1$, where $d \mid mp$. By Proposition 4.8, M'' is contained in a principal homogeneous space P_F for the p-torsion $J_F[p]$. We apply Theorem 8.3: If P_F is nontrivial then it is irreducible of degree p^2 and $p \mid m$, a contradiction. If P_F is trivial then $d = p^2 - 1$, whence $(p^2 - 1) \mid m$, a contradiction. ∎

Remark 8.11. For the application to Theorem 8.8, we only really need M' to be nontorsion. Then we may apply Proposition 4.13 to X' and potential density for X follows from Proposition 3.1. (If X is defined over a number field then X' and M' are as well.)

The precise description of the multisection M does give interesting consequences:

Corollary 8.12. *Every nonisotrivial elliptic K3 surface contains an infinite number of irreducible rational curves.*

Corollary 8.13. *For each nonisotrivial Jacobian elliptic K3 surface with Néron–Severi group of rank two, there exists an infinite sequence of irreducible rational curves with decomposible divisor classes.*

These cannot be (-2)-curves [27] §2.

Proof of corollary. A Jacobian elliptic K3 surface with Néron–Severi rank two has effective cone generated by the classes of a fiber E and the zero section Σ [34]

	E	Σ
E	0	1
Σ	1	-2

Thus any multisection of relative degree > 1 is decomposible. ∎

9. Approach II: Genus Estimates

Before diving into technical details, we indicate the main idea of this approach. Consider the modular curve

$$X_1(n) = \{ (E,P) \ : \ E \text{ an elliptic curve, } P \in E \text{ of order } n\} / \simeq .$$

Any family of elliptic curves $\mathcal{E} \to B$ with n-torsion section gives a morphism $B \to X_1(n)$. For large primes $X_1(p)$ has positive genus [28] §4.2, so there are no nonisotrivial families of elliptic curves with p-torsion section over $\mathbb{P}^1$. We shall prove similar results for torsion multisections: If they have large torsion order they must have positive genus.

9.1. The Riemann–Hurwitz formula

Let F be the function field of a smooth projective curve B, M_F a finite reduced scheme over F, M the normalization of B in the fraction ring of M_F, and $f : M \to B$ the induced finite morphism. We do not assume M is irreducible. The *genus* $g(M_F)$ is defined as the arithmetic genus of M, and is computed by the Riemann–Hurwitz formula

$$2g(M) - 2 = \deg(f)\big(2g(B) - 2\big) + \sum_{m \in M} e_m,$$

where e_m is the local ramification order at m.

Let Γ be the absolute Galois group of F (or the fundamental group of the open subset $U \subset B$ over which f is unbranched). For each $b \in B$, write

$$f^{-1}(b) = \{m_1, \ldots, m_{r(b)}\}.$$

Let $\gamma_b \in \Gamma$ represent the conjugacy class of a generator for the Galois group of the completion $\hat{F}_b$ of F at b (or the class of a small loop around b.) The cover f yields a representation into the symmetric group

$$\varphi : \Gamma \to \mathfrak{S}_{\deg f}$$

where $\varphi(\gamma_b)$ gives the local monodromy at b. If we represent this as a product of disjoint cycles

$$\varphi(\gamma_b) = \sigma_1 \ldots \sigma_{r(b)},$$

then the cycles are in one-to-one correspondence with the points

$$\{m_1, \ldots, m_{r(b)}\}$$

and

$$\text{length}\,(\sigma_i) = e_{m_i} + 1.$$

This allows us to rewrite the Riemann–Hurwitz formula

$$2g(M) - 2 = \deg f\big(2g(B) - 2\big) + \sum_{b \in B} \big(\deg(f) - r(b)\big),$$

where $r(b)$ is the number of orbits of $\varphi(\gamma_b) \in \mathfrak{S}_{\deg f}$.

9.2. Genera of principal homogeneous spaces

Theorem 9.1. *Let F be the function field of a smooth projective curve B; for each $b \in B$, let $\hat{F}_b$ denote the completion of F at b. Let H_F be a finite group scheme over F and P_F a H_F-principal homogeneous space over F. Then $g(H_F) \leq g(P_F)$, with equality only when $P_{\hat{F}_b}$ is a trivial $H_{\hat{F}_b}$-principal homogeneous space for each $b \in B$.*

Proof of theorem. We use the notation of the Appendix and §9.1, and use H and P to denote the normalizations of B in H_F and P_F respectively. Let Γ denote the absolute Galois group of F, $G = H_{F^a}$ the group of points of H_F over the algebraic closure, and G' the associated twisted form, with classifying cocycle

$$\alpha \in H^1(\Gamma, \mathrm{Aut}(G)) = \mathrm{Hom}(\Gamma, \mathrm{Aut}(G)).$$

By Proposition A.2, P_F is classified by a representation

$$\phi : \Gamma \to \mathrm{Aff}(G)$$

so that $q \circ \phi = \alpha$, where $q : \mathrm{Aff}(G) \to \mathrm{Aut}(G)$ is the quotient map.

For each $b \in B$ let $\gamma_b \in \Gamma$ correspond to the generator of the absolute Galois group of $\hat{F}_b$ (or the class of a loop around b in the fundamental group.) As we saw in §9.1, the ramification of $H \to B$ (resp. $P \to B$) over b is determined by the number of orbits of $\alpha(\gamma_b)$ (resp. $\phi(\gamma_b)$) on G. By Proposition A.2, the $H_{\hat{F}_b}$-principal homogeneous space $P_{\hat{F}_b}$ is trivial only if $\phi(\gamma_b)$ is conjugate to $\alpha(\gamma_b)$ by some $\tau_g \in G$.

To apply the Riemann–Hurwitz formula, it suffices to establish the following:

Proposition 9.2. *Let G be a finite group, $a \in \sigma(\mathrm{Aut}(G))$, and $\psi \in \mathrm{Aff}(G)$ so that $q(\psi) = q(a)$. Then we have*

$$\#\ \mathit{orbits}(\psi) \leq \#\ \mathit{orbits}(a),$$

with equality only if $\psi = \tau_g a \tau_g^{-1}$ for some $g \in G$.

Proof of proposition. Let $\mathrm{Aut}(G)$ and $\mathrm{Aff}(G)$ act on G from the left. Since $\mathrm{Aff}(G)$ is a semidirect product of G by $\mathrm{Aut}(G)$ and $q(\psi) = q(a)$, we may represent $\psi = \tau_h a$ for some τ_h, with $h \in G$ acting on G by left translation.

For each $\beta \in \mathrm{Aff}(G)$ and $N \in \mathbb{N}$, define

$$\mathrm{Fix}(\beta, N) = \{g \in G : \beta^N(g) = g\},$$

the elements fixed by the Nth power of β; $\mathrm{Fix}(\beta, 1)$ is the set of β-fixed points. We write $O(\beta, N) = |\mathrm{Fix}(\beta, N)|$. Observe the following properties:

1. $\mathrm{Fix}(\beta, N_1) \subset \mathrm{Fix}(\beta, N_2)$ whenever $N_1 \mid N_2$.
2. $\mathrm{Fix}(\beta, N)$ is the elements contained in a β-orbit with cardinality dividing N.
3. When $a \in \sigma(\mathrm{Aut}(G)) \subset \mathrm{Aff}(G)$, $\mathrm{Fix}(a, N)$ is a subgroup of G.

4. When $\psi = \tau_h a$ for $a \in \sigma(\operatorname{Aut}(G))$, $\operatorname{Fix}(\psi, N)$ is either empty or is a principal homogeneous space for $\operatorname{Fix}(a, N)$. In particular, either $O(\psi, N) = 0$ or $O(\psi, N) = O(a, N)$.

We verify the last assertion: Take $g_1, g_2 \in \operatorname{Fix}(\psi, N)$ and $g \in \operatorname{Fix}(a, N)$, so that $g = a^N g$ and $g_i = \psi^N(g_i)$, $i = 1, 2$. Since $\psi^N = (\tau_h a)^N = \tau_{h'} a^N$ for some $h' \in G$, we have $g_1 = \tau_{h'} a^N(g_1)$, $\tau_{h'} a^N(g_1 g) = \tau_{h'} a^N(g_1) a^N(g) = g_1 g$, and $g_1 g \in \operatorname{Fix}(\psi, N)$. Furthermore,

$$a^N(g_1^{-1} g_2) = a^N(g_1)^{-1} a^N(g_2) = \left(\tau_{h'}^{-1} g_1\right)^{-1} \tau_{h'}^{-1} g_2 = g_1^{-1} g_2$$

so that $g_1^{-1} g_2 \in \operatorname{Fix}(a, N)$.

Let $o(\beta, d)$ denote the number of β-orbits of G with cardinality d, so that

$$O(\beta, N) = \sum_{d|N} o(\beta, d) d$$

$$o(\beta, d) = 1/d \sum_{j|d} O(\beta, d/j) \mu(j)$$

where μ is the Mobius function. Then the total number of β-orbits is

$$\begin{aligned}
\#\operatorname{orbits}(\beta) &= \sum_{d|\operatorname{order}(\beta)} o(\beta, d) = \sum_{d} 1/d \sum_{j|d} O(\beta, d/j) \mu(j) \\
&= \sum_{j|\operatorname{order}(\beta)} O(\beta, j)/j \sum_{k|\operatorname{order}(\beta)/j} \mu(k)/k \\
&= \sum_{j|\operatorname{order}(\beta)} O(\beta, j)/j \prod_{\ell \text{ prime}|\operatorname{order}(\beta)/j} (\ell - 1)/\ell.
\end{aligned}$$

This is a positive linear combination of the $O(\beta, j)$. We have seen that $O(\psi, j) \leq O(a, j)$ for each j, hence

$$\#\operatorname{orbits}(\psi) \leq \#\operatorname{orbits}(a).$$

If equality holds then $O(\psi, j) = O(a, j)$ for each j so, in particular, $O(\psi, 1) \neq 0$. If ψ fixes an element $g \in G$ then $\tau_g^{-1} \psi \tau_g$ fixes the identity. However, any element of $\operatorname{Aff}(G)$ fixing the identity is in $\sigma(\operatorname{Aut}(G))$ and

$$q(\tau_g^{-1} \psi \tau_g) = q(\psi) = q(a),$$

hence $\psi = \tau_g a \tau_g^{-1}$. ∎

9.3. Genus of the p-torsion

We organize our computation by the Kodaira type of the degenerate fiber (see [21] §9, Table 1 or [2] pp. 159).

Theorem 9.3. *Let $\overline{J} \to B$ be a relatively minimal Jacobian elliptic fibration with generic fiber J_F. Suppose there are n_0 degenerate fibers of type I_0^*, n_1' degenerate fibers of type $I_a, a > 0$, n_1'' degenerate fibers of type $I_a^*, a > 0$, n_2 fibers of type II or II^*, n_3 fibers of type III or III^*, and n_4 fibers of type IV or IV^*.*

If p is a sufficiently large prime number then

$$2g\big(J_F[p]\big) - 2 = p^2\big(2g(B) - 2\big) + n_1'(p-1)^2 + n_1''(p^2 - p)$$
$$+ (p^2 - 1)(1/2n_0 + 5/6n_2 + 3/4n_3 + 2/3n_4)$$
$$2g\big(J_F[p] - \{0\}\big) - 2 = n_1'(p-1)^2 + n_1''(p^2 - p) + (p^2 - 1)\big(2g(B) - 2)$$
$$+ 1/2n_0 + 5/6n_2 + 3/4n_3 + 2/3n_4\big).$$

Remark 9.4. $J_F[p]$ does not generally extend to a group scheme finite and flat over B. Any such group scheme would be smooth over B, but the proof of the theorem shows the existence of ramification at singular fibers.

Corollary 9.5. *Retain the notation and assumptions of Theorem 9.3. Suppose $B = \mathbb{P}^1$ and assume*

$$c(\overline{J}) := 1/2n_0 + n_1' + n_1'' + 5/6n_2 + 3/4n_3 + 2/3n_4 - 2 > 0.$$

Then for $p \gg 0$ both $J_F[p]$ and $J_F[p] - \{0\}$ have positive genus, as does any $J_F[p]$-principal homogeneous space.

The last assertion follows from Theorem 9.1.

Proof of theorem. To compute the full torsion we take $M_F = J_F[p]$, so the resulting morphism $f : M \to B$ has degree p^2.

In order to apply the Riemann–Hurwitz formula as in §9.1, we record the local monodromy $\rho(\gamma_b) \in \mathrm{SL}_2(\mathbb{Z})$, the number of orbits of $(\mathbb{Z}/p\mathbb{Z})^{\oplus 2}$ under $\rho(\gamma_b)$, and the contribution to the total ramification. In the table below we assume that $p \neq 2, 3$, so that for fibers of types I_0^*, II, II^*, III, III^*, IV, IV^* the mod p reduction of $\rho(\gamma_b)$ is semisimple and fixes only the origin. If there are fibers of type I_a (resp. I_a^*) we also assume $(p,a)=1$, so the mod p reduction is conjugate to

$$\begin{pmatrix} 1 & 1 \\ 0 & 1 \end{pmatrix} \left(\text{resp.} \; - \begin{pmatrix} 1 & 1 \\ 0 & 1 \end{pmatrix} \right).$$

Table 1. Ramification contributions at degenerate fibers

Kodaira type	monodromy matrix $\rho(\gamma_b)$	# orbits mod p $r(b)$	ramification $\sum_{m\in f^{-1}(b)} e_m$
I_0	$\begin{pmatrix} 1 & 0 \\ 0 & 1 \end{pmatrix}$	p^2	0
I_0^*	$-\begin{pmatrix} 1 & 0 \\ 0 & 1 \end{pmatrix}$	$1/2(p^2+1)$	$1/2(p^2-1)$
I_a	$\begin{pmatrix} 1 & a \\ 0 & 1 \end{pmatrix}$	$2p-1$	$(p-1)^2$
I_a^*	$-\begin{pmatrix} 1 & a \\ 0 & 1 \end{pmatrix}$	p	p^2-p
II	$\begin{pmatrix} 1 & 1 \\ -1 & 0 \end{pmatrix}$	$1/6(p^2+5)$	$5/6(p^2-1)$
II^*	$\begin{pmatrix} 0 & -1 \\ 1 & 1 \end{pmatrix}$	$1/6(p^2+5)$	$5/6(p^2-1)$
III	$\begin{pmatrix} 0 & 1 \\ -1 & 0 \end{pmatrix}$	$1/4(p^2+3)$	$3/4(p^2-1)$
III^*	$\begin{pmatrix} 0 & -1 \\ 1 & 0 \end{pmatrix}$	$1/4(p^2+3)$	$3/4(p^2-1)$
IV	$\begin{pmatrix} 0 & 1 \\ -1 & -1 \end{pmatrix}$	$1/3(p^2+2)$	$2/3(p^2-1)$
IV^*	$\begin{pmatrix} -1 & -1 \\ 1 & 0 \end{pmatrix}$	$1/3(p^2+2)$	$2/3(p^2-1)$

To compute the nonzero torsion we take $M_F = J_F[p] - \{0\}$, so the resulting morphism $f : M \to B$ has degree p^2-1. On the other hand, removing the origin involves eliminating an orbit of size one, which does not change the ramification contribution. ∎

We classify the K3 surfaces where Corollary 9.5 *fails* to apply:

Proposition 9.6. *Let $X \to \mathbb{P}^1$ be an elliptic K3 surface, with Jacobian $\overline{J} \to \mathbb{P}^1$ and generic fiber J_F. If $c(\overline{J}) \leq 0$ (cf. Corollary 9.5) then X is isotrivial and takes one of the following forms*

1. *four degenerate fibers of type* I_0^*, *and* $\operatorname{rank} \mathrm{NS}(X) = 18, 19, 20$;
2. *three degenerate fibers of types* I_0^*, II^*, IV^*, *and* $\operatorname{rank} \mathrm{NS}(X) = 20$;
3. *one degenerate fiber of type* I_0^*, *two of type* III^*, *and* $\operatorname{rank} \mathrm{NS}(X) = 20$;
4. *three degenerate fibers of type* IV^*, *and* $\operatorname{rank} \mathrm{NS}(X) = 20$.

Proof. Retain the notation of Theorem 9.3 and §7.4. If $c(\overline{J}) \leq 0$ then

$$2n_0 + 4n_1' + 4n_1'' + 10/3n_2 + 3n_3 + 8/3n_4 \leq 8$$

and subtracting from the formula of Corollary 7.15

$$\operatorname{rank} \mathrm{NS}(X) + n_1' + 2(n_0 + n_1'' + n_2 + n_3 + n_4) = 26 + r(J_F)$$

gives

$$\operatorname{rank} \mathrm{NS}(X) \geq 18 + r(J_F) + 3n_1' + 2n_1'' + 4/3n_2 + n_3 + 2/3n_4.$$

Since $\operatorname{rank} \mathrm{NS}(X) \leq 20$, it follows that $n_1' = 0$.

The expressions above admit the following solutions:

	n_0	n_1'	n_1''	n_2	n_3	n_4	$\operatorname{rank} \mathrm{NS}(X)$	$r(J)$
1	4	0	0	0	0	0	18, 19, 20	2, 1, 0
2	3	0	0	0	0	0	20	0
3	2	0	1	0	0	0	20	0
4	2	0	0	1	0	0	20	0
5	2	0	0	0	1	0	20	0
6	2	0	0	0	0	1	20	0
7	1	0	0	1	0	1	20	0
8	1	0	0	0	2	0	20	0
9	1	0	0	0	1	1	20	0
10	1	0	0	0	0	2	20	0
11	0	0	0	0	0	3	20	0

Several of these can be excluded. Solutions 2, 4, 5, 6, 9, and 10 are inconsistent with the Euler characteristic computation of Proposition 7.14. As for solution 3, suppose $X \to \mathbb{P}^1$ is an elliptic surface with one fiber of type I_a^* for $a > 0$ and two fibers of type I_0^*. Such a fiber has one rational curve with multiplicity two, intersected transversally by four rational curves with multiplicity two (see Figure 1). Let $B' \to \mathbb{P}^1$ be a double cover branched at the two points $b_1, b_2 \in \mathbb{P}^1$ with fibers of type I_0^*. The fiber product $X \times_{\mathbb{P}^1} B'$ is non-normal along the preimages of the rational curves with

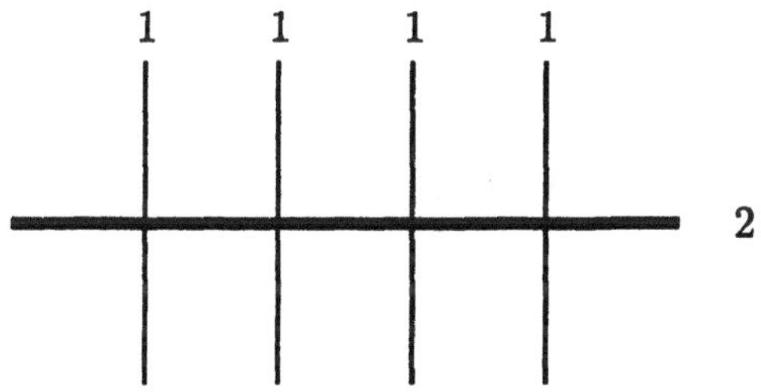

Fig. 1. Figure 1. Degenerate fiber of type I_0^*

multiplicty two. Let Y be the normalization and $\psi : Y \to B'$ the induced elliptic fibration. The fibers $\psi^{-1}(b_i)$ consist of a smooth elliptic curve, along with four (-1)-curves intersecting it transversally, i.e., the preimages of the rational curves of multiplicity one. Thus ψ is not relatively minimal over the points b_1, b_2, so we blown down the eight (-1)-curves. Let $\psi' : Y' \to B'$ be the resulting fibration, which has only two degenerate fibers, both with potentially multiplicative reduction, i.e., j-invariant infinity (see [21], Part II, pp. 604). We obtain a nonisotrivial elliptic fibration with abelian monodromy, an impossibility (see Lemma 8.5.)

The remaining solutions are consistent with Proposition 7.14 and are enumerated above. These correspond to isotrivial elliptic fibrations because there are no fibers of potentially multiplicative reduction. ∎

Remark 9.7. Ad hoc arguments give potential density for the degenerate cases enumerated in Proposition 9.6. The first case is a Kummer surface: Suppose that $X \to \mathbb{P}^1$ is an elliptic K3 surface with type I_0^* degenerate fibers at the points $b_1, b_2, b_3, b_4 \in \mathbb{P}^1$. The local and global monodromy is multiplication by -1. Let $E \to \mathbb{P}^1$ be the ramified double cover branched over b_1, b_2, b_3, b_4, an elliptic curve. Let $\psi' : Y' \to E$ be the minimal elliptic fibration obtained from $X \times_{\mathbb{P}^1} E \to E$, which is now smooth and has trivial monodromy. Such fibrations are classified (see [2] §V.5): ψ' is isotrivial with constant fiber C, is classified by the induced representation $\pi_1(E)$ into the torsion subgroup of C, and Y' is an abelian surface. By Proposition 4.2, Y' has potentially dense rational points. Since Y' dominates X, it also has potentially dense rational points.

The remaining cases have rank twenty, and thus have infinite automorphism group [40] Theorem 5. Potential density follows from Theorem 6.3.

9.4. Second proof of Theorem 8.8

In light of Remark 8.11, it suffices to establish:

Proposition 9.8. *Let $X \to \mathbb{P}^1$ be a nonisotrivial elliptic K3 surface, with Jacobian $\overline{J}$ and generic fiber J_F. Then there exists an elliptic K3 surface $X' \to \mathbb{P}^1$ with Jacobian $\overline{J}$, a dominant rational map $X' \dashrightarrow X$ over $\mathbb{P}^1$, and a nontorsion rational multisection $M' \subset X'$.*

Proof. We retain the notation of the proof of Theorem 8.7, so that X' is an order-p twist of X. If M' has order N then the image of M' under the multiplication map

$$\mu_N : X' \dashrightarrow J^N(X')$$

is a section of $J^N(X') \to \mathbb{P}^1$. Since X' has order divisible by p, Proposition 7.5 implies $p \mid N$. Write $N = pn$ and let $Y = J^n(X')$, an elliptic K3 surface with Jacobian J_F, so that $[Y] = n[X']$. We have a factorization of μ_N

$$X' \overset{\mu_n}{\dashrightarrow} Y \overset{\mu_p}{\dashrightarrow} J^N(X') = J^p(Y).$$

The image $M := \mu_n(M')$ is a *rational* multisection in Y of order exactly p and $\Sigma := \mu_p(M)$ is a section in $J^p(Y)$, with $M \subset \mu_p^{-1}(\Sigma)$. Furthermore, $P_F := \mu_p^{-1}(\Sigma)_F$ is a principal homogeneous space for $J_F[p]$.

If $P_F \simeq J_F[p]$ then $M_F \simeq J_F[p] - \{0\}$ because Theorem 8.3 implies $J_F[p] - \{0\}$ is irreducible when $p \gg 0$. It has positive genus by Proposition 9.6 and Corollary 9.5. As our family is nonisotrivial, this can also be extracted from standard computations of the genera of modular curves (e.g., [28] §4.2).

If P_F is nontrivial then Theorem 8.3 guarantees it is irreducible for $p \gg 0$. Then $M = \mu_p^{-1}(\Sigma)$ and Corollary 9.5 implies M has positive genus when $p \gg 0$, a contradiction. ■

9.5. Completing Theorem 6.4: the isotrivial case

After Theorem 8.8 and Remark 9.7, the only case where potential density remains open is isotrivial elliptic K3 surfaces with $c(\overline{J}) > 0$. Can the argument of §9.4 be applied in this situation? The main complication is that $J_F[p] - \{0\}$, and even nontrivial $J_F[p]$-principal homogeneous spaces, may fail to be irreducible. Indeed, the monodromy group is a subgroup of the automorphism group of the geometric generic fiber; the irreducible components of $J_F[p]$ correspond to orbits of p-torsion points under these automorphisms. Unfortunately, Theorems 9.3 and 9.1 shed little light on the genus of an *irreducible component* of a principal homogeneous space.

With some extra bookkeeping, one can still prove the following:

Theorem 9.9. *Let $\overline{J} \to \mathbb{P}^1$ be an isotrivial Jacobian elliptic surface, with generic fiber J_F. Suppose that $c(\overline{J}) > 0$. For prime numbers $p \gg 0$, each irreducible component of $J_F[p] - \{0\}$ has positive genus. If P_F is a nontrivial $J_F[p]$-principal homogeneous space then each irreducible component of P_F has positive genus.*

Repeating the argument for Proposition 9.8, we obtain

Proposition 9.10. *Let $X \to \mathbb{P}^1$ be an isotrivial elliptic K3 surface, with Jacobian $\overline{J} \to \mathbb{P}^1$ and generic fiber J_F, so that $c(\overline{J}) > 0$. Then there exists an elliptic K3 surface $X' \to \mathbb{P}^1$ with Jacobian $\overline{J}$, a dominant rational map $X' \dashrightarrow X$ over $\mathbb{P}^1$, and a nontorsion rational multisection $M' \subset X'$.*

Given this, the argument of §9.4 gives potential density for isotrivial elliptic K3 surfaces with $c(\overline{J}) > 0$.

Before establishing Theorem 9.9, we generalize Proposition 8.1 to classify the irreducible components of principal homogeneous spaces for the p-torsion of an isotrivial elliptic fibration. We use the notation of the Appendix, in particular, the exact sequence

$$(2) \qquad 1 \to (\mathbb{Z}/p\mathbb{Z})^{\oplus 2} \to \mathrm{AffSL}_2(\mathbb{Z}/p\mathbb{Z}) \xrightarrow{q} \mathrm{SL}_2(\mathbb{Z}/p\mathbb{Z}) \to 1,$$

with canonical splitting σ.

Proposition 9.11. *Let $T \in \mathrm{SL}_2(\mathbb{Z})$ be an element of finite order n generating a subgroup H, $p \neq 2, 3$ a prime, and $H' \subset \mathrm{AffSL}_2(\mathbb{Z}/p\mathbb{Z})$ so that $q(H') = H$. Then there is a split exact sequence*

$$(3) \qquad 1 \to V \to H' \to H \to 1, \quad V := H' \cap (\mathbb{Z}/p\mathbb{Z})^{\oplus 2}.$$

For each splitting σ', $\sigma'(H)$ is conjugate to a subgroup of $\sigma(\,\mathrm{SL}_2(\mathbb{Z}/p\mathbb{Z}))$, where σ is the canonical splitting of exact sequence 2.

The orbit decomposition of $(\mathbb{Z}/p\mathbb{Z})^{\oplus 2}$ under the action of H' is one of the following:

1. *If* $\dim_{\mathbb{Z}/p\mathbb{Z}}(V) = 0$, *$H'$ has one fixed point and* $(p^2-1)/n$ *orbits with* n *elements.*
2. *If* $\dim_{\mathbb{Z}/p\mathbb{Z}}(V) = 1$, *$H'$ has one orbit with p elements (the subspace V) and* $(p-1)/n$ *orbits with* pn *elements.*
3. *If* $\dim_{\mathbb{Z}/p\mathbb{Z}}(V) = 2$, *$H'$ has one orbit with p^2 elements.*

Proof. If $T \in \mathrm{SL}_2(\mathbb{Z})$ has finite order n, it is semisimple and its eigenvalues are primitive nth roots of unity. The characteristic polynomial of T is quadratic, so $n = 2, 3, 4$, or 6. As $p \neq 2, 3$, the reduction of $T(\mathrm{mod}\ p)$ still has eigenvalues which are nth roots of unity, and $T(\mathrm{mod}\ p)$ has order n.

The exact sequence 3 is clearly induced from exact sequence 2; it is split because $|V|$ is prime to $n = |H|$. Now $\sigma'(H)$ is conjugate to a subgroup of $\sigma(\,\mathrm{SL}_2(\mathbb{Z}/p\mathbb{Z}))$, provided $\sigma'(T)$ fixes some point $v \in (\mathbb{Z}/p\mathbb{Z})^{\oplus 2}$: Consider the action of $\sigma'(T)$ on polynomials over $(\mathbb{Z}/p\mathbb{Z})^{\oplus 2}$ of degree ≤ 1

$$c_0 + c_1x_1 + c_2x_2, \quad c_0, c_1, c_2 \in \mathbb{Z}/p\mathbb{Z}.$$

We know $\sigma'(T)$ fixes the constants c_0 and has order prime to p, so its action decomposes as a direct sum of irreducibles

$$\langle 1 \rangle \oplus \langle x_1 - v_1, x_2 - v_2 \rangle,$$

and the induced action on the second factor is semisimple. The fixed point is $v = (v_1, v_2)$; the orbit analysis in the next paragraph will show that $v \in V$.

It remains to analyze the orbit decomposition. In each case, we first conjugate so that $\sigma'(H) \subset \mathrm{SL}_2(\mathbb{Z}/p\mathbb{Z})$. If $V = 0$, $H' \subset \mathrm{SL}_2(\mathbb{Z}/p\mathbb{Z})$, generated by a semisimple matrix $\sigma'(T)$ of order $n > 1$. The fixed point is the origin and every other orbit has n elements. If $V = (\mathbb{Z}/p\mathbb{Z})^{\oplus 2}$ then H' contains the full translation group, so the action is transitive. Now assume V is one dimensional. Of course, V is an eigenspace for $\sigma'(T)$. The group H' is generated by translations by elements of V and the action of $\sigma'(T)$. Again, the only fixed point under the action of $\sigma'(T)$ is the origin, so any orbit not containing the origin has order divisible by n. No element of $(\mathbb{Z}/p\mathbb{Z})^{\oplus 2}$ is fixed under translation by V, so each orbit has order divisible by p. The description of the orbits follows. ■

Proof of Theorem 9.9. Let M denote a component of the normalization of $\mathbb{P}^1$ in $J_F[p] - \{0\}$ or P_F, corresponding to an orbit of H' on $(\mathbb{Z}/p\mathbb{Z})^{\oplus 2}$. In the first case of Proposition 9.11, M corresponds to an H'-orbit with n elements and Riemann–Hurwitz takes the form

$$2g(M) - 2 = c(\overline{J})n > 0.$$

In the last case of Proposition 9.11 P_F is irreducible, so Theorems 9.3 and 9.1 apply as in the nonisotrivial case.

For the remaining cases, we analyze more closely the behavior at fibers of additive reduction:

Lemma 9.12. *Let $\hat{F} = \mathbb{C}((t))$ with absolute Galois group $\hat{\Gamma}$, and $J_{\hat{F}}$ an elliptic curve over $\hat{F}$. Assume that the closed fiber is not of type I_a. For primes $p > 3$, $H^1(\hat{\Gamma}, J_{\hat{F}}[p]) = 0$.*

Proof of lemma. Take the associated Néron/relatively minimal model $\overline{J}$ over $\mathbb{C}[[t]]$. The locus where the fibers of $\overline{J}$ are smooth and reduced is called the 'group-like' part. In the case of additive reduction, the identity component of the group-like part is the additive group, which has no torsion. Hence, the order of the torsion group is equal to the number of connected components of the group-like part. The Kodaira classification ([21] §9, Table 1 or Figure 2 of §10) gives

Kodaira type	I_a^*	II	II^*	III	III^*	IV	IV^*
torsion order	4	0	0	2	2	3	3

By Proposition 7.2, in these cases any local $J_{\hat{F}}[p]$-principal homogeneous space (with $p > 3$) is trivial. ■

We complete the proof of Theorem 9.9. Only the second case of Proposition 9.11 remains. Lemma 9.12 says that at each point $b \in \mathbb{P}^1$ of additive reduction, the local monodromy $\phi(\gamma_b) \in H'$ has a fixed point (see Proposition A.2.) If M corresponds to an H'-orbit with p elements then one of the elements is fixed at each point of additive reduction, and Riemann–Hurwitz takes the form

$$2g(M)-2 \geq -2p+(1/2n_0+5/6n_2+3/4n_3+2/3n_4)(p-1) = (p-1)c(\overline{J})-2.$$

We have inequality because $f : M \to \mathbb{P}^1$ may ramify at smooth fibers. If M corresponds to an orbit with pn elements then none of the elements is fixed and

$$2g(M) - 2 \geq -2pn + (1/2n_0 + 5/6n_2 + 3/4n_3 + 2/3n_4)(np) = npc(\overline{J}).$$

This completes the proof. ■

10. Approach III: Elliptic Multisections (based on correspondence with J. Kollár)

10.1. Cubical curves

Definition 10.1. A *cubical curve* C is a reduced curve which may be imbedded as a plane cubic $C \subset \mathbb{P}^2$, with at most two irreducible components.

Example 10.2. 1. an elliptic curve is cubical;

2. an irreducible rational curve with a single node is cubical with equation $y^2z = x^2z + x^3$, and is called the 'nodal cubic';
3. an irreducible rational curve with a single cusp is cubical with equation $y^2z = x^3$, and is called the 'cuspidal cubic';
4. a curve with two smooth rational components intersecting in two nodes is cubical with equation $x(x^2 - yz) = 0$, and is called the 'secant cubic';
5. a curve with two smooth rational components intersecting in one point tangentially is cubical with equation $z(x^2 - yz) = 0$, and is called the 'tangential cubic'.

Proposition 10.3. 1. *Cubical curves are connected of arithmetic genus one and admit flat deformations to smooth curves.*

2. *Any isomorphism $C_1^\nu \to C_2^\nu$ between the normalizations of cubical curves that respects their conductors descends to an isomorphism $C_1 \to C_2$ of the cubical curves.*
3. *Let R be a projective integral singular rational curve. Then there exists a cubical curve C and birational morphism $f : C \to R$.*
4. *Let R be a projective connected curve with two rational irreducible components, at least one of which is smooth. Assume that the smooth component intersects the second component in at least two distinct points, or intersects a smooth branch of the second component with multiplicity greater than one. Then there exists a cubical curve C and birational morphism $f : C \to R$.*

Proof. We leave the proof of the first part to the reader. One way to establish the isomorphism assertion is to observe that Example 10.2 gives a complete classification of cubical curves.

For the remaining claims, consider the *seminormalization* $\sigma : R^\sigma \to R$ [23] I.7.2. This is a finite, birational, *bijective* morphism, and is maximal with these properties; it is obtained from the normalization R^ν by identifying points which are identified by $\nu : R^\nu \to R$.

For the third assertion, $R^\nu \simeq \mathbb{P}^1$ and either $R^\nu \to R^\sigma$ or $R^\sigma \to R$ fails to be an isomorphism, because R is singular. If $R^\nu \to R^\sigma$ fails to be an isomorphism then there exist distinct $r_1, r_2 \in R^\nu$ that are identified in R. Let C_0 be the nodal cubic obtained from R^ν by identifying r_1 and r_2. The induced $C_0 \to R$ is the desired morphism from a cubical curve. Otherwise, $R^\sigma \simeq \mathbb{P}^1$ and we choose $r \in R^\sigma$ at which σ is not an isomorphism. Consider the local rings $\mathcal{O}_{R,\sigma(r)} \subset \mathcal{O}_{R^\sigma,r}$, a finite extension of $\mathcal{O}_{R^\sigma,r}$-modules. If $t \in \mathfrak{m}_{R^\sigma,r}$ is a local uniformizer then $t^n \in \mathfrak{m}_{R,\sigma(r)}$ for $n \gg 0$ but not $n = 1$. The intermediate ring

$$\mathcal{O}_{R,\sigma(r)} \subset \mathcal{O}_{R,\sigma(r)}[t^2, t^3, t^4, \ldots] \subset \mathcal{O}_{R^\sigma,r}$$

corresponds to a factorization $R^\sigma \to C \to R$ through a cuspidal curve C.

For the last assertion, we have $R^\nu = R_1^\nu \cup R_2^\nu \simeq \mathbb{P}^1 \cup \mathbb{P}^1$ with R_1^ν mapped isomorphically onto its image in R. Suppose we have distinct $r_1, r_1' \in R_1^\nu$ and $r_2, r_2' \in R_2^\nu$ so that $\nu(r_1) = \nu(r_2)$ and $\nu(r_1') = \nu(r_2')$. Let C be the curve obtained by gluing the R_i^ν so that r_1 and r_2 (resp. r_1' and r_2') are identified. This is a 'secant cubic' and we obtain a factorization $R^\nu \to C \to R$. Now suppose we have $r_i \in R_i^\nu$ so that $\nu(r_1) = \nu(r_2) = r$, and ν maps an open neighborhood of r_2 isomorphically onto its image, which intersects $\nu(R_1^\nu)$ with multiplicity at least two at r. Algebraically, $\mathcal{O}_{R,r}$ is contained in the subring of elements $(f_1, f_2) \in \mathcal{O}_{R_1^\nu,r_1} \times \mathcal{O}_{R_2^\nu,r_2}$ with $f_1(r_1) = f_2(r_2)$ and $f_1 \in \mathfrak{m}^2_{R_1^\nu,r_1}$ if and only if $f_2 \in \mathfrak{m}^2_{R_2^\nu,r_2}$. This is the ring of functions of a curve C of arithmetic genus one consisting of two smooth rational components meeting tangentially at a single point, i.e., a tangential cubic. Thus we get the desired factorization $R^\nu \to C \to R$. ∎

10.2. Production of cubical curves

Proposition 10.4. *Let $\eta : X \to \mathbb{P}^1$ be an elliptic K3 surface with Jacobian fibration $\overline{J} \to \mathbb{P}^1$ and generic fiber J_F. There exist the following*

1. *an elliptic K3 surface $\eta' : X' \to \mathbb{P}^1$ with Jacobian $\overline{J} \to \mathbb{P}^1$;*
2. *a dominant rational map $X' \dashrightarrow X$;*
3. *a morphism $f_0 : C_0 \to X'$ from a cubical curve that is birational onto its image; this image is not contained in any fiber of η'.*

Proof. Corollary 7.13 yields an elliptic K3 surface $X' \to \mathbb{P}^1$ with Jacobian $\overline{J}$ so that

$$\text{order}\,[X'_F] = p \cdot \text{order}\,[X_F] \quad \text{and} \quad p[X'_F] = [X_F]$$

as well as a dominant rational map

$$\mu_p : X' \dashrightarrow J^p(X') = X.$$

Let $M' \subset X'$ be the rational multisection of degree $p \cdot \text{order}\,[X_F]$ guaranteed by Proposition 7.12. If M' is singular then we obtain the desired map directly from the third part of Proposition 10.3. If M' is nonsingular, we apply the fourth part of Proposition 10.3 to the union of M' and a suitable irreducible component of a degenerate fiber. It remains to show there exists a component satisfying the hypotheses of Proposition 10.3 provided M' has sufficiently large degree:

Lemma 10.5. *Let $\eta' : X' \to \mathbb{P}^1$ be an elliptic K3 surface and M' a nonsingular multisection of degree $d \geq 35$. Then M' intersects some irreducible component of a degenerate fiber in at least two points, or intersects one smooth reduced branch of the degenerate fiber with multiplicity greater than one.*

Proof of Lemma. Recall the multiplicities of irreducible components of degenerate fibers in the Kodaira classification [2], pp. 150, displayed in Figure 2.

First consider the fibers where all the reduced irreducible components are nonsingular. We choose d so that it is greater than the sum of the multiplicities over all the components:

type	I_a, $a > 1$	III	IV	I_a^*	II^*	III^*	IV^*
sum of multiplicities	a	2	3	$2a+6$	30	18	12

Then M' intersects some component twice, or perhaps at one point with multiplicity greater than one. Proposition 7.14 implies that an elliptic K3 can only have fibers of type I_a (resp. I_a^*) for $a \leq 19$ (resp. $a \leq 14$). Thus it suffices to take $d \geq 35$.

Now consider the remaining fibers. For fibers of type I_1, any smooth curve intersecting the nodal point with multiplicity ≥ 3 intersects one of the branches with multiplicity ≥ 2. It suffices that $d \geq 3$. In case II, any smooth curve intersects the cuspidal point in at most multiplicity three, so it suffices that $d \geq 4$. ∎

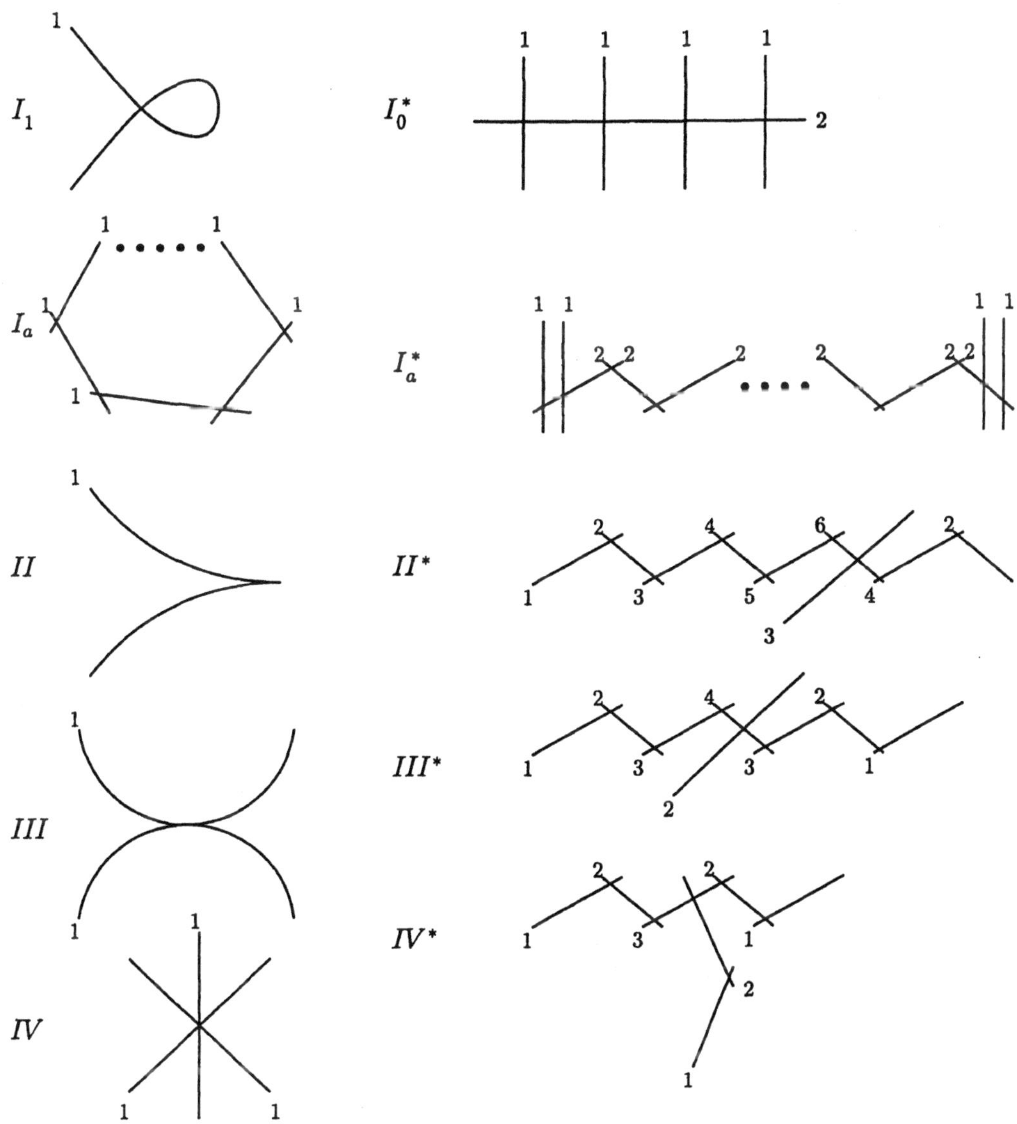

Fig. 2. Figure 2. Multiplicities of degenerate fibers

10.3. Production of elliptic curves

Proposition 10.6. *Let X be a K3 surface, C_0 a cubical curve, and $f_0 : C_0 \to X$ a morphism birational onto its image. Then $\{f_0 : C_0 \to X\}$ deforms to a morphism $\{f : C \to X\}$, where C is a smooth elliptic curve. The resulting family of elliptic curves dominates X.*

Proof. Let $v : \mathcal{C} \to S$ be a versal deformation of C_0, with C_0 the fiber over the distinguished point $s_0 \in S$. This can be realized as a linear series of plane cubic curves containing C_0 (cf. the first part of Proposition 10.3). The generic fiber is a smooth elliptic curve and $\dim(S) = \dim \operatorname{Aut}(C_0)$ [23] II.1.11.

Consider the functor of morphisms of flat S-schemes

$$\operatorname{Hom}(\mathcal{C}, X \times S)(T) = \left\{T\text{-morphisms } g : \mathcal{C} \times_S T \to (X \times S) \times_S T\right\},$$

where T is an S-scheme. Assigning to each morphism its graph, we may represent this functor by an S-scheme $\operatorname{Hom}(\mathcal{C}, X \times S)$, an open subset of the relative Hilbert scheme of $\mathcal{C} \times_S (X \times S)$. The morphism f_0 yields a point of $\operatorname{Hom}(\mathcal{C}, X \times S)$ over the basepoint s_0.

Our next task is to bound the dimension of $\operatorname{Hom}(\mathcal{C}, X \times S)$ at f_0 from below. General theory [23] I.2.17, II.1.13 guarantees that

$$\begin{aligned}\dim_{[f_0]} \operatorname{Hom}(\mathcal{C}, X \times S) &\geq \dim \operatorname{Hom}(f_0^*\Omega^1_X, \mathcal{O}_{C_0}) \\ &\quad - \dim \operatorname{Ext}^1(f_0^*\Omega^1_X, \mathcal{O}_{C_0}) + \dim_{s_0} S \\ &= -K_X \cdot C_0 + (\dim(X) - 3)\chi(\mathcal{O}_{C_0}) + \dim \operatorname{Aut}(C_0) \\ &= \dim \operatorname{Aut}(C_0).\end{aligned}$$

In our particular situation this can be improved (cf. [23] II.1.13.1 and [32]): We claim that

$$\dim_{[f_0]} \operatorname{Hom}(\mathcal{C}, X \times S) \geq \dim \operatorname{Aut}(C_0) + 1.$$

Let $\mathcal{X} \to \Delta$ be a one-parameter complex-analytic deformation of X, so that the generic fiber contains no algebraic curves. This can be achieved using the Local Torelli Theorem [27], by choosing a deformation for which none of the classes in $H^2(X, \mathbb{Z})$ remains of type $(1,1)$. Now consider the new mapping functor over $S \times \Delta$

$$\operatorname{Hom}(\mathcal{C} \times \Delta, \mathcal{X} \times S)$$

at the point

$$g_0 : C_0 \xrightarrow{f_0} X \hookrightarrow \mathcal{X}.$$

The general theory now gives

$$\dim_{[g_0]} \operatorname{Hom}(\mathcal{C} \times \Delta, \mathcal{X} \times S) \geq \dim \operatorname{Hom}(g_0^*\Omega^1_{\mathcal{X}}, \mathcal{O}_{C_0})$$

$$- \dim \operatorname{Ext}^1(g_0^*\Omega^1_{\mathcal{X}}, \mathcal{O}_{C_0}) + \dim_{(s_0,0)}(S \times \Delta)$$

$$= \dim \operatorname{Aut}(C_0) + 1.$$

The generic fiber of $\mathcal{X}$ contains no algebraic curves, so the family $\operatorname{Hom}(\mathcal{C} \times \Delta, \mathcal{X} \times S)$ must parametrize curves contained in X.

Consequently, there are deformations $\{f : C \to X\}$ of $\{f_0 : C_0 \to X\}$ that are not obtained by composing f_0 with an automorphism of C_0. We claim that $f(C) \not\subset f_0(C_0)$. Indeed, suppose that $f(C) = f_0(C_0)$. Since f_0 and f are birational onto their images and normalizations are unique, the composed morphisms

$$C^\nu \to C \to f(C) \quad C_0^\nu \to C_0 \to f_0(C_0)$$

agree up to an isomorphism $C^\nu \to C_0^\nu$ which preserves conductors. By part two of Proposition 10.3, this descends to an isomorphism $C \to C_0$, a contradiction. Thus the images $f(C)$ dominate X, and since X is not ruled we conclude that the generic domain curve C is elliptic. ■

10.4. Proof of Theorem 6.4

As in our previous approaches to Theorem 6.4, the key point is that each elliptic K3 surface $X \to \mathbb{P}^1$ is dominated by an elliptic K3 surface $X' \to \mathbb{P}^1$ admitting a nontorsion multisection with dense rational points, so that Proposition 4.13 applies. What is new here is that the multisection is an elliptic rather than a rational curve:

Theorem 10.7. *Let $\eta : X \to \mathbb{P}^1$ be an elliptic K3 surface with Jacobian fibration $\overline{J} \to \mathbb{P}^1$ and generic fiber J_F. There exist the following*

1. *an elliptic K3 surface $\eta' : X' \to \mathbb{P}^1$ with Jacobian $\overline{J} \to \mathbb{P}^1$;*
2. *a dominant rational map $X' \dashrightarrow X$;*
3. *a nontorsion elliptic multisection $\hat{M} \subset X'$.*

Furthermore, if X is defined over a number field then X' and $\hat{M}$ can be chosen so they are defined over a finite extension of that field.

This is an immediate consequence of Proposition 10.4 and the following:

Proposition 10.8. *Let $\eta' : X' \to \mathbb{P}^1$ be an elliptic K3 surface. Assume there exists a cubical curve C_0 and a morphism $f_0 : C_0 \to X'$ birational onto its image, so that $f_0(C_0)$ is not contained in any fiber of η' Then η' admits an elliptic multisection of infinite order. If X' is defined over a number field then the multisection can be defined over a finite extension of that field.*

In particular, any elliptic multisection deforms to a nontorsion elliptic multisection.

Proof of Proposition 10.8. Proposition 10.6 implies there exists a smooth projective elliptic surface $S \to B$, and a surjective morphism $g : S \to X'$ so that the generic fiber is mapped birationally onto its image. Under our assumptions, this image is not contained in a fiber of η'. Moreover, since X' is a K3 surface the images of any two fibers are linearly equivalent

$$[C] \equiv g_*[S_b] \quad \text{for each} \quad b \in B.$$

We distinguish some special fibers of $S \to B$. We have the singular fibers and the fibers that fail to be mapped birationally onto their image. The image of these fibers maps to a closed subset $Z \subset X'$.

Choose a smooth fiber $X'_p = \eta'^{-1}(p)$ not lying in Z, defined over a number field if X' is. After passage to a larger number field L, we can find a point $x \in X'_p(L)$, $x \notin Z$, so that $t_x = -\deg(C.X'_p)x + [C]|_{X'_p} \in J(X'_p)$ has infinite order in the Jacobian (by the arithmetic result, Proposition 4.2). Let S_b be a fiber of $S \to B$ so that $x \in g(S_b)$, which by assumption is smooth and maps birationally onto its image $\hat{M} := g(S_b)$. As S_b maps birationally onto $\hat{M}$, $[\hat{M}] = [g_*S_b] = [C]$. Now t_x has infinite order in $J(X'_p)$ and coincides with the restriction of $\tau_{\hat{M}}$ to X'_p. Hence $\tau_{\hat{M}}$ itself has infinite order and the multisection $\hat{M}$ is nontorsion. ∎

Corollary 10.9. *Every elliptic K3 surface admits a nontorsion elliptic multisection.*

Proof. Retain the notation of Theorem 10.7. The proof above yields an elliptic multisection $\hat{M}$ for η' through the generic point of X'. Its image $M \subset X$ is a multisection for η through the generic point of X; M has positive genus because X is not ruled and thus is elliptic. Proposition 10.8 shows M deforms to a nontorsion elliptic multisection. ∎

11. Symmetric Products of K3 Surfaces

11.1. Do generic K3 surfaces have potentially dense rational points?

In the previous sections, we have seen many special examples of K3 surfaces with potentially dense rational points. However, we know very little about the density of rational points on generic K3 surfaces. Indeed, all the K3 surfaces considered up to this point are very special in moduli:

Proposition 11.1. *A K3 surface X admits an elliptic fibration if and only if there exists a class $D \in \mathrm{NS}(X)$ with $D.D = 0$. If a K3 surface admits an elliptic fibration or an infinite automorphism group, then* $\operatorname{rank} \mathrm{NS}(X) \geq 2$.

Proof. If X admits an elliptic fibration $\eta : X \to \mathbb{P}$, then the generic fiber $[\eta^{-1}(b)] \in \mathrm{NS}(X)$ has self-intersection zero. The class of the fiber is independent from the polarization of X, so $\operatorname{rank} \mathrm{NS}(X) \geq 2$. Conversely, suppose X admits a class $[D] \in \mathrm{NS}(X)$ with $D.D = 0$. By Riemann Roch, either D or $-D$ is effective. Since the effective cone is the union of images of the nef cone under reflections by (-2)-curves ([27] §2), we may asssume D is nef. Basic results on linear series on K3 surfaces [34] imply that $|D|$ is basepoint free and defines an elliptic fibration on X.

If X has Néron–Severi rank one, then the polarization generates the Néron–Severi group and every automorphism fixes the polarization, so the automorphism group is finite (cf. the proof of Lemma 6.9). ∎

Question 11.2. Does there exist any K3 surface X over a number field with $\operatorname{rank} \mathrm{NS}(X) = 1$ and dense rational points?

11.2. Symmetric powers of surfaces

To obtain potential density theorems for generic K3 surfaces, we consider auxilliary varieties, the *symmetric powers*. We refer the reader to [18] for further details. The n-fold symmetric power of a surface X is the quotient

$$X^{(n)} = \underbrace{X \times X \times \cdots \times X}_{n \text{ times}} / \mathfrak{S}_n.$$

This variety admits a natural desingularization by the Hilbert scheme of length-n, zero-dimensional subschemes of X

$$\sigma : X^{[n]} \to X^{(n)}.$$

For a curve C, the behavior of rational points on the symmetric products $C^{(n)}$ bears little resemblance to the behavior of rational points on C. Indeed, for $n > 2g(C) - 2$, $C^{(n)}$ is a projective bundle over $J^n(C)$, and thus always has potentially dense rational points. For a surface X, the rational points on $X^{(n)}$ might very well behave similarly to rational points on X. There is a simple formula relating the Kodaira dimension of a surface and its symmetric product

$$\kappa\big(X^{(n)}\big) = n\kappa(X).$$

Moreover, the quotient morphism $q : X^n \to X^{(n)}$ is almost étale; it is unramified away from a codimension-two subset, the diagonal. If q *were* étale then the Chevalley–Weil Theorem (Proposition 3.4) would apply: potential density of rational points on $X^{(n)}$ implies potential density of rational points on X^n, and thus on X.

11.3. Main Theorem

Theorem 11.3. *Let X be a K3 surface defined over a number field. Assume that X admits a polarization f of degree $2N - 2$. Then rational points on $X^{(n)}$ are potentially dense for some $n \leq N$.*

The proof of Theorem 11.3 divides into cases, depending on the geometry of the elliptic curves contained in X. We first need a geometric result:

Proposition 11.4. *Let X be a K3 surface and f a divisor with $h^0(X, f) > 1$. Then there exists an irreducible, possibly singular, curve of genus one $C \subset X$ so that $f - [C]$ is effective.*

This is proved in detail in §4 of [18]. The main ingredient is [29], where it is shown that for a generic polarized K3 surface there is a one-parameter family of singular curves of genus one in the polarizing class. As in Lemma 6.10, these specialize to the desired curves.

We apply Proposition 11.4 to the polarization class f. The resulting genus-one curve C has self-intersection

$$C.C \leq f.f.$$

If $C.C > 0$ then $[C]$ is big by Riemann–Roch and Theorem 11.3 follows from:

Proposition 11.5. *Let X be a K3 surface defined over a number field, and g the class of a big line bundle of degree $2(n-1)$. Assume there is an irreducible, possibly singular, curve of genus one $C \subset X$ with $[C] = g$. Then rational points on $X^{(n)}$ are potentially dense.*

Sketch Proof. Again, we only give the main ideas of the argument; see [18] for more details. The linear series $|g|$ contains an irreducible curve and so is basepoint free. Indeed, linear series on a K3 surface have no isolated fixed points, only fixed components [34] §2. It follows that g is numerically effective and has no higher cohomology, and thus defines a morphism $j : X \to \mathbb{P}^n$. Consider the incidence correspondence

$$\begin{array}{ccccc} & \mathcal{H} := \{(x, H) \,:\, j(x) \in H\} & \subset & X \times \check{\mathbb{P}}^n \\ \pi_1 \swarrow & & & \searrow \pi_2 \\ X & & & & \check{\mathbb{P}}^n \end{array}$$

where $\check{\mathbb{P}}^n$ is the dual projective space. Over an open subset $V \subset \check{\mathbb{P}}^n$, the fibers of π_2 are smooth curves of genus n.

Consider the degree n component of the relative Albanese

$$\eta \,:\, J^n(\mathcal{H})_V \to V.$$

The Jacobi inversion formula says that the degree n component of the Albanese of a smooth projective curve C of genus n is birational to the symmetric product $C^{(n)}$. Globalizing, we find that $J^n(\mathcal{H})_V$ is birational to the symmetric product $X^{(n)}$. Indeed, for any generic $x_1 + \cdots + x_n \in X^{(n)}$, there is a unique hyperplane $H \supset \{j(x_1), \ldots, j(x_n)\}$, and we obtain an element $\mathcal{O}_{j^{-1}(H)}(x_1 + \cdots + x_n) \in J^n(j^{-1}(H))$.

By Proposition 4.13, it suffices to find a nondegenerate multisection M of η, defined over a number field, with dense rational points. Suppose we have n distinct irreducible curves $C_1, \ldots, C_n$ in $|g|$. Then we can map the product

$$C_1 \times C_2 \times \cdots \times C_n \dashrightarrow X^{(n)}$$

and let $C_1 * \cdots * C_n \subset X^{(n)}$ denote the closed image. The induced morphism $M = C_1 * \cdots * C_n \to V$ is generically finite: There are a finite number of points in the product supported in a generic hyperplane section of X.

The remainder of the argument follows Proposition 10.8. As before, there exists an elliptic fibration $S \to B$ and a surjective morphism $S \to X$ so that the generic fiber is mapped birationally onto a genus-one curve with class g. Pick a smooth fiber $A = \eta^{-1}(H)$ of $J^n(\mathcal{H})_V \to V$ and a point $\alpha \in A(K)$, where K is some number field, so that $-\deg(M/V)\alpha + [M]|_A$ is nondegenerate (see Proposition 4.2.) Using Jacobi inversion, we can express

$$\alpha = x_1 + \cdots + x_n, \quad x_1, \ldots, x_n \in X, \quad j(x_i) \in H.$$

After perhaps choosing a more general α, we may assume that each $x_i \in S_{b_i}$, where the S_{b_i} are distinct, irreducible fibers of $S \to B$, mapped birationally onto their image in X, and defined over a number field. It follows that α lies in the multisection $S_{b_1} * \cdots * S_{b_n}$, which is necessarily nondegenerate. ■

It remains to deal with the case $C.C \leq 0$. Since C is irreducible, adjunction implies C is smooth and $C.C = 0$. In particular, the linear series $|C|$ yields an elliptic fibration $\eta : X \to \mathbb{P}^1$. Then rational points on X are potentially dense by Theorem 6.4. ■

Appendix: Galois Cohomology and Principal Homogeneous Spaces

Let G and Γ be groups and let $\mathrm{Aut}(G)$ denote the automorphism group of G. When G and Γ admit topological structures, all the maps described below are tacitly assumed to be continuous with respect to the relevant topologies.

Let G' be a twisted form of G with respect to Γ. This means we have an action

$$\Gamma \times G \to G$$

$$(\gamma, g) \to \gamma(g),$$

respecting the group structure, i.e., a homomorphism

$$\alpha : \Gamma \to \mathrm{Aut}\,(G).$$

Remark A.1. In applications, Γ is often the Galois group $\mathrm{Gal}\,(F^a/F)$ of the algebraic closure of a field of characteristic zero, G the group of F^a-points of a group scheme over F, and G' the associated Galois module. In this context, we will use the same notation for G' and the group scheme. We are mainly interested in the case where G' is an abelian variety or its N-torsion subgroup. The relation between Galois cohomology and principal homogeneous spaces in this context is developed in [26].

The zeroth cohomology group of G' is the subgroup of invariant elements

$$H^0(\Gamma, G') = \{g \in G' \,:\, \gamma(g) = g \quad \text{for each} \quad \gamma \in \Gamma\}.$$

A cocycle with values in G' is a map

$$\xi : \Gamma \to G'$$

$$\gamma \to \xi(\gamma)$$

satisfying the cocycle condition

$$\xi(\gamma\gamma') = \xi(\gamma)\gamma(\xi(\gamma')).$$

Two cocycles ξ and θ are cohomologous if there exists a $g \in G'$ so that

$$g\theta(\gamma) = \xi(\gamma)\gamma(g), \quad \text{for each} \quad \gamma \in \Gamma.$$

The first cohomology set $H^1(\Gamma, G')$ is the set of equivalence classes of cocycles under the cohomology relation.

If the Γ-action on G' is trivial, we have

$$H^1(\Gamma, G') = \mathrm{Hom}\,(\Gamma, G),$$

the group homomorphisms from Γ to G. In particular, a twisted form of G is governed by an element $\alpha \in H^1(\Gamma, \mathrm{Aut}(G))$.

A G'-principal homogeneous space is a set P with two actions

$$\begin{aligned} \Gamma \times P &\to P \\ (\gamma, p) &\to \gamma(p) \\ P \times G' &\to P \\ (p, g) &\to p \cdot g \end{aligned}$$

satisfying

1. the action of G' is compatible with the Γ-action
$$\gamma(p \cdot g) = \gamma(p) \cdot \gamma(g);$$
2. for each $p_1, p_2 \in P$, there is a unique $g \in G'$ with $p_1 \cdot g = p_2$.

G' acts on itself by multiplication: This is called the *trivial* principal homogeneous space.

Choose $p \in P$. For each $\gamma \in \Gamma$ there exists a unique $\xi(\gamma) \in G'$ so that $\gamma(p) = p \cdot \xi(\gamma)$. We have

$$\begin{aligned} \gamma(\gamma'(p)) &= \gamma(p \cdot \xi(\gamma')) = \gamma(p) \cdot \gamma(\xi(\gamma')) = p \cdot \xi(\gamma)\gamma(\xi(\gamma')) \\ (\gamma\gamma')p &= p \cdot \xi(\gamma\gamma') \end{aligned}$$

so $\xi(\gamma\gamma') = \xi(\gamma)\gamma(\xi(\gamma'))$ and ξ is a cocycle for G'. Changing the basepoint p, we may write

$$\gamma(p \cdot g) = (p \cdot g)\theta(\gamma)$$

where $g\theta(\gamma) = \xi(\gamma)\gamma(g)$, so θ is cohomologous to ξ. Thus every G'-principal homogeneous space determines an element of $H^1(\Gamma, G')$ and conversely.

Now let G be a group and $\mathrm{Aff}(G)$ the semidirect product of G by $\mathrm{Aut}(G)$, so we have an exact sequence

$$1 \to G \to \mathrm{Aff}(G) \xrightarrow{q} \mathrm{Aut}(G) \to 1 \tag{4}$$

admitting a splitting $\sigma : \mathrm{Aut}(G) \hookrightarrow \mathrm{Aff}(G)$. We interpret $\mathrm{Aff}(G)$ as the permutations of G generated by left translations

$$\tau_g : x \to gx \quad g \in G$$

and automorphisms $a \in \mathrm{Aut}\,(G)$. Given $g_1, g_2 \in G$ and $a_1, a_2 \in \mathrm{Aut}\,(G)$, we have

$$\tau_{g_1} a_1 \tau_{g_2} a_2 = \tau_{g_1}(a_1 \tau_{g_2} a_1^{-1}) a_1 a_2 = \tau_{g_1} \tau_{a_1(g_2)} a_1 a_2.$$

Proposition A.2. *Let G' be a Γ-twisted form of a group G, with classifying cocycle $\alpha \in H^1(\Gamma, \mathrm{Aut}\,(G)) = \mathrm{Hom}\,(\Gamma, \mathrm{Aut}\,(G))$. Then $H^1(\Gamma, G')$ corresponds to G-conjugacy classes of homomorphisms*

$$\phi : \Gamma \to \mathrm{Aff}\,(G), \quad \textit{with} \quad q \circ \phi = \alpha.$$

The trivial element corresponds to $\sigma \circ \alpha$.

Proof. Given a cocycle $\xi(\gamma)$, we define $\phi(\gamma) = \tau_{\xi(\gamma)} \alpha(\gamma)$ so that

$$\begin{aligned}
\phi(\gamma\gamma') &= \tau_{\xi(\gamma\gamma')} \alpha(\gamma\gamma') \\
&= \tau_{\xi(\gamma)} \tau_{\gamma(\xi(\gamma'))} \alpha(\gamma) \alpha(\gamma') \\
&= \tau_{\xi(\gamma)} \alpha(\gamma) \tau_{\xi(\gamma')} \alpha(\gamma') \\
&= \phi(\gamma) \phi(\gamma').
\end{aligned}$$

Conversely, each homomorphism $\phi : \Gamma \to \mathrm{Aff}\,(G)$ with $q \circ \phi = \alpha$ yields a cocycle $\xi(\gamma)$. Now suppose that θ and ξ are cohomologous, so that $\theta(\gamma) = g^{-1}\xi(\gamma)\gamma(g)$ for some $g \in G'$, and let ϕ_θ and ϕ_ξ be the corresponding homomorphisms. Then we have

$$\begin{aligned}
\phi_\theta(\gamma) &= \tau_{g^{-1}\xi(\gamma)\gamma(g)} \alpha(\gamma) = \tau_g^{-1} \tau_{\xi(\gamma)} \tau_{\gamma(g)} \alpha(\gamma) \\
&= \tau_g^{-1} \tau_{\xi(\gamma)} \alpha(\gamma) \tau_g = \tau_g^{-1} \phi_\xi(\gamma) \tau_g,
\end{aligned}$$

and the homomorphisms are conjugate. ∎

Remark A.3. For each normal subgroup $H \subset \mathrm{Aut}\,(G)$, the split exact sequence (4) restricts to a split exact sequence

$$1 \to G \to \mathrm{Aff}\, H(G) \xrightarrow{q} H \to 1.$$

Assume that G' is governed by a coycle with values in H,

$$\alpha \in H^1(\Gamma, H) \simeq \mathrm{Hom}\,(\Gamma, H).$$

Then $H^1(\Gamma, G')$ corresponds to homomorphisms

$$\phi : \Gamma \to \mathrm{Aff}\, H(G), \quad \text{with} \quad q \circ \phi = \alpha.$$

References

[1] S. A. Amitsur, Generic splitting fields of central simple algebras, *Ann. of Math. (2)*, **62** (1955), 8–43.

[2] W. Barth, C. Peters and A. Van de Ven, *Compact complex surfaces*, Ergebnisse der Mathematik und ihrer Grenzgebiete (3), vol. 4, Springer-Verlag (Berlin, 1984).

[3] F. Bogomolov and Y. Tschinkel, Density of rational points on Enriques surfaces, *Math. Res. Letters*, **5** (1998), 623-628.

[4] F. Bogomolov and Y. Tschinkel, On the density of rational points on elliptic fibrations. *J. reine angew. Math.*, **511** (1999), 87–93.

[5] F. Bogomolov and Y. Tschinkel, Density of rational points on elliptic K3 surfaces, *Asian J. of Math.*, **4** (2000), 351–368.

[6] F. Campana, Remarques sur le revêtement universel des variétés kählériennes compactes, *Bull. Soc. Math. France,* **122** (1994), 255–284.

[7] F. Campana, Special varieties and classification theory, math.AG/0110051.

[8] J. L. Colliot-Thélène, A. N. Skorobogatov and P. Swinnerton-Dyer, Double fibres and double covers: paucity of rational points, *Acta Arith.*, **79** (1997), 113–135.

[9] A. Corti. A. Pukhlikov, M. Reid, Fano 3-fold hypersurfaces, *Explicit birational geometry of 3-folds*, London Math. Soc. Lecture Note Ser., vol. 281, Cambridge Univ. Press (Cambridge, 2000), pp. 175–258.

[10] O. Debarre, Fano varieties. this volume.

[11] G. Faltings, Endlichkeitssätze für abelsche Varietäten über Zahlkörpern, *Invent. Math.*, **73** (1983). 349–366; Erratum: *Invent. Math.*, **75** (1984), 381.

[12] G. Faltings. Diophantine approximation on abelian varieties, *Ann. of Math. (2)*, **133** (1991), 549–576.

[13] G. Frey and M. Jarden, Approximation theory and the rank of Abelian varieties over large algebraic fields. *Proc. London Math. Soc.*, **28** (1974), 112–128.

[14] G. Grant and E. Manduchi, Root numbers and algebraic points on elliptic surfaces with base P^1, *Duke Math. J.*, **89** (1997), 413–422.

[15] G. Grant and E. Manduchi. Root numbers and algebraic points on elliptic surfaces with elliptic base. *Duke Math. J.*, **93** (1998), 479–486.

[16] A. Grothendieck, Éléments de géométrie algébrique IV: Étude locale des schémas et des morphismes de schémas IV, *Publ. Math. IHES*, vol. 32 (1967).

[17] J. Harris and Y. Tschinkel, Rational points on quartics, *Duke Math. J.*, **104** (2000), 477–500.

[18] B. Hassett and Y. Tschinkel, Abelian fibrations and rational points on symmetric products. *Intern. J. of Math.*, **11** (2000), 1163–1176.

[19] B. Hassett and Y. Tschinkel, Density of integral points on algebraic varieties, in: *Rational Points on Algebraic Varieties* (E. Peyre, Yu. Tschinkel, eds.) Progress in mathematics, vol. 199, Birkhhäuser (2001), pp. 165–197.

[20] V. A. Iskovskikh and Yu. G. Prokhorov, Fano varieties, *Algebraic geometry V*, Encyclopaedia Math. Sci., vol. 47, Springer (1999), pp. 1–247.

[21] K. Kodaira, On compact analytic surfaces I, II, III, *Ann. of Math. (2)*, **71** (1960), 111–152; **77** (1963), 563–626; **78** (1963), 1–40.

[22] J. Kollár, Higher direct images of dualizing sheaves II, *Ann. of Math. (2)*, **124** (1986), no. 1, 171–202.

[23] J. Kollár, *Rational curves on algebraic varieties*, Ergebnisse der Mathematik und ihrer Grenzgebiete (3), vol. 32, Springer (1996).

[24] S. J. Kovács, The cone of curves of a $K3$ surface, *Math. Ann.*, **300** (1994), 681–691.

[25] S. Lang, *Abelian Varieties*, Interscience Tracts in Pure and Applied Mathematics, Number 7 (1959).

[26] S. Lang and J. Tate, Principal homogeneous spaces over abelian varieties, *Amer. J. Math.*, **80** (1958), 659–684.

[27] E. Looijenga and C. Peters, Torelli theorems for Kähler $K3$ surfaces, *Compositio Math.*, **42** (1980/81), 145–186.

[28] T. Miyake, *Modular forms*, Springer (1989).

[29] S. Mori and S. Mukai, The uniruledness of the moduli space of curves of genus 11, *Algebraic geometry (Tokyo/Kyoto, 1982)*, LNM 1016, Springer (1983), pp. 334–353.

[30] S. Mukai, On the moduli space of bundles on $K3$ surfaces I, *Vector bundles on algebraic varieties (Bombay, 1984)*, Tata Inst. Fund. Res. Stud. Math., 11, Tata Inst. Fund. Res. (Bombay, 1987), pp. 341–413.

[31] A. P. Ogg, Cohomology of abelian varieties over function fields, *Ann. of Math. (2)*, **76** (1962), 185–212.

[32] Z. Ran, Hodge theory and deformations of maps, *Compositio Math.*, **97** (1995), 309–328.

[33] P. Roquette, On the Galois cohomology of the projective linear group and its applications to the construction of generic splitting fields of algebras, *Math. Ann.*, **150** (1963), 411–439.

[34] B. Saint-Donat, Projective models of K3 surfaces, *Amer. J. Math.*, **96** (1974), 602–639.

[35] J. P. Serre, *A course in arithmetic*, Graduate Texts In Mathematics, vol. 7, Springer (1973).

[36] J. P. Serre, *Lectures on the Mordell-Weil Theorem*, 2nd edition, Vieweg (Braunschweig, 1990).

[37] I. R. Shafarevich, Principal homogeneous spaces defined over a function field (Russian) *Trudy Mat. Inst. Steklov.*, **64** (1961), 316–346; English translation: *Amer. Math. Soc. Transl. (2)*, **37** (1964), 85–114.

[38] I. R. Shafarevich (ed.) Algebraic surfaces *Proceedings of the Steklov Institute of Mathematics, No. 75* (1965) American Mathematical Society (Providence, R.I., 1965).

[39] T. Shioda, On elliptic modular surfaces, *J. Math. Soc. Japan*, **24** (1972), 20–59.

[40] T. Shioda and H. Inose, On singular $K3$ surfaces, *Complex analysis and algebraic geometry*, Iwanami Shoten (Tokyo, 1977), pp. 119–136.

[41] J. Silverman, Rational points on $K3$ surfaces: a new canonical height, *Invent. Math.*, **105** (1991), 347–373.

[42] H. Sterk, Finiteness results for algebraic $K3$ surfaces, *Math. Z.*, **189** (1985), 507–513.

[43] A. Weil, Arithmétique et géométrie sur les variétés algébriques, in: *Actualités Sci. Indust.*, **206**, Hermann (Paris, 1936), pp. 3–16; reprinted in: *Oeuvres Scientifiques* Vol. I, Springer (1980), pp. 87–100.

Brendan Hassett
Department of Mathematics
Rice University, MS 136
6100 South Main Street
Houston, Texas 77005-1892
U.S.A

hassett@math.rice.edu

BOLYAI SOCIETY
MATHEMATICAL STUDIES, 12

Higher Dimensional Varieties and Rational Points, pp. 283–310.

Fujita's Program and Rational Points

YURI TSCHINKEL

1. Introduction

A classical theme in mathematics is the study of integral solutions of diophantine equations, that is, equations with integral coefficients. The main problems are

— decide the existence (or nonexistence) of solutions;

— find (some or all) solutions;

— describe (qualitatively or quantitatively) the set of solutions.

Even the first of these questions is difficult, in full generality. As we know from logic, it is, in a sense, equivalent to *all* (formal) mathematics: for a given formal language, there exists *one* (non-homogeneous) polynomial $f(t, x_1, \ldots, x_n)$ (with $\mathbb{Z}$-coefficients and ≤ 9 variables x_j) such that the Statement $\#t$ is provable in this language iff the equation $f(t, x_1, \ldots, x_n) = 0$ has a solution with $(x_1, \ldots, x_n) \in \mathbb{Z}^n$ (see [34], [35] for more details). Thus, at least theoretically, one can convert a problem in any field of mathematics, e.g. topology, to a problem in number theory. This is convincing evidence that general diophantine equations are extraordinarily complex.

Nevertheless, there exist classes of equations, for which the above questions may have a satisfactory answer. Below I will show how some of these classes arise in practice. But first notice that the questions are "invariant" under coordinate changes. This central observation signals the entry of algebraic geometry into the subject. Algebraic geometry does not only simplify and streamline manipulations of bulky equations (providing "models"). More importantly, it opens the field to geometric intuition and geometric constructions.

From now on, let $X/\mathbb{Q}$ be a projective algebraic variety. Thus X is defined by a system of homogeneous equations with $\mathbb{Q}$-coefficients. Rational points on X are (equivalence classes of) rational solutions to these equations. It is a natural idea that in the *stable* range, that is, after passing to a finite extension of $\mathbb{Q}$ and restricting to some Zariski open subset X^0 of X there should be a relation between the set of rational points of X^0 and geometric invariants of X. This idea not only explains and unifies "statistical" data (numerical experiments, theoretical results) but also arms us with predictive power. It is a source of inspiration to both fields: arithmetic and geometry.

The focus of this survey is the relationship between the asymptotic distribution of rational points of bounded height on varieties with many rational points and their global geometric invariants: the cone of (classes of) effective divisors and the position of the anticanonical class with respect to this cone. We will be mostly concerned with *rational* varieties so that the points are already Zariski dense. In fact, most of the varieties are compactifications of algebraic groups or homogeneous varieties and the proofs of asymptotics of rational points rely on harmonic analysis on the corresponding adelic spaces. My main goal in these notes is to show where the geometric invariants enter the analysis and how the analysis forces us to introduce new geometric tools.

Here is a brief outline of the paper. In Section 2 we introduce the main geometric invariants of interest to us. In Section 3 we explain the problem and state known results. In Section 4 we define Tamagawa numbers which appear in asymptotics of rational points of bounded height. Finally, in Section 5 we illustrate the interaction between geometry and arithmetic in several examples.

Acknowledgments. I would like to thank the organizers of the conference "Higher Dimensional Varieties and Rational Points" (Budapest, 2001) for the invitation. I gratefully acknowledge the collaboration with V. Batyrev, A. Chambert-Loir, J. Franke, B. Hassett, Yu. Manin, E. Peyre, J. Shalika, M. Strauch and R. Takloo-Bighash which led to the development of the ideas and techniques discussed in this paper. My research was partially supported by the NSF and the Clay Foundation.

2. Geometry

In this section we introduce some geometric background material and define certain geometric invariants which will be used in arithmetic investigations later on. Here we work over an algebraically closed field of characteristic zero.

2.1. Main invariants

Let X be a smooth projective algebraic variety. The main global geometric invariants of interest to us are

— the Picard group $\mathrm{Pic}\,(X)$ and the Néron–Severi group $\mathrm{NS}\,(X)$;
— the (closed) cone of effective divisors

$$\Lambda_{\mathsf{eff}}(X) \subset \mathrm{Pic}\,(X)_{\mathbb{R}} = \mathrm{Pic}\,(X) \otimes \mathbb{R}$$

and (the closure of) the ample cone $\Lambda_{\mathsf{ample}}(X) \subset \Lambda_{\mathsf{eff}}(X)$;
— the canonical class K_X.

One can think of these as being combinatorial data encoding the geometry of X (fibration structures etc.). In all of our applications, $\mathrm{Pic}\,(X) = \mathrm{NS}\,(X)$, $\Lambda_{\mathsf{eff}}(X)$ is finitely generated and $-K_X$ is contained in the interior of $\Lambda_{\mathsf{eff}}(X)$.

Example 2.1.1. Let $X = X_d \subset \mathbb{P}^n$ be the hypersurface defined by a (smooth) homogeneous form f of degree d in $n+1$ variables (with $n \geq 4$). Then

— $\mathrm{Pic}\,(X) = \mathbb{Z} \cdot H$ (where H is the hyperplane class);
— $\Lambda_{\mathsf{eff}}(X) = \Lambda_{\mathsf{ample}}(X)$ is generated by H;
— $-K_X = (n+1-d)H$.

Example 2.1.2. Let $X = \bar{M}_{0,5}$ be the moduli space of stable curves of genus 0 with 5 punctures, classically known as the Del Pezzo surface of degree 5. It can be realized as a blow-up of $\mathbb{P}^2$ in four points. We write H for the (proper transform) of the hyperplane class on $\mathbb{P}^2$ and $E_1, \ldots, E_4$ for the classes of the exceptional curves. Then

— $\mathrm{Pic}\,(X) = \mathbb{Z}^5 = \langle H, E_1, \ldots, E_4 \rangle$;

— $\Lambda_{\mathrm{eff}}(X)$ is generated by the classes $E_i, H - E_i - E_j$, and $\Lambda_{\mathrm{ample}}(X)$ is dual to it (with respect to the intersection pairing);

— $-K_X = 3H - (E_1 + \cdots + E_4)$.

For more information, see [33] or [24].

Example 2.1.3. Let $X = \bar{M}_{0,6}$ be the moduli space of stable curves of genus 0 with 6 punctures. It can be realized as a blow-up of $\mathbb{P}^3$ in 5 points and 10 lines joining pairs of these points. We write H for the hyperplane class, $E_1, \ldots, E_5$ for the proper transforms of the 5 points and E_{ij} for the proper transforms of lines through the points i and j. By [24], we have:

— $\mathrm{Pic}\,(X) = \mathbb{Z}^{16} = \langle H, E_i, E_{ij}\rangle$ with $i \neq j \in [1, \ldots, 5]$, $E_{ij} = E_{ji}$;

— $\Lambda_{\mathrm{eff}}(X)$ is generated by

$$E_i, E_{ij}, H - (E_i + E_j + E_k + E_{ij} + E_{ik} + E_{jk}),$$

$$2H - (E_1 + \cdots + E_5) - (E_{ik} + E_{i\ell} + E_{jk} + E_{j\ell})$$

(altogether 40 generators);

— $-K_X = 4H - 2(E_1 + \cdots + E_5) - (\sum_{i<j} E_{ij})$.

The $\mathfrak{S}_6$-invariant ample cone has two generators, denoted by L_3 and L_4. The associated line bundles give maps

$$f_3 : \bar{M}_{0,6} \to S_3 \subset \mathbb{P}^4,$$

$$f_4 : \bar{M}_{0,6} \to I_4 \subset \mathbb{P}^4,$$

with images the *Segre* cubic:

$$S_3 : \sum_{i=0}^{5} x_i^3 = \sum_{i=0}^{5} x_i = 0.$$

respectively, the *Igusa* quartic:

$$I_4 : (x_0x_1 + x_0x_2 + x_1x_2 - x_3x_4)^2 - 4x_0x_1x_2(x_0 + x_1 + x_2 + x_3 + x_4) = 0,$$

(see [26] for more information about these varieties).

Example 2.1.4. Let X be an equivariant compactification of a unipotent algebraic group. Then

— Pic $(X) = \oplus_{j\in\mathcal{J}} D_j$, where D_j are classes of the irreducible components of the boundary;
— $\Lambda_{\mathrm{eff}}(X)$ is generated by the D_j;
— $-K_X = \sum_{j\in\mathcal{J}} \kappa_j D_j$, with $\kappa_j \geq 2$ for all $j \in \mathcal{J}$ (see [23] for the case of a commutative group).

Example 2.1.5. Let $X = \mathrm{B} \setminus \mathrm{G}$ be a generalized flag variety: $\mathbb{G}$ is a semi-simple algebraic group and $\mathrm{B} \subset \mathrm{G}$ a Borel subgroup. Then
— Pic $(X)_{\mathbb{Q}} = \mathfrak{X}^*(\mathrm{T})_{\mathbb{Q}}$ (the characters of the maximal torus $\mathrm{T} \subset \mathrm{B}$);
— $\Lambda_{\mathrm{eff}}(X) = \Lambda_{\mathrm{ample}}(X)$ is generated by the fundamental weights ω_j, $j = 1, \ldots, \mathrm{rk}\, \mathrm{G}$;
— $-K_X = 2\rho$ (sum of roots occurring in the unipotent radical of B).

Example 2.1.6. Let X be the *wonderful* compactification of a semi-simple group G of adjoint type (see [15]). Then
— Pic $(X)_{\mathbb{Q}} = \mathfrak{X}^*(\mathrm{T})_{\mathbb{Q}}$ (where T is the maximal torus in G);
— $\Lambda_{\mathrm{eff}}(X)$ is generated by the positive simple roots, α_j, $j = 1, \ldots, \mathrm{rk}\, \mathrm{G}$;
— $\Lambda_{\mathrm{ample}}(X)$ is generated by the fundamental weights ω_j, $j = 1, \ldots, \mathrm{rk}\, \mathrm{G}$;
— $-K_X = 2\rho + \sum_j \alpha_j$.

2.2. Rough classification

Let me recall the very rough classification of smooth algebraic varieties (over an algebraically closed field) according to the position of the anticanonical class with respect to the ample cone. There are three main cases:

— **Fano:** $-K_X$ ample;
— **general type:** K_X ample;
— **intermediate type:** none of the above.

In finer classification schemes and in the minimal model program it is important to include varieties with certain "mild" singularities (see [29]). Similarly, in many arithmetic questions, the passage to singular varieties is inevitable (see Section 4). Some examples of varieties in each group are contained in the following table:

dim	Fano	Intermediate type	General type
1	$\mathbb{P}^1$	elliptic curves	C, $g(C) \geq 2$
2	$\mathbb{P}^2$, $\mathbb{P}^1 \times \mathbb{P}^1$, $\bar{M}_{0,5}$, $X_3 \subset \mathbb{P}^3, \ldots$	$K3$ surfaces: $X_4 \subset \mathbb{P}^3, \ldots$ abelian surfaces,...	...
3	smooth: ~ 120 families singular: $\bar{M}_{0,6}, \ldots$.	Calabi–Yau varieties ...	...

Smooth Fano surfaces, also known as Del Pezzo surfaces, have been classified by the Italian school. The classification of smooth Fano threefolds, initiated by Fano, advanced by Iskovskikh [27] and completed by Mori–Mukai [37], is a major achievement of modern algebraic geometry. After Kollár and Mori [31], one knows that there are only finitely many families of smooth Fano varieties in each dimension. It seems plausible, that certain classes of varieties of intermediate type will also admit a good description (for example, holomorphic symplectic varieties). For a recent approach to the classification of algebraic varieties see [8].

In our arithmetic applications we will mostly encounter varieties on the left side of the table.

2.3. Cones

Let (A, Λ) be a pair consisting of a lattice and a strictly convex (closed) cone in $A_{\mathbb{R}}$: $\Lambda \cap -\Lambda = 0$. Let $(\check{A}, \check{\Lambda})$ the pair consisting of the dual lattice and the dual cone defined by

$$\check{\Lambda} := \left\{ \check{\lambda} \in \check{A}_{\mathbb{R}} \mid \langle \lambda', \check{\lambda} \rangle \geq 0, \ \forall \lambda' \in \Lambda \right\}.$$

The lattice $\check{A}$ determines the normalization of the Lebesgue measure $d\check{a}$ on $\check{A}_{\mathbb{R}}$ (covolume =1). For $a \in A_{\mathbb{C}}$ define

$$\mathcal{X}_{\Lambda}(a) := \int_{\check{\Lambda}} e^{-\langle a, \check{a} \rangle} d\check{a}. \tag{2.1}$$

The integral converges absolutely and uniformly for $\Re(a)$ in compacts contained in the interior Λ° of Λ.

Definition 2.3.1. Assume that X is smooth, $\mathrm{NS}(X) = \mathrm{Pic}(X)$ and that $-K_X$ is in the interior of $\Lambda_{\mathrm{eff}}(X)$. We define

$$\gamma(X) := \mathcal{X}_{\Lambda_{\mathrm{eff}}(X)}(-K_X).$$

Remark 2.3.2. This constant measures the volume of the polytope obtained by intersecting the affine hyperplane $\langle -K_X, \cdot \rangle = 1$ with the *dual* to the cone of effective divisors $\Lambda_{\mathrm{eff}}(X)$ in the dual to the Néron–Severi group. The explicit determination of $\gamma(X)$ can be a serious problem in linear programming. For example, let X be the moduli space $\bar{M}_{0,6}$ (see Example 2.1.3). The dual to the cone $\Lambda_{\mathrm{eff}}(X)$ has 3905 generators (in a 16-dimensional vector space), forming 25 orbits under the action of the symmetric group $\mathbb{S}_6$.

Let $(A, \Lambda, -K)$ be a triple consisting of a (torsion free) lattice $A = \mathbb{Z}^n$, a (closed) strictly convex polyhedral cone in $A_{\mathbb{R}}$ generated by finitely many vectors in A and a vector $-K \subset \Lambda^\circ$ (the interior of Λ). For $L \in \Lambda^\circ$ we define

$$a(\Lambda, L) = \inf \{a \mid aL + K \in \Lambda\}$$

and $b(\Lambda, L)$ as the codimension of the minimal face of Λ containing the class $a(\Lambda, L)L + K$. Obviously, for $L = -K$ we get that $a(\Lambda, -K) = 1$ and that $b(\Lambda, -K)$ equals the rank of A.

2.4. Cones in geometry

In this section we assume that X is a smooth projective Fano variety ($-K_X \in \Lambda_{\mathrm{ample}}(X)$). We have a sequence of inclusions:

$$\Lambda_{\mathrm{ample}}(X) \subset \Lambda_{\mathrm{eff}}(X) \subset \mathrm{Pic}\,(X)_{\mathbb{R}}.$$

Projectivity implies that $\Lambda_{\mathrm{ample}}(X)$ contains an open subset. The finer classification theories of Mori (resp. Fujita) are based on the study of $\Lambda_{\mathrm{ample}}(X)$ (resp. $\Lambda_{\mathrm{eff}}(X)$). We now give a simplified picture of these theories by stating some basic theorems and conjectures. We refer to [29] and [20] for more details. The conjectures are proved for all Fano varieties of dimension ≤ 3 and certain classes of varieties, like toric varieties, in arbitrary dimension [31], [1].

Theorem 2.4.1. *The cone $\Lambda_{\mathrm{ample}}(X)$ is finitely generated.*

Conjecture 2.4.2. *The cone $\Lambda_{\mathrm{eff}}(X)$ is finitely generated.*

Remark 2.4.3. The ample cone $\Lambda_{\mathrm{ample}}(X)$ can be finitely generated even when $-K_X$ is not ample (or even effective). For example, $\Lambda_{\mathrm{ample}}(X)$ is finitely generated for *every* toric variety or every variety with $\mathrm{rk}\,\mathrm{NS}\,(X) = 1$.

Let $\Lambda \subset \mathrm{Pic}\,(X)_{\mathbb{R}}$ be a finitely generated rational cone. For $L \in \Lambda^{\circ}$, with $a(\Lambda, L) \in \mathbb{Q}$, we define

$$\mathrm{R}(\Lambda, L) := \oplus_{\nu \geq 0} H^0\Big(X, \big(L^{\otimes a(\Lambda,L)} \otimes K_X\big)^{\otimes k\nu}\Big)$$

(where k is the denominator of $a(\Lambda, L)$).

Conjecture 2.4.4. *Let X be a smooth Fano variety and L (the class of) a very ample line bundle on X. Then the rings $\mathrm{R}\big(\Lambda_{\mathrm{ample}}(X), L\big)$ and $\mathrm{R}\big(\Lambda_{\mathrm{eff}}(X), L\big)$ are finitely generated.*

Moreover, there should be diagrams:

Mori	**Fujita**
$X \quad L\vert_{X_y} \sim -K_{X_y}$	$X \quad L\vert_{X_y} \sim -K_{X_y} + E$
$\vdots$	$\vdots$
$\downarrow$	$\downarrow$
$Y = \mathrm{Proj}\,\big(\mathrm{R}\big(\Lambda_{\mathrm{ample}}(X), L\big)\big)$	$Y = \mathrm{Proj}\,\big(\mathrm{R}\big(\Lambda_{\mathrm{eff}}(X), L\big)\big)$

where L is ample and E a rigid effective divisor (and $\sim$ means proportional). Recall that a divisor E is called *rigid* if $h^0\big(\mathcal{O}(\nu E)\big) = 1$ for $\nu \gg 0$. Intuitively, the following is happening: at $a = a\big(\Lambda_{\mathrm{ample}}(X), L\big)$ (resp. $a = \big(\Lambda_{\mathrm{ample}}(X), L\big)$) the divisor $aL + K_X$ just fails to be ample (resp. *big*). But a divisor in this class could still move and define a fibration, rather than an embedding.

Remark 2.4.5. Fujita's program follows from Mori's Minimal Model Program [1]. We formulated it independently since in arithmetic applications we need the *effective* cone. Moreover, in our examples the anticanonical class is in the interior of the effective cone but not always ample (for example, toric varieties) and the cone $\Lambda_{\mathrm{ample}}(X)$ is not always finitely generated (for example, equivariant compactifications of unipotent groups).

3. Arithmetic

3.1. Asymptotics

Let F be a finite extension of $\mathbb{Q}$. The conjectures of Bombieri, Lang and Vojta predict that on a variety X (over F) of general type the set $X(F)$ of F-rational points is not Zariski dense (see [52]). Faltings proved this for subvarieties of abelian varieties [18]. On the other hand, one expects that Fano varieties should have a Zariski dense set of rational points, at least after a finite extension of F. See [24], [7] for some partial results and [22] for a recent survey. Most of the time, we will be interested in *rational* varieties, so that rational points are *a priori* Zariski dense. We will study the asymptotic distribution of rational points of bounded height.

For $X = \mathbb{P}^n$ one can define a *height*

$$\begin{aligned} H : \quad \mathbb{P}^n(F) &\to \mathbb{R}_{>0}, \\ (x_0 : \cdots : x_n) &\mapsto \prod_v \max_j \left(|x_j|_v \right), \end{aligned}$$

where the product is over the set $\mathrm{Val}\,(F)$ of all valuations of F. The choice of the norm $\max_j \left(|x_j|_v \right)$ in the vector space F^{n+1} is to some extent arbitrary. Namely, replacing the norms by comparable (homogeneous) norms at *finitely* many places one obtains a new height. For example, at the infinite places one could replace

$$\max \left(|x_j|_v \right) \quad \text{by} \quad \left(\sum_j |x_j|_v^2 \right)^{1/2}.$$

One refers to such choices of norms as fixing the *metrization* (see Section 4 for more details).

Let X be projective, L a very ample line bundle on X and

$$f_L : X \to \mathbb{P}^n = \mathbb{P}\big(H^0(X, L)\big)$$

the corresponding embedding into a projective space. We have an induced (exponential) height:

$$\begin{aligned} H_{\mathcal{L}} : X(F) &\to \quad \mathbb{R}_{>0}, \\ x \quad &\mapsto H\big(f_L(x)\big). \end{aligned}$$

We wrote $\mathcal{L}$ to stress the dependence of the height on the metrization. The set of points of $\mathcal{L}$-height bounded by $B > 0$ is finite. We define the *counting function*

$$N(U, \mathcal{L}, B) := \#\{ x \in U(F) \mid H_{\mathcal{L}}(x) \leq B \},$$

where $U \subset X$ is a Zariski open subset.

The following theorem summarizes most of the known results concerning asymptotics of rational points of bounded height on algebraic varieties.

Theorem 3.1.1. *Let X/F be one of the following varieties:*

— *smooth complete intersection of small degree (for example,* [6]*);*

— *flag variety* [19];

— *smooth toric variety* [5];

— *smooth equivariant compactification of* G/U*-horospherical variety, where* G *is a semi-simple group and* $\mathrm{U} \subset \mathrm{G}$ *a maximal unipotent subgroup* [49];

— *smooth equivariant compactification of* $\mathbb{G}_a^n$ [12];

— *smooth bi-equivariant compactification of a unipotent group* [46], [47];

— *a wonderful compactification of an anisotropic form of a simple algebraic group of adjoint type* [44], [45].

Let $H_{\mathcal{L}}$ be an appropriate height such that the corresponding class $L \in$ Pic(X) is contained in the interior of the cone of effective divisors and

$$a(L) = a\big(\Lambda_{\mathrm{eff}}(X), L\big), \quad b(L) = b\big(\Lambda_{\mathrm{eff}}(X), L\big).$$

There exist a dense Zariski open subset $U \subset X$ and a and a constant $c(U, \mathcal{L}) > 0$ such that

$$N(U, \mathcal{L}, B) = \frac{c(U, \mathcal{L})}{a(L)(b(L)-1)!} B^{a(L)} \big(\log(B)\big)^{b(L)-1} \big(1 + o(1)\big),$$

as $B \to \infty$.

Remark 3.1.2. The constant $c(U, \mathcal{L})$ depends, of course, not only on the geometric data (U, L) but also on the metrization. The definitions are postponed until Section 4.

Remark 3.1.3. The theorem holds for *singular* varieties as well. One has to apply the methods to some (equivariant) desingularization and use the functoriality of heights: if $\pi : X' \to X$ is a birational map, inducing an isomorphism $\pi^{-1}(U) \xrightarrow{\sim} U \subset X$ then

$$H_{\pi^*\mathcal{L}}(\pi^{-1}(x)) = H_{\mathcal{L}}(x).$$

Consequently,

$$N(U, \mathcal{L}, B) = N(\pi^{-1}(U), \pi^*\mathcal{L}, B).$$

Remark 3.1.4. The only other classes of varieties (that I am aware of), for which similar asymptotic formulas have been proved, is the class of certain twisted products of flag varieties considered in [48] and $\bar{M}_{0,5}$ in its anticanonical embedding [16].

Remark 3.1.5. Notice that with the exception of complete intersections the varieties from Theorem 3.1.1 have a rather simple "cellular" structure. In particular, we can parametrize all rational points in some dense Zariski open subset. The theorem is to be understood as a statement about *heights*: even the torus $\mathbb{G}_m^2$ has very nontrivial embeddings into projective spaces and in each of these embeddings we have a different counting problem.

Remark 3.1.6. The restriction to Zariski open subvarieties is necessary, as X may contain proper subvarieties Z such that

$$\lim_{B\to\infty} \frac{N(X \setminus Z, \mathcal{L}, B)}{N(Z, \mathcal{L}, B)} = 0.$$

It is futile to try to relate the asymptotic of rational points on Z to geometric invariants of X. Such varieties are called $\mathcal{L}$-*accumulating*. The simplest example is the blow-up of $\mathbb{P}^2$ in one point: the exceptional curve is accumulating with respect to $-K_X$.

Stratifications by accumulating subvarieties have been considered in [36].

4. Tamagawa Numbers

In this section we show how to associate to a metrized ample line bundle $\mathcal{L}$ a constant $c(\mathcal{L})$ which should appear in the leading term of asymptotic expansions for the number of rational points of bounded $\mathcal{L}$-height. If L is proportional to $-K_X$ then the main ingredient of $c(\mathcal{L})$ is an adelic integral. Otherwise. we use Fujita's program to reduce to this situation.

The fact that a metrized canonical line bundle determines a measure is well known in differential geometry. The same reasoning applies over the adeles. Integrals of such (suitably regularized) measures give Tamagawa numbers which appear as factors in leading coefficients in Laurent expansions of zeta functions.

4.1. Metrizations of line bundles

Notations 4.1.1. Let F be a number field and disc (F) the discriminant of F (over $\mathbb{Q}$). The set of places of F will be denoted by Val (F). We shall write $v \mid \infty$ if v is archimedean and $v \nmid \infty$ if v is nonarchimedean. For any place v of F we denote by F_v the completion of F the completion of F at v and by $\mathfrak{o}_v$ the ring of v-adic integers (for $v \nmid \infty$). Let q_v be the cardinality of the residue field $\mathbb{F}_v$ of F_v for nonarchimedean valuations. The local absolute value $|\cdot|_v$ on F_v is the multiplier of the Haar measure, i.e., $d(ax_v) = |a|_v dx_v$ for some Haar measure dx_v on F_v. We denote by $\mathbb{A} = \mathbb{A}_F = \prod_v' F_v$ the adele ring of F.

Definition 4.1.2. Let X be an algebraic variety over F and L a line bundle on X. A v-adic metric on L is a family $\left(\|\cdot\|_x\right)_{x\in X(F_v)}$ of v-adic Banach norms on L_x such that for every Zariski open $U \subset X$ and every section $g \in H^0(U, L)$ the map

$$U(F_v) \to \mathbb{R}, \quad x \mapsto \|g\|_x,$$

is continuous in the v-adic topology on $U(F_v)$.

Example 4.1.3. Assume that L is generated by global sections. Choose a basis $(g_j)_{j\in[0,\ldots,n]}$ of $H^0(X, L)$ (over F). If g is a section such that $g(x) \neq 0$ then define

$$\|g\|_x := \max_{0\le j\le n} \left(\left|\frac{g_j}{g}(x)\right|_v\right)^{-1},$$

otherwise $\|0\|_x := 0$. This defines a v-adic metric on L. Of course, this metric depends on the choice of $(g_j)_{j\in[0,\dots,n]}$.

Definition 4.1.4. Assume that L is generated by global sections. An adelic metric on L is a collection of v-adic metrics, for every $v \in \mathrm{Val}(F)$, such that for all but finitely many $v \in \mathrm{Val}(F)$ the v-adic metric on L is defined by means of some *fixed* basis $(g_j)_{j\in[0,\dots,n]}$ of $H^0(X, L)$.

We shall write $\|\cdot\| := (\|\cdot\|_v)$ for an adelic metric on L and call a pair $\mathcal{L} = (L, \|\cdot\|)$ an adelically metrized line bundle. Metrizations extend naturally to tensor products and duals of metrized line bundles, which shows a way to define adelic metrizations on arbitrary line bundles L (on projective X): represent L as $L = L_1 \otimes L_2^{-1}$ with very ample L_1 and L_2. Assume that L_1, L_2 are adelically metrized. An adelic metrization of L is any metrization which for all but finitely many v is induced from the metrizations on L_1, L_2.

Definition 4.1.5. Let $\mathcal{L} = (L, \|\cdot\|)$ be an adelically metrized line bundle on X and g an F-rational section of L. Let $U \subset X$ be the maximal Zariski open subset of X where g is defined and is $\neq 0$. For all $x = (x_v)_v \in U(\mathbb{A})$ we define the local

$$H_{\mathcal{L},g,v}(x_v) := \|g\|_{x_v}^{-1}$$

and the global *height function*

$$H_{\mathcal{L}}(x) := \prod_{v\in\mathrm{Val}(F)} H_{\mathcal{L},g,v}(x_v).$$

By the product formula, the restriction of the global height to $U(F)$ does not depend on the choice of g.

Example 4.1.6. For $X = \mathbb{P}^1 = (x_0 : x_1)$ one has $\mathrm{Pic}(X) = \mathbb{Z}$, spanned by the class $L = [(1:0)]$. For all $g = x_0/x_1 \in \mathbb{G}_a(\mathbb{A})$ we define

$$H_{\mathcal{L},g,v}(g_v) = \max(1, |g|_v).$$

The restriction of $H_{\mathcal{L}} = \prod_v H_{\mathcal{L},g,v}$ to $\mathbb{G}_a(F) \subset \mathbb{P}^1$ is the usual height on $\mathbb{P}^1$ (with respect to the usual metrization of $\mathcal{L} = \mathcal{O}(1)$).

4.2. Geometry

I follow closely the exposition in [5]. Let E/F be some finite Galois extension such that all of the following constructions are defined over E. Let $(V, \mathcal{L})$ be a smooth *quasi-projective* d-dimensional variety together with a metrized very ample line bundle $\mathcal{L}$ which embeds V in some projective space $\mathbb{P}^n$. We denote by $\overline{V}^{\mathcal{L}}$ the normalization of the projective closure of $V \subset \mathbb{P}^n$. In general, $\overline{V}^{\mathcal{L}}$ is singular. We will introduce several notions relying on a resolution of singularities

$$\rho : X \to \overline{V}^{\mathcal{L}}.$$

Naturally, the defined objects will be independent of the choice of the resolution. For $\Lambda \subset \mathrm{NS}\,(X)_{\mathbb{R}}$ we define

$$a(\Lambda, \mathcal{L}) := a(\Lambda, \rho^*\mathcal{L}).$$

We will always assume that $a\big(\Lambda_{\mathrm{eff}}(X), \mathcal{L}\big) > 0$.

Definition 4.2.1. A pair $(V, \mathcal{L})$ is called *primitive* if $a\big(\Lambda_{\mathrm{eff}}(X), \mathcal{L}\big) \in \mathbb{Q}_{>0}$ and if there exists a resolution of singularities

$$\rho : X \to \overline{V}^{\mathcal{L}}$$

such that for some $k \in \mathbb{N}$

$$\Big((\rho^*\mathcal{L})^{\otimes a(\Lambda_{\mathrm{eff}}(X), \mathcal{L})} \otimes K_X\Big)^{\otimes k} = \mathcal{O}(D),$$

where D is a *rigid* effective divisor $(h^0\big(X, \mathcal{O}(\nu D)\big) = 1$ for all $\nu \gg 0)$.

Example 4.2.2. of a primitive pair: $(V, -\mathcal{K}_V)$, where V is a smooth projective Fano variety and $-\mathcal{K}_V$ is a metrized anticanonical line bundle.

Let $k \in \mathbb{N}$ be such that $a(\Lambda, \mathcal{L})k \in \mathbb{N}$ and consider

$$\mathrm{R}(\Lambda, \mathcal{L}) := \oplus_{\nu \geq 0} H^0\Big(X, \Big(\big((\rho^*\mathcal{L})^{a(\Lambda, \mathcal{L})} \otimes K_X\big)^{\otimes k}\Big)^{\otimes \nu}\Big).$$

As explained in Section 2, in both cases ($\Lambda = \Lambda_{\mathrm{ample}}$ or $\Lambda = \Lambda_{\mathrm{eff}}$) it is expected that $\mathrm{R}(\Lambda, \mathcal{L})$ is finitely generated and that we have a fibration

$$\pi = \pi_{\mathcal{L}} : X \to Y^{\mathcal{L}},$$

where $Y^{\mathcal{L}} = \mathrm{Proj}\left(\mathrm{R}(\mathcal{L}, \Lambda)\right)$. For $\Lambda = \Lambda_{\mathrm{eff}}(X)$ the generic fiber of π is (expected to be) a primitive variety in the sense of Definition 4.2.1. More precisely, there should be a diagram:

$$\begin{array}{c} \rho : X \to \overline{V}^{\mathcal{L}} \supset V \\ \downarrow \\ Y^{\mathcal{L}} \end{array}$$

such that:

- $\dim(Y^{\mathcal{L}}) < \dim(X)$;
- there exists a Zariski open $U \subset Y^{\mathcal{L}}$ such that for all $y \in U(\mathbb{C})$ the pair $(V_y, \mathcal{L}_y)$ is primitive (here $V_y = \pi^{-1}(y) \cap V$ and $\mathcal{L}_y$ is and $\mathcal{L}_y$ is the restriction of $\mathcal{L}$ to V_y);
- for all $y \in U(\mathbb{C})$ we have $a\left(\Lambda_{\mathrm{eff}}(X), \mathcal{L}\right) = a\left(\Lambda_{\mathrm{eff}}(V_y), \mathcal{L}_y\right)$;
- For all $k \in \mathbb{N}$ such that $a\left(\Lambda_{\mathrm{eff}}(X), \mathcal{L}\right) k \in \mathbb{N}$ the vector bundle

$$\mathcal{L}_k := R^0\pi_*\left(\left((\rho^*\mathcal{L})^{\otimes a(\Lambda_{\mathrm{eff}}(X),\mathcal{L})} \otimes K_X\right)^{\otimes k}\right)$$

is in fact an ample invertible sheaf on $Y^{\mathcal{L}}$.

Such a fibration will be called an *$\mathcal{L}$-primitive fibration*. A variety may admit several primitive fibrations.

Example 4.2.3. Let $X \subset \mathbb{P}_1^n \times \mathbb{P}_2^n$ $(n \geq 2)$ be a hypersurface given by a bi-homogeneous form of bi-degree (d_1, d_2). Both projections $X \to \mathbb{P}_1^n$ and $X \to \mathbb{P}_2^n$ are $\mathcal{L}$-primitive, for appropriate $\mathcal{L}$. In particular, for $n = 3$ and $(d_1, d_2) = (1, 3)$ there are *two* distinct $-\mathcal{K}_X$-primitive fibrations: one onto a point and another onto $\mathbb{P}_1^3$.

4.3. Tamagawa numbers

For smooth projective Fano varieties X with an adelically metrized anticanonical line bundle Peyre defined in [41] a Tamagawa number, generalizing the classical construction for linear algebraic groups. We need to further generalize this to primitive pairs.

Abbreviate $a(\mathcal{L}) = a\left(\Lambda_{\mathrm{eff}}(X), \mathcal{L}\right)$ and let $(V, \mathcal{L})$ be a primitive pair such that

$$\mathcal{O}(D) := \left((\rho^*\mathcal{L})^{\otimes a(\mathcal{L})} \otimes K_X\right)^{\otimes k},$$

where k is such that $a(\mathcal{L})k \in \mathbb{N}$ and D is a rigid effective divisor as in Definition 4.2.1. Choose an F-rational section $g \in H^0(X, \mathcal{O}(D))$; it is unique up to multiplication by F^*. Choose local analytic coordinates $x_{1,v}, \ldots, x_{d,v}$ in a neighborhood U_x of $x \in X(F_v)$. In U_x the section g has a representation

$$g = f^{ka(\mathcal{L})}(dx_{1,v} \wedge \cdots \wedge dx_{d,v})^k,$$

where f is a local section of L. This defines a local v-adic measure in U_x by

$$\omega_{\mathcal{L},g,v} := \|f\|_{x_v}^{a(\mathcal{L})} dx_{1,v} \cdots dx_{d,v},$$

where $dx_{1,v} \cdots dx_{d,v}$ is the selfdual Haar measure on F_v^d normalized as in [50]. A standard argument shows that $\omega_{\mathcal{L},g,v}$ glues to a v-adic measure on $X(F_v)$. The restriction of this measure to $V(F_v)$ does not depend on the choice of the resolution $\rho : X \to \overline{V}^{\mathcal{L}}$. Thus we have a measure on $V(F_v)$.

Denote by $(D_j)_{j\in\mathcal{J}}$ the irreducible components of the support of D and by

$$\mathrm{Pic}\,(V, \mathcal{L}) := \mathrm{Pic}\,(X \setminus \cup_{j\in\mathcal{J}} D_j).$$

The Galois group Γ acts on $\mathrm{Pic}\,(V, \mathcal{L})$. Let S be a finite set of valuations of bad reduction for the data (ρ, D_j, etc.), including the archimedean valuations. Put $\lambda_v = 1$ for $v \in S$, $\lambda_v = L_v(1, \mathrm{Pic}\,(V, \mathcal{L}))$ (for $v \notin S$) and

$$\omega_{\mathcal{L}} := L_S^*(1, \mathrm{Pic}\,(V, \mathcal{L}))\,|\,\mathrm{disc}\,(F)|^{-d/2} \prod_v \lambda_v^{-1} \omega_{\mathcal{L},g,v}.$$

(Here L_v is the local factor of the Artin L-function associated to the Γ-module $\mathrm{Pic}\,(V, \mathcal{L})$ and $L_S^*(1, \mathrm{Pic}\,(V, \mathcal{L}))$ is the leading coefficient at $s = 1$ of the partial Artin L-function.) By the product formula, the measure does not depend on the choice of the F-rational section g. Define

$$\tau_{\mathcal{L}}(V) := \int_{\overline{X(F)}} \omega_{\mathcal{L}},$$

where $\overline{X(F)} \subset X(\mathbb{A})$ is the closure of $X(F)$ in the direct product topology. The convergence of the Euler product follows from

$$h^1(X, \mathcal{O}_X) = h^2(X, \mathcal{O}_X) = 0$$

(which holds for all varieties in Theorem 3.1.1 and, conjecturally, for all X arising from primitive pairs $(V, \mathcal{L})$). We have a map

$$\tilde{\rho} : \mathrm{Pic}\,(X)_{\mathbb{R}} \to \mathrm{Pic}\,(V, \mathcal{L})_{\mathbb{R}}$$

and we denote by

$$\Lambda_{\mathrm{eff}}(V,\mathcal{L}) := \tilde{\rho}\big(\Lambda_{\mathrm{eff}}(X)\big) \subset \mathrm{Pic}\,(V,\mathcal{L})_{\mathbb{R}}.$$

Definition 4.3.1. Let $(V,\mathcal{L})$ be a primitive pair as above. Define

$$c(V,\mathcal{L}) := \mathcal{X}_{\Lambda_{\mathrm{eff}}(V,\mathcal{L})}\big(\tilde{\rho}([-K_X])\big)\cdot\big|H^1(\Gamma,\mathrm{Pic}\,(V,\mathcal{L}))\big|\cdot\tau_{\mathcal{L}}(V).$$

Example 4.3.2. Let us return to the Example 2.1.3. For the image of $\bar{M}_{0,6}$ under f_3 (the Segre cubic) one knows an upper and lower bound:

$$c'B^2\log\,(B)^5 \le N(U,L_3,B) \le c''B^2\log\,(B)^5,$$

for an appropriate Zariski open U and some constants $c',c'' \geq 0$ [53]. An *asymptotic* formula for $N(U,L_3,B)$ of this shape would be compatible with the description in Theorem 3.1.1. Moreover, we are now in the position to specify a constant c which should appear in this asymptotic (answering a question in [53]). Indeed, the Segre cubic threefold is singular (it has 10 isolated double points). The blow-up $X = \tilde{S}_3$ of 5 points on $\mathbb{P}^3$ is a desingularization of S_3. Its Picard group $\mathrm{Pic}\,(X)$ is freely generated by the classes H, E_j (for $j = 1,\ldots,5$). The effective cone is generated by the classes E_j, $H-(E_i+E_j+E_k)$ and the anticanonical class is given by

$$-K_X = 4H - 2(E_1+\cdots+E_5).$$

The line bundle on S_3 giving the Segre embedding pulls back to $L = -1/2\cdot K_X$. Clearly, $a(L) = 2$ and $b(L) = \mathrm{rk}\,\mathrm{Pic}\,(X) = 6$ (see also [5], Section 5.2). The predicted leading constant $c = \gamma\cdot\tau$, where

$$\gamma = \frac{2^5}{5!}\mathcal{X}_{\Lambda}(-K_X)$$

and

$$\tau = \tau_\infty\cdot\prod_p (1-1/p)^6(1+6/p+6/p^2+1/p^3),$$

(τ_∞ is the "singular integral" — the archimedean density of X).

If $(V,\mathcal{L})$ is not primitive then, by Section 4, some Zariski open subset $U \subset V$ admits a primitive fibration: there is a diagram

$$\begin{array}{ccc} X & \to & \overline{V}^{\mathcal{L}} \\ \downarrow & & \\ Y^{\mathcal{L}} & & \end{array}$$

such that for all $y \in Y^{\mathcal{L}}(F)$ the pair $(U_y, \mathcal{L}_y)$ is primitive. Then

$$c(U, \mathcal{L}) := \sum_{y \in Y^0} c(U_y, \mathcal{L}_y), \tag{4.1}$$

where the right side is a (possibly infinite converging (!)) sum over the subset $Y^0 \subset Y^{\mathcal{L}}(F)$ of all those fibers U_y where

$$a(\mathcal{L}) = a(\mathcal{L}_y) \quad \text{and} \quad \mathrm{rk}\,\mathrm{Pic}\,(V, \mathcal{L})^{\Gamma} = \mathrm{rk}\,\mathrm{Pic}\,(V_y, \mathcal{L}_y)^{\Gamma}.$$

Remark 4.3.3. A conceptual (conjectural) explanation of the convergence of the sum (4.1) is as follows: the Tamagawa type number $c(U_y, \mathcal{L}_y)$ is comparable to a *height* of the corresponding point y on the base Y, with respect to a line bundle whose class is contained in the interior of the cone of effective divisors shifted by the anticanonical class (in $\mathrm{Pic}\,(Y)$) and the (possibly infinite) sum is estimated from above by a *height zeta function* for Y in its domain of absolute convergence. This is exactly what happens in all examples in Theorem 3.1.1 (where it is actually an easy matter to give a direct proof of convergence).

In Section 5 we will see that even if we start with pairs $(V, \mathcal{L})$ where V is a smooth projective variety and $\mathcal{L}$ is a very ample adelically metrized line bundle on V we still need to consider singular varieties.

5. Analysis

5.1. Height zeta functions

We illustrate our approach to asymptotics of rational points in the particular case of equivariant compactifications X of $\mathrm{G} = \mathbb{G}_a^n$. The geometry of these varieties can be quite complicated already in dimension three (see [23] for more details). Nevertheless, we are able to prove asymptotic formulas for *every* such compactification and *every* (big) line bundle on it.

The idea is to use harmonic analysis on the adelic points $\mathrm{G}(\mathbb{A})$. We may assume that X is projective and that the boundary $X \setminus \mathrm{G}$ is a strict normal crossing divisor (using equivariant resolution of singularities, if necessary).

Recall that $\text{Pic}(X) = \text{NS}(X)$ is generated by the irreducible boundary components (see Example 2.1.4). Next we define a height pairing

$$H = \prod_v H_v : \text{Pic}(X)_{\mathbb{C}} \times \mathrm{G}(\mathbb{A}) \to \mathbb{C}$$

such that its restriction to $L \in \text{Pic}(X) \times \mathrm{G}(F)$ is the usual height (with respect to some metrized line bundle $\mathcal{L}$ in this class). Indeed, $\text{Pic}(X)$ has a basis of (classes of) ample line bundles with sections which do not vanish in G and we can apply the constructions in Section 4. The equivariance of X implies that H is invariant under some compact subgroup $\mathbf{K} \subset \mathrm{G}(\mathbb{A}_f)$ (this may fail for general compactifications of the affine space).

Now we consider the *height zeta function*

$$\mathcal{Z}(\mathrm{G}, \mathrm{s}) = \sum_{x \in \mathrm{G}(F)} H(\mathrm{s}; x)^{-1}.$$

The projectivity of X implies that $\mathcal{Z}(\mathrm{G}, \mathrm{s})$ converges *a priori* for $\Re(\mathrm{s})$ in some (shifted) open cone in $\text{Pic}(X)_{\mathbb{R}}$ (see Lemma 3.2 in [9]). Tauberian theorems relate analytic properties of $\mathcal{Z}$ to asymptotics of rational points of bounded height. Thus we need to determine the domain of holomorphy of $\mathcal{Z}$, find its poles and obtain its meromorphic continuation beyond those poles.

To achieve this we pass to a "spectral" expansion of $\mathcal{Z}$ (Poisson formula) to obtain a representation

$$\mathcal{Z}(\mathrm{G}, \mathrm{s}) = \sum_{(\mathrm{G}(\mathbb{A})/\mathrm{G}(F)\mathbf{K})^*} \hat{H}(\mathrm{s}; \psi), \tag{5.1}$$

where the sum is over the group of unitary characters of $\mathrm{G}(\mathbb{A})$ which are trivial on $\mathrm{G}(F)\mathbf{K}$. The triviality of ψ on $\mathbf{K}$ follows from the invariance of H under $\mathbf{K}$. This is crucial for the subsequent analysis — the right side in (5.1) is a summation over a *lattice* (and the trivial representation is isolated). The next step is the determination of the contribution of the trivial representation to $\mathcal{Z}$, achieved in the following section.

5.2. Height integrals

Consider more generally a smooth d-dimensional equivariant compactification X of a linear algebraic group G over F. Assume that over some finite Galois extension E/F with Galois group $\Gamma = \mathrm{Gal}(E/F)$ the boundary is a *strict normal crossing* divisor consisting of geometrically irreducible components

$$X \setminus \mathrm{G} = \cup_{j\in\mathcal{J}} D_j.$$

Then Γ acts by permutations on the set of boundary components. Denote by $\mathcal{J}/\Gamma$ (resp. $\mathcal{J}/\Gamma_v$, where Γ_v is the decomposition group at v) the set of orbits of Γ (resp. Γ_v) on $\mathcal{J}$. We put $D_\emptyset = \mathrm{G}$ and define for every subset $J \subset \mathcal{J}/\Gamma_v$

$$D_J = \cap_{j\in J} D_j$$

$$D_J^0 = D_J \setminus \cup_{J'\supsetneq J} D_{J'}.$$

For every $j \in \mathcal{J}/\Gamma$ (resp. $j \in \mathcal{J}/\Gamma_v$) we denote by F_j/F (resp. $F_{j,v}/F_v$) the minimal extension over which the orbit j splits completely and by $f_{j,v}$ the degree $[F_{j,v} : F_v]$. Choose for each v a Haar measure dg_v on $\mathrm{G}(F_v)$ such that for almost all v

$$\int_{\mathrm{G}(\mathfrak{o}_v)} dg_v = 1,$$

(to define $\mathrm{G}(\mathfrak{o}_v)$ one fixes a model for G over the integers). As in Section 4, one can define a *height* pairing between

$$\mathrm{Div}_{\mathbb{C}} := \oplus_j \mathbb{C} D_j$$

and $\mathrm{G}(\mathbb{A})$ (for *unipotent* groups, $\mathrm{Div}_{\mathbb{C}} = \mathrm{Pic}\,(X)_{\mathbb{C}}$, by Example 2.1.4). In the above basis, we have coordinates $\mathbf{s} = (s_j)_{j\in\mathcal{J}/\Gamma}$ on $\mathrm{Div}_{\mathbb{C}}$. Choose an F-rational (bi-)invariant differential form d-form on G. Then it has poles along each boundary component, and we denote by κ_j the corresponding multiplicities. For all but finitely many nonarchimedean valuations v, one has (see [12])

(5.2)

$$\int_{\mathrm{G}(F_v)} H_v(\mathbf{s}; g_v)^{-1} dg_v = \tau_v(\mathrm{G})^{-1}\Big(\sum_{J\subset\mathcal{J}/\Gamma_v} \frac{\# D_J^0(\mathbb{F}_v)}{q_v^d} \prod_{j\in J} \frac{q_v^{f_{j,v}} - 1}{q_v^{f_{j,v}(s_j-\kappa_j+1)} - 1}\Big),$$

where $\tau_v(\mathrm{G})$ is the local Tamagawa number of G (see [54], [38]).

Example 5.2.1. Let $X = \mathbb{P}^1$, $H_v(g_v)$ the local height as in Example 4.1.6. Then

$$\int_{\mathbb{G}_a(F_v)} H_v(g_v)^{-s} dg_v = \frac{1 - q_v^{-s}}{1 - q_v^{-(s-2)}}. \tag{5.3}$$

Corollary 5.2.2. *Let X be an equivariant compactification of a unipotent group. We have a height pairing*

$$\mathrm{Pic}\,(X)_{\mathbb{C}} = \oplus_{j \in \mathcal{J}/\Gamma} \mathbb{C} D_j \times \mathbb{G}_a^n(\mathbb{A}) \to \mathbb{C},$$

generalizing the usual heights. There exists a function $\varphi(\mathbf{s})$ (with $\mathbf{s} = (s_j)_{j \in \mathcal{J}/\Gamma}$) which is holomorphic for $s_j \geq \kappa_j - \delta$ (for some $\delta > 0$) such that

$$\prod_v \int_{\mathrm{G}(F_v)} H_v(\mathbf{s}; g_v)^{-1} dg_v = \prod_{j \in \mathcal{J}/\Gamma} \zeta_{F_j}(s_j - \kappa_j + 1) \cdot \varphi(\mathbf{s}) \tag{5.4}$$

(here we used that for unipotent groups $\tau_v(\mathrm{G}) = 1$).

Remark 5.2.3. As often in harmonic analysis on *compact* spaces, the trivial representation contributes the main term in "asymptotics". Indeed, the Euler product above will provide the poles of "highest" order in the expansion of the zeta function $\mathcal{Z}(\mathbf{s})$ around the point $-K_X$. Recall that (in the basis D_j) the effective cone is the positive octant (see Example 2.1.4). We observe that the polar hyperplanes of $\mathcal{Z}$ cut out the shifted, by the anticanonical class, effective cone. Even if we didn't know from geometric considerations (see 2.1.4) what the effective cone or the anticanonical class are — the adelic integral (5.4) is "telling" us the answer.

Example 5.2.4. Let us consider a wonderful compactification X of a (split) semi-simple group G of adjoint type (see 2.1.6). By construction, X is a bi-equivariant compactification of G (the group G acts on both sides). For almost all v, the local heights H_v are invariant under $\mathbf{K}_v = \mathrm{G}(\mathfrak{o}_v)$. This implies that $H_v(\mathbf{s}; g_v)$ depends only on the a_v^+-component in the Cartan decomposition

$$g_v = k_v a_v^+ k_v'$$

of g_v (here $\mathbf{s} \in \mathfrak{X}^*(\mathrm{T})_{\mathbb{C}} = \mathrm{Pic}\,(X)_{\mathbb{C}}$ and $a_v^+ \in \mathrm{T}(F_v)$). One can check that the height pairing

$$H_v : \mathfrak{X}^*(\mathrm{T})_{\mathbb{C}} \times \mathrm{G}(F_v) \to \mathbb{C}$$

is given by

$$(\mathbf{s}, g_v) \mapsto q_v^{\langle \mathbf{s}, \overline{a}_v^+ \rangle},$$

where $\overline{a}_v^+$ is the image of $a_v^+ \in \mathrm{T}(F_v)$ in the Lie algebra of T (under the logarithmic map). The local height integral can be computed geometrically as in (5.2) or directly

$$\int_{\mathrm{G}(F_v)} H_v(\mathbf{s}; g_v)^{-1} dg_v = \sum_{\overline{a}_v^+} q_v^{-\langle \mathbf{s}, \overline{a}_v^+ \rangle} \operatorname{vol}(\mathbf{K}_v a_v^+ \mathbf{K}_v).$$

Comparing these two expressions one obtains closed formulas for the volumes

$$\operatorname{vol}(\mathbf{K}_v a_v^+ \mathbf{K}_v).$$

Notice that the boundary strata $D_{\mathcal{J}}$ have a nice inductive description: they are fibrations over $(\mathrm{P}_{\mathcal{J}} \setminus \mathrm{G}) \times (\mathrm{P}_{\mathcal{J}} \setminus \mathrm{G})$ with fibers wonderful compactifications of the Levi components of the parabolic Levi components of the parabolic $\mathrm{P}_{\mathcal{J}}$ (where $\mathcal{J} \subset [1, \ldots, \operatorname{rk} \mathrm{G}]$). This allows a transparent inductive computation of the number of $\mathbb{F}_v$-points in each stratum. A combinatorial approach to these volumes for *split* G can be found in [32]. I am not aware of a treatment of the general case in the literature.

Finally. as in Remark 5.2.3, the local computation is telling us that the effective cone $\Lambda_{\mathrm{eff}}(X) \subset \mathfrak{X}^*(\mathrm{T})$ "has to be" the cone dual to the cone A_v^+ (parametrizing the $\mathbf{K}_v a_v^+ \mathbf{K}_v$-cosets in the Cartan decomposition), that is exactly the cone spanned by simple roots. Similarly, we can read off the anticanonical class.

5.3. Height zeta functions continued

We return to the study of analytic properties of the height zeta function $\mathcal{Z}(\mathbf{s}; g)$ for equivariant compactifications of $\mathbb{G}_a^n$ and consider now nontrivial characters ψ, occurring in the expansion (5.1). For each ψ we compute the Fourier transforms $\hat{H}_v$ at almost all nonarchimedean places and find estimates at the remaining places. At good reduction places, this generalizes the computation in Section 5. A character $\psi = \psi_{\mathbf{a}}$ of $\mathbb{G}_a^n(\mathbb{A})/\mathbb{G}_a^n(F)$ is defined by a linear form $f_{\mathbf{a}} := \langle \mathbf{a}, \cdot \rangle$, with $\mathbf{a} \in \mathbb{G}_a^n(F)$. The divisor of the corresponding rational function on X can be written as

$$\operatorname{div}(f_{\mathbf{a}}) = E(\mathbf{a}) - \sum_{j \in \mathcal{J}/\Gamma} d_j(\mathbf{a}) D_j,$$

where all $d_j \geq 0$ (by Example 2.1.4). Denote by $\mathcal{J}_0(\mathbf{a})$ the set of all $j \in \mathcal{J}/\Gamma$ such that the corresponding $d_j(\mathbf{a}) = 0$. For nontrivial $\mathbf{a}$ the set $\mathcal{J}_0(\mathbf{a})$ is a *proper* subset of $\mathcal{J}/\Gamma$. "Motivic" integration (combined with estimates at bad reduction places) yields

$$\hat{H}(\mathbf{s};\psi_{\mathbf{a}}) = \prod_{j\in\mathcal{J}_0(\mathbf{a})} \zeta_{F_j}(s_j - \kappa_j + 1)\varphi(\mathbf{s};\mathbf{a})$$

where φ is holomorphic in the neighborhood of $-K_X$. At this stage we observe that *each* term in (5.1) admits a meromorphic continuation and that the poles of each term are contained in the faces of the (shifted) cone $\Lambda_{\text{eff}}(X) - K_X$. To obtain a meromorphic continuation for the *sum* we need a bound

$$|\varphi(\mathbf{s};\mathbf{a})| \ll (1 + \|\mathbf{a}\|)^{-n}$$

(for sufficiently large n). For this we use integration by parts at the archimedean places. Once again, the equivariance of X is essential.

In the neighborhood of $-K_X$ the pole of $\mathcal{Z}$ of highest order is contributed by the trivial character and the leading constant at this pole is an adelic integral (a Tamagawa number) times a rational number (the γ-constant). Let us consider a line bundle L which is *not* proportional to $-K_X$ and restrict $\mathcal{Z}(\mathbf{s})$ to sL. Denote by $\mathcal{J}(L)$ the set of boundary divisors which do not lie in the face of the cone $\Lambda_{\text{eff}}(X)$ containing $a(L)L + K_X$. The one-parameter zeta function $\mathcal{Z}(sL)$ is holomorphic for $\Re(s) > a(L)$, admits a meromorphic continuation to $\Re(s) > a(L) - \delta$ (for some $\delta > 0$) and has a pole of order *at most* $b(L)$ at $s = a(L)$. Notice that *every* term in the sum

$$\sum_{\mathbf{a}:\mathcal{J}_0(\mathbf{a})\supseteq\mathcal{J}(L)} \hat{H}(\mathbf{s};\psi_{\mathbf{a}}) \tag{5.5}$$

contributes to this pole (the other terms in $\mathcal{Z}$ have poles of smaller order). We need to show that the (in general, infinite) sum of leading coefficients of these terms (complex numbers) converges to a *non-zero* real number. Further, we need to identify this real number.

One possible approach, suggested by harmonic analysis, is to use the Poisson formula one more time to convert the sum over Fourier transforms of the height to an integral over some subgroup of $\mathrm{G}(\mathbb{A})$. The common kernel of the characters appearing in (5.5) is a subgroup

$$\mathrm{G}^L(F) \cdot \mathrm{G}_L(\mathbb{A}) \subset \mathrm{G}(\mathbb{A}),$$

where G_L is the intersection of kernels of $f_{\mathbf{a}}$, for all $\mathbf{a}$ occurring in (5.5). Applying the Poisson formula we find that the leading term equals

$$(5.6) \qquad \lim_{s\to a(L)} \left(s-a(L)\right)^{b(L)} \cdot \left(\sum_{g\in \mathrm{G}(F)/\mathrm{G}_L(F)} \int_{\mathrm{G}_L(\mathbb{A})} H(s;g+g')^{-1} dg' \right).$$

Now we want to find a geometric explanation of this formula. First of all, the outer sum indicates that we are dealing with a fibration. It turns out that this fibration is exactly of the type appearing in Fujita's program (the map onto the base of the fibration is given by the *sections* $f_{\mathbf{a}}$ as in (5.5)). Secondly, the adelic integrals are very similar to those already treated in Section 5. In fact, they are height integrals for the induced equivariant compactifications of translates of G_L. There is one important difference: previously, we started with a smooth equivariant compactification X, with boundary a strict normal crossing divisor. We cannot guarantee, however, that the compactification of G_L (and its shifts) inside X has the same property! This explains the technical setup of Section 4: analysis tells us that we need to deal with fibrations and singular varieties. And finally, the (possibly infinite) regularized sum in (5.6) converges for s sufficiently close to $a(L)$ (it can be estimated by an integral of type (5.2)).

Example 5.3.1. Consider the simplest case $X = \mathbb{P}^n \times \mathbb{P}^n$, $\mathcal{L}$ the product of the line bundles $\mathcal{O}(a)$ and $\mathcal{O}(b)$. The Picard group $\mathrm{Pic}\,(X) = \mathbb{Z}\oplus\mathbb{Z}$, so that the class of $\mathcal{L} = (a,b)$ and $-K_X = (n+1, n+1)$. The height zeta function factors as well

$$\mathcal{Z}_X(s_1, s_2) = \mathcal{Z}_{\mathbb{P}^n}(s_1)\mathcal{Z}_{\mathbb{P}^n}(s_2).$$

Its restriction to the line $s(n+1, n+1)$ has a pole at $s=1$ of order 2, with leading coefficient an adelic integral (the Tamagawa number τ^2) times a rational number (the γ-constant). The restriction to the line $s(a,b)$, for $a > b$, has a pole at $s = (n+1)/b$ with residue $\tau \cdot \gamma \cdot \mathcal{Z}_{\mathbb{P}^n}(a(n+1)/b)$. The (converging) sum

$$\sum_{x\in\mathbb{P}^n(F)} \tau\cdot\gamma\cdot H_{\mathcal{O}(1)}(x)^{a(n+1)/b}$$

is exactly the constant $c(X, \mathcal{L})$ defined in Section 4.

REFERENCES

[1] V. Batyrev, The cone of effective divisors of threefolds, in: *Proc. of the Intern. Conf. on Algebra (Novosibirsk, 1989)*, Contemp. Math., **131**, Part 3, Amer. Math. Soc. (Providence, 1992), pp. 337–352.

[2] V. Batyrev and Yu. Tschinkel, Rational points of bounded height on compactifications of anisotropic tori, *Intern. Math. Res. Notices,* **12** (1995), 591–635.

[3] V. Batyrev and Yu. Tschinkel, Rational points on some Fano cubic bundles, *C.R. Acad. Sci.*, **323** (1996), 41–46.

[4] V. Batyrev and Yu. Tschinkel, Manin's conjecture for toric varieties, *J. Algebraic Geom.*, **7** (1998), 15–53.

[5] V. Batyrev and Yu. Tschinkel, Tamagawa numbers of polarized algebraic varieties, in: *Nombre et répartition des points de hauteur bornée* (E. Peyre, ed.), Astérisque **251** (1998), 299–340.

[6] B. Birch, Forms in many variables, *Proc. Roy. Soc. Ser. A*, **265** (1961–62), 245–263.

[7] F. Bogomolov and Yu. Tschinkel, On the density of rational points on elliptic fibrations, *J. reine angew. Math.*, **511** (1999), 87–93.

[8] F. Campana, *Special varieties and classification theory*, e-print math.AG/0110051 (2001).

[9] A. Chambert-Loir and Yu. Tschinkel, Points of bounded height on equivariant compactifications of vector groups I, *Compositio Math.*, **124** (2000), 65–93.

[10] A. Chambert-Loir and Yu. Tschinkel, Torseurs arithmétiques et espaces fibrés, in: *Rational points on algebraic varieties* (E. Peyre, Yu. Tchinkel, eds.), Progress in Math., vol. 199, Birkhäuser (2001), pp. 37–69.

[11] A. Chambert-Loir and Yu. Tschinkel, Fonctions zêta de hauteurs des espaces fibrés, in: *Rational points on algebraic varieties* (E. Peyre, Yu. Tchinkel, eds.), Progress in Math., vol. 199, Birkhäuser (2001), pp. 71–116.

[12] A. Chambert-Loir and Yu. Tschinkel, On the distribution of rational points on equivariant compactifications of vector groups, *Invent. Math.*, **148** (2002), 421–452.

[13] J.-L. Colliot-Thélène and J.-J. Sansuc, La R-équivalence sur les tores, *Ann. Sci. ENS,* **10** (1977), 175–230.

[14] J.-L. Colliot-Thélène, P. Swinnerton-Dyer and A. Skorobogatov, Double fibres and double covers: paucity of rational points, *Acta Arith.*, **79** (1997), 113–135.

[15] C. De Concini and C. Procesi, *Complete symmetric varieties,* Invariant theory (Montecatini, 1982), LNM **996**, Springer (1983), pp. 1–44.

[16] R. de la Bretèche, Nombre de points de hauteur bornée sur les surfaces de del Pezzo de degré 5, to appear in *Duke Math. J.* (2002).

[17] J. Denef and F. Loeser, Motivic Igusa zeta functions, *J. Algebraic Geom.*, **7(3)** (1998), 505–537.

[18] G. Faltings, Diophantine approximation on abelian varieties, *Ann. of Math.*, **133(2)** (1991), 549–576.

[19] J. Franke, Yu. Manin and Yu. Tschinkel, Rational points of bounded height on Fano varieties, *Invent. Math.*, **95** (1989), 421–435.

[20] T. Fujita, On Kodaira energy of polarized varieties, *J. Math. Soc. Japan*, **48(1)** (1996), 1–12.

[21] J. Harris and Yu. Tschinkel, Rational points on quartics, *Duke Math. J.*, **104** (2000), 477–500.

[22] B. Hassett, *Potential density of rational points on algebraic varieties*, this volume.

[23] B. Hassett and Yu. Tschinkel, Geometry of equivariant compactifications of $\mathbb{G}_a^n$, *Intern. Math. Res. Notices*, **22** (1999), 1211–1230.

[24] B. Hassett and Yu. Tschinkel, On the effective cone of the moduli spaces of pointed rational curves, to appear in *SISTAG Commemorative Volume: Singapore Intern. Symp. in Topology and Geometry* (A. J. Berrick et al., eds.) (2001).

[25] R. Heath-Brown, The density of rational points on cubic surfaces, *Acta Arith.*, **272** (1997), 17–30.

[26] B. Hunt, *The geometry of some special arithmetic quotients*, LNM **1637**, Springer (1996).

[27] V. A. Iskovskih, Fano threefolds I, II, *Izv. Akad. Nauk SSSR Ser. Mat.*, **41** (1977), 516–562 and **42** (1978), 506–549.

[28] J. Igusa, *Forms of higher degree*, Tata Institute of Fundamental Research Lectures on Mathematics and Physics **59** (Bombay, 1978).

[29] Y. Kawamata, K. Matsuda and K. Matsuki, Introduction to the Minimal Model Program, in: *Algebraic Geometry, Sendai, 1985*, Adv. Studies in Pure Math., vol. 10, North-Holland (1987), pp. 283–360.

[30] J. Kollár, Low degree polynomial equations: arithmetic, geometry and topology, in: *European Congress of Mathematics* (Budapest, 1996), Progress in Math., vol. 168, Birkhäuser, vol. 1 (1998), pp. 255–288.

[31] J. Kollár and Sh. Mori, *Birational geometry of algebraic varieties*, With the collaboration of C. H. Clemens and A. Corti. Cambridge Tracts in Mathematics **134**, Cambridge University Press (1998).

[32] I. G. MacDonald, *Spherical functions on a group of p-adic type*, Publications of the Ramanujan Institute, No. 2. Ramanujan Institute, Centre for Advanced Study in Mathematics, University of Madras (1971).

[33] Yu. I. Manin, *Cubic forms. Algebra, geometry, arithmetic*, Second edition, North-Holland Mathematical Library, **4**, North-Holland (Amsterdam, 1986).

[34] Yu. Matiyasevich, Hilbert's tenth problem: what was done and what is to be done, in: *Hilbert's tenth problem: relations with arithmetic and algebraic geometry* (Ghent, 1999), Contemp. Math. vol. 270, Amer. Math. Soc. (Providence, 2000), pp. 1–47.

[35] Yu. Matiyasevich, Le dixième problème de Hilbert: que peut-on faire avec les équations diophantiennes?, in: *La recherche de la vérité*, ACL-Éd. Kangourou (Paris, 1999), pp. 281–305.

[36] D. McKinnon, Counting rational points on $K3$ surfaces, *J. Number Theory*, **84** (2000), 49–62.

[37] Sh. Mori and Sh. Mukai, *On Fano 3-folds with $B_2 \geq 2$*, Algebraic varieties and analytic varieties (Tokyo, 1981), 101–129, Adv. Stud. Pure Math., vol. 1, North-Holland (Amsterdam, 1983).

[38] T. Ono, On the Tamagawa number of algebraic tori, *Ann. of Math.*, **78(1)** (1963), 47–73.

[39] T. Ono, On the relative theory of Tamagawa numbers, *Ann. of Math.*, **82(1)** (1965), 88–111.

[40] T. Ono, *On Tamagawa numbers*, Proc. Symp. Pure Math., vol. 9, Amer. Math. Soc. (Providence, 1966), pp. 122–132.

[41] E. Peyre, Hauteurs et nombres de Tamagawa sur les variétés de Fano, *Duke Math. J.*, **79** (1995), 101–218.

[42] E. Peyre, Terme principal de la fonction zêta des hauteurs et torseurs universels, in: *Nombre et répartition des points de hauteur bornée* (E. Peyre, ed.), Astérisque, **251** (1999), 259–298.

[43] E. Peyre and Yu. Tschinkel, Tamagawa numbers of cubic surfaces, *Math. Comp.*, **70** (2001), 367–387.

[44] J. Shalika, R. Takloo-Bighash and Yu. Tschinkel, *Rational points and automorphic forms*, preprint (2001).

[45] J. Shalika, R. Takloo-Bighash and Yu. Tschinkel, *Distribution of rational points on equivariant compactifications of inner forms of reductive groups*, preprint (2001).

[46] J. Shalika and Yu. Tschinkel, *Height zeta functions of equivariant compactifications of the Heisenberg group*, preprint (2001).

[47] J. Shalika and Yu. Tschinkel, *Height zeta functions of equivariant compactifications of unipotent groups*, in preparation.

[48] M. Strauch, Arithmetic stratifications and partial Eisenstein series, in: *Rational points on algebraic varieties* (E. Peyre, Yu. Tchinkel, eds.), Progress in Math., vol. 199, Birkhäuser (2001), pp. 335–355.

[49] M. Strauch and Yu. Tschinkel, Height zeta functions of toric bundles over flag varieties, *Selecta Math.*, **5** (1999), 325–396.

[50] J. Tate, Fourier analysis in number fields, and Hecke's zeta functions, in: *Algebraic Number theory* (J. W. S. Cassels, A. Fröhlich, eds.), Thompson (Washington, D.C., 1967), pp. 305–347.

[51] E. B. Vinberg, The theory of convex homogeneous cones, *Trans. Moscow Math. Soc.*, **12** (1963), 340–403.

[52] P. Vojta, *Diophantine approximations and value distribution theory*, LNM **1239**, Springer (1987).

[53] R. C. Vaughan and T. D. Wooley, On a certain nonary cubic form and related equations, *Duke Math. J.*, **80** (1995), 669–735.

[54] A. Weil, *Adeles and algebraic groups,* Progress in Math., vol. 23, Birkhäuser (1982).

Yuri Tschinkel

Department of Mathematics
Princeton University
Princeton, NJ 08544-1000
U.S.A.

ytschink@math.princeton.edu

MIX
Papier aus verantwortungsvollen Quellen
Paper from responsible sources
FSC® C105338

If you have any concerns about our products,
you can contact us on
ProductSafety@springernature.com

In case Publisher is established outside the EU,
the EU authorized representative is:
Springer Nature Customer Service Center GmbH
Europaplatz 3, 69115 Heidelberg, Germany

Printed by Libri Plureos GmbH
in Hamburg, Germany